T0358301

Invariants
and Pictures

Low-dimensional Topology and
Combinatorial Group Theory

K&E Series on Knots and Everything — Vol. 66

Invariants and Pictures

Low-dimensional Topology and Combinatorial Group Theory

Vassily Olegovich Manturov
Bauman Moscow State Technical University, Russia

and

Denis Fedoseev
Moscow State University, Russia

Seongjeong Kim
Bauman Moscow State Technical University, Russia
& Moscow Institute of Physics and Technology, Russia

Igor Nikonov
Moscow State University, Russia

 World Scientific

NEW JERSEY · LONDON · SINGAPORE · BEIJING · SHANGHAI · HONG KONG · TAIPEI · CHENNAI · TOKYO

Published by

World Scientific Publishing Co. Pte. Ltd.

5 Toh Tuck Link, Singapore 596224

USA office: 27 Warren Street, Suite 401-402, Hackensack, NJ 07601

UK office: 57 Shelton Street, Covent Garden, London WC2H 9HE

Library of Congress Cataloging-in-Publication Data

Names: Manturov, V. O. (Vasiliĭ Olegovich), author. | Fedoseev, Denis, author. |
Kim, Seongjeong, author. | Nikonov, Igor (Igor Mikhailovich), author.
Title: Invariants and pictures : low-dimensional topology and combinatorial group theory /
Vassily Olegovich Manturov and Denis Fedoseev, Seongjeong Kim, Igor Nikonov.
Description: New Jersey : World Scientific, [2020] | Series: Series on knots and everything,
0219-9769 ; vol. 66 | Includes bibliographical references and index.
Identifiers: LCCN 2020012113 | ISBN 9789811220111 (hardcover) |
ISBN 9789811220128 (ebook for institutions) | ISBN 9789811220135 (ebook for individuals)
Subjects: LCSH: Low-dimensional topology. | Combinatorial group theory. | Invariants.
Classification: LCC QA612.14 .M36 2020 | DDC 514/.22--dc23
LC record available at https://lccn.loc.gov/2020012113

British Library Cataloguing-in-Publication Data

A catalogue record for this book is available from the British Library.

For any available supplementary material, please visit
https://www.worldscientific.com/worldscibooks/10.1142/11821#t=suppl

Printed in Singapore

Preface

A long time ago, when I first encountered knot tables and started unknotting knots "by hand", I was quite excited with the fact that some knots may have more than one minimal representative. In other words, in order to make an object simpler, one should first make it more complicated. For example, see Fig. 0.1 [Kauffman and Lambropoulou, 2012]: this diagram represents the trivial knot, but in order to simplify it, one needs to perform an *increasing* Reidemeister move first.

Fig. 0.1   Culprit knot

Being a first year undergraduate student (in Moscow State University), I first met free groups and their presentation. The power and beauty, and simplicity of these groups for me were their exponential growth and extremely easy solution to the word problem and conjugacy problem by means of a gradient descent algorithm in a word (in a cyclic word, respectively).

Also, I was excited with the Diamond lemma: a simple condition which guarantees the uniqueness of the minimal objects, and hence, solution to many problems (Chapter 1.4).

Being a last year undergraduate and teaching a knot theory course for the first time, I thought: "Why do not we have it (at least partially) in knot theory?"

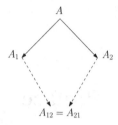

Fig. 0.2   The Diamond lemma

By that time I knew about the Diamond lemma and solvability of many problems like word problem in groups by gradient descent algorithm. The van Kampen lemma and Greendlinger's theorem came to my knowledge much later.

I spent a lot of time working with virtual knot theory; my doctoral (habilitation) thesis [Manturov, 2007] was devoted to various problems in that theory: from algorithmic recognition of virtual knots to the construction of Khovanov homology for virtual knots with arbitrary coefficients.

Virtual knot theory (Chapter 3), which can be formally defined via Gauß diagrams which are not necessarily planar, is a theory about knots in thickened surfaces $S_g \times I$ considered up to addition/removal of nugatory handles. It contains classical knot theory as a proper part: classical knots can be thought of as knots in the thickened sphere. Hence, virtual knots have a lot of additional information coming from the topology of the ambient space (S_g).

By playing with formal Gauß diagrams, I decided to drop any arrow and sign information and called such objects *free knots* [Manturov, 2009].

It was an interesting puzzle for me in December 2008 in Heidelberg to construct invariants of free knots as I had never heard of any. What one should pay attention to is that all chords of a Gauß diagram of a virtual knots can be *odd* and *even*. Gauß himself knew that Gauß diagrams of planar curves and knots have no odd chords. Hence, odd chords can be the key point of non-triviality and non-classicality. When looking at Reidemeister moves, one can see that the chord taking part in a first Reidemeister move is even; two chords taking part in a second Reidemeister move are of the same parity, and the sum of parities of the three chords taking part in a third Reidemeister move is 0 modulo 2 if we count odd chords as 1 and even chords as 0 (Definition 5.8). Hence, odd chords can only cancel with "neighbouring" odd chords by the second Reidemeister moves; otherwise they persist.

In January 2009, I constructed a state-sum invariant of free knots valued in diagrams of free knots, i.e., framed four-valent graphs[1] [Manturov, 2010]. For states, I was taking all possible smoothings at even crossings, imposing some diagrams to be zero. This invariant was constructed in such a way that all odd chords persisted and I got the formula

$$[K] = K$$

whenever K is a diagram of a free knot with all chords where no two chords can be cancelled by a second Reidemeister move (see Section 5.1).

The deep sense of this formula can be expressed as follows:

If a virtual diagram is complicated enough, then it realises itself.

Namely, K on the left hand side is *some* free knot diagram K, and K on the right hand side is a *concrete graph*. Hence, if K' is another diagram of the same knot, we shall have $[K'] = K$ meaning that K is obtained as a result of smoothing of K'. This principle is depicted on Fig. 0.3.

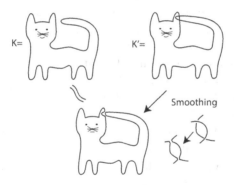

Fig. 0.3   A picture which is its own invariant

This is an example of what we miss in classical knot theory: *local minimality yields global minimality* or, in other words, *if a diagram is odd and irreducible* then it is minimal in a very strong sense (Lemma 5.1, Section 5.1). Not only one can say that K' has larger crossing number than K, we can say that K "lives" inside K'. Having these "graphical" invariants,

---

[1] After I constructed such invariant, I learnt that free knots were invented by Turaev five years before that and thought to be trivial [Turaev, 2007], hence, I disproved Turaev's conjecture without knowing that.

we get immediate consequences about many characteristics of K. The main problem is to construct invariants of similar nature in the case of classical knot theory.

Something similar can be seen in other situations: free groups or free products of cyclic groups, cobordism theory for free knots, or while considering other geometrical problems. For example, if we want to understand the genus of a surface where the knot K can be realised, it suffices for us to look at the minimal genus where the concrete diagram K can be realised for the genus of any other K' is a priori larger than that of K [Manturov, 2012b], see also [Ilyutko, Manturov and Nikonov, 2011].

Actually, having lectured knot theory over many years and having published several knot theory books by that time [Manturov, 2018; Ilyutko and Manturov, 2013; Ilyutko, Manturov and Nikonov, 2011], I knew that a knot group (fundamental group of the complement to a knot) may be very complicated, hence, it may contain powerful information inside.

What is the main difference between classical knot theory and virtual knot theory? In my opinion, the existence of a large ambient group $(\pi_1(S_g))$ in the virtual knot case. By extracting some information out of it, one can get parities and other enhancements for knots, hence, leading us to the above "picture-valued" invariants (Section 5.1).

But how to extract nice group information out of classical knots? For my purposes (say, for picture-valued brackets), there is a lack of "canonical coordinate system" *how* to make the knot see these nodes (crossings) visible in a way to get some parity.

As a topologist, I was always interested in *cobordisms* and *concordance*: much more subtle equivalence relations than isotopy or homotopy. As a knot theorist, I knew from my childhood about the Fox conjecture: all slice knots are conjectured to be ribbon.

Now, let us look at free knots: very coarse objects; it looks like if we impose a coarse equivalence relation, like cobordism, it will kill everything. However, it turned out not only that cobordism classes are non trivial, but for odd irreducible free knots the Fox conjecture is true (it is my joint work with D. A. Fedoseev [Fedoseev and Manturov, 2019b; Fedoseev and Manturov, 2018]).

Again we can say that some local information allows one to judge about some global *dynamics*: if it is not possible to pair chord ends and cap an odd framed 4-graph at once without singularities, just with double lines, then there is no chance for it to be capped after any long sequence of Reidemeister moves, maxima and minima, triple points and cusps (*statics*).

This takes me back to the time of my habilitation thesis. Once writing a knot theory paper and discussing it with Oleg Yanovich Viro, I wrote "a classical knot is an equivalence class of classical knot diagrams modulo Reidemeister moves". Well, — said Viro, — you are restricting yourself very much. Staying at this position, how can you prove that a non-trivial knot has a quadrisecant?

Reflecting this after several years, I understood that it is not quite necessary to consider knots by using Reidemeister planar projections, one can look for other "nodes".

This led me to my initial preprint [Manturov, 2015a] and to an extensive study of braids and dynamical systems. This happened around New Year 2015.

Namely, I looked at usual Artin braids as everybody does: as dynamical systems of points in $\mathbb{R}^2$, but, instead of creating Reidemeister's diagram by projecting braids to a screen (say, the plane Oxz), I decided to look at those moments when some three points are *collinear*. This is quite a good property of a "node" which behaves nicely under generic isotopy. Let us denote such situations by letters a_{ijk} where i, j, k are numbers of points (this triple of numbers is unordered).

When considering four collinear points, we see that *the tetrahedron (Zamolodchikov,* see, for example, [Etingof, Frenkel and Kirillov, 1998]) *equation* emerges. Namely, having a dynamics with a quadruple point and slightly perturbing it, we get a dynamics, where this quadruple point splits into four triple points.

Writing it algebraically, we get:

$$a_{ijk}a_{ijl}a_{ikl}a_{jkl} = a_{jkl}a_{ikl}a_{ijl}a_{ijk}.$$

Definition 0.1. The groups G_n^k are defined as follows.

$$G_n^k = \langle a_m | (1), (2), (3) \rangle,$$

where the generators a_m are indexed by all k-element subsets of $\{1, \ldots, n\}$, the relation (1) means

$$(a_m)^2 = 1 \text{ for any unordered sets } m \subset \{1, \ldots, n\}, Card(m) = k;$$

(2) means

$$a_m a_{m'} = a_{m'} a_m, \text{ if } Card(m \cap m') < k - 1;$$

and, finally, the relation (3) looks as follows. For every set $U \subset \{1, \ldots, n\}$ of cardinality $(k + 1)$, let us order all its k-element subsets arbitrarily and denote them by $m^1, \ldots, m^{k+1}$. Then (3) is:

$$a_{m^1} \ldots a_{m^{k+1}} = a_{m^{k+1}} \ldots a_{m^1}.$$

This situation with the Zamolodchikov equation happens almost everywhere, hence, I formulated the following principle:

If dynamical systems describing the motion of n particles possess a nice codimension one property depending on exactly k particles, then these dynamical systems admit a topological invariant valued in G_n^k.

In topological language, it means that we get a certain homomorphism from some fundamental group of a topological space to the groups G_n^k.

Collecting all results about the G_n^k groups, I taught a half-year course of lectures in the Moscow State University entitled "Invariants and Pictures" and a 2-week course in Guangzhou. The notes taken by my colleagues I. M. Nikonov[2], D. A. Fedoseev and S. Kim were the starting point for the present book.

Since that time, my seminar in Moscow, my students and colleagues in Moscow, Novosibirsk, Beijing, Guangzhou, and Singapore started to study the groups G_n^k, mostly from two points of view:

From the topological point of view, which spaces can we study?

Besides the homomorphisms from the pure braid group PB_n to G_n^3 and G_n^4 (Sections 8.1 and 8.2), I just mention that I invented braids for higher-dimensional spaces (or projective spaces).

Of course, the configuration space $C(\mathbb{R}^{k-1}, n)$ is simply connected for $k > 3$ but if we take some *restricted* configuration space $C'(\mathbb{R}^{k-1}, n)$, it will not be simply connected any more and leads to a meaningful notion of higher dimensional braid (Chapter 11).

What sort of the restriction do we impose? On the plane, we consider just braids, so we say that no two points coincide. In $\mathbb{R}^3$ we forbid collinear triples, in $\mathbb{R}^4$ we forbid coplanar quadruple of points.

When I showed the spaces I study to my coauthor, Jie Wu, he said: look, these are k-regular embeddings, they go back to Carol Borsuk. Indeed, after looking at some papers by Borsuk, I saw similar ideas were due to P. L. Chebysheff ([Borsuk, 1957; Kolmogorov, 1948]).

---

[2]I.M. Nikonov coauthored my first published paper about G_n^k.

By the way, once Wu looked at the group G_n^k, he immediately asked about the existence of simplicial group structure on such groups, the joint project we are working on now with S. Kim, J. Wu, F. Li.

An interested reader may ask whether such braids exist not only for $\mathbb{R}^k$ (or $\mathbb{R}P^k$), but also for other spaces. This question we shall touch on later.

From the algebraic point of view, why are these groups good, how are they related to other groups, how to solve the word and the conjugacy problems, etc.?

It is impossible to describe all directions of the G_n^k group theory in the preface, the reader will find many directions in the unsolved problem list; I just mention some of them.

For properties of G_n^k, we can think of them as n-strand braids with k-fold strand intersection.

Like $\binom{n}{k} = \binom{n-1}{k} + \binom{n-1}{k-1}$, there are nice "strand forgetting" and "strand deletion" maps to G_{n-1}^k and G_{n-1}^{k-1}, see Fig. 0.4.

The groups G_n^k have lots of epimorphisms onto free products of cyclic groups; hence, invariants constructed from them are powerful enough and easy to compare.

For example, the groups G_n^2 are commensurable with some Coxeter groups of special type, see Fig. 0.5, which immediately solves the word problem for them.

As Diamond lemma works for Coxeter groups, it works for G_n^2, and in many other places throughout the book.

After a couple of years of study of G_n^k, I understood that I was not completely free and this approach is still somewhat restrictive. Well, we can study braids, we can invent braids in $\mathbb{R}^n$ and $\mathbb{R}P^n$, but what if we consider just braids on a 2-surface? What can we study then? The property "three points belong to the same line" is not quite good even in the metrical case because even if we have a Riemannian metric on a 2-surface of genus g, there may be infinitely many geodesics passing through two points. Irrational cables may destroy the whole construction.

Then I decided to transform the "G_n^k-point of view" to make it more local and more topological. Assume we have a collection of points in a 2-surface and seek G_n^4-property: four points belong to the same circle.

Consider a 2-surface of genus g with N points on it. We choose N to be sufficiently large and put points in a position to form the centers of Voronoï cells. It is always possible for the sphere $g = 0$, and for the plane we may think that all our points live inside a triangle forming a Voronoï tiling of the latter.

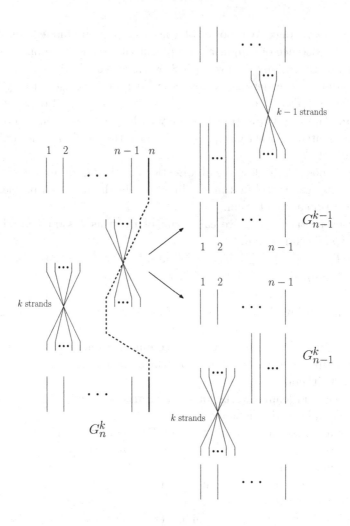

Fig. 0.4   Maps from G_n^k to G_{n-1}^k and G_{n-1}^{k-1}

We are interested in those moments when the combinatorics of the Voronoï tiling changes, see Fig. 0.7.

This corresponds to a flip, the situation when four *nearest points belong to the same circle*. This means that no other point lies inside the circle passing through these four, see [Gelfand, Kapranov and Zelevinsky, 1994].

The most interesting situation of codimension 2 corresponds to five points belonging to the same circle.

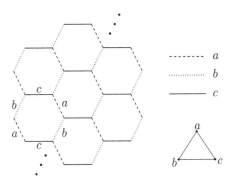

Fig. 0.5   The Cayley graph of the group G_3^2 and the Coxeter group $C(3,2)$

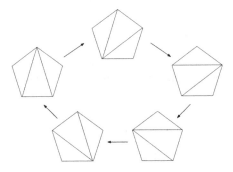

Fig. 0.6   Flips on a pentagon

This leads to the relation:

$$d_{1234}d_{1245}d_{2345}d_{1235}d_{1345} = 1.$$

Note that unlike the case of G_n^4, here we have five terms, not ten. What is the crucial difference? The point is that if we have five points in the neighbourhood of a circle, then every quadruple of them appears to be on the same circle *twice*, but one time the fifth point is outside the circle, and one time it is inside the circle. We denote the set $\{1, \ldots, n\}$ by $\bar{n}$ and introduce the following

Definition 0.2. The group Γ_n^4 is the group given by group presentation generated by $\{d_{(ijkl)} \mid \{i, j, k, l\} \subset \bar{n}, |\{i, j, k, l\}| = 4\}$ subject to the following relations:

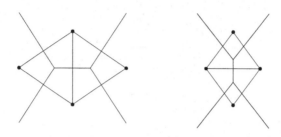

Fig. 0.7 Voronoï tiling change

(1) $d^2_{(ijkl)} = 1$ for $\{i, j, k, l\} \subset \bar{n}$,

(2) $d_{(ijkl)}d_{(stuv)} = d_{(stuv)}d_{(ijkl)}$, for $|\{i, j, k, l\} \cap \{s, t, u, v\}| < 3$,

(3) $d_{(ijkl)}d_{(ijlm)}d_{(jklm)}d_{(ijkm)}d_{(iklm)} = 1$ for distinct i, j, k, l, m,

(4) $d_{(ijkl)} = d_{(kjil)} = d_{(ilkj)} = d_{(klij)} = d_{(jkli)} = d_{(jilk)} = d_{(lkji)} = d_{(lijk)}$, for distinct i, j, k, l, m.

Just like we formulated the G^k_n principle, here we could formulate the Γ^k_n principle in whole generality, but we restrict ourselves with several examples.

It turns out that groups Γ^4_n have nice presentation coming from *the Ptolemy relation* and the cluster algebras. The Ptolemy relation

$$xy = ac + bd$$

says that the product of diagonals of an inscribed quadrilateral equals the sum of products of its opposite faces, see Fig. 0.8.

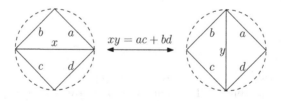

Fig. 0.8 The Ptolemy relation

We can use it when considering triangulations of a given surface: when performing a flip, we replace one diagonal (x) with the other diagonal (y) by using this relation. It is known that if we consider all five triangulations of the pentagon and perform five flips all the way around, we return to the initial triangulation with the same label, see Fig. 0.6.

This well known fact gives rise to presentations of Γ_n^4.

Thus, by analysing the groups Γ_n^4, we can get

(1) Invariants of braids on 2-surfaces valued in polytopes;
(2) Invariants of knots;
(3) Relations to groups G_n^4;
(4) Braids on $\mathbb{R}^3$.

Going slightly beyond, we can investigate braids in $\mathbb{R}^3$ and the configuration space of polytopes.

We will not say much about the groups Γ_n^k for $k > 4$. The main idea is:

(1) Generators (codimension 1) correspond to simplicial $(k-2)$-polytopes with k vertices;
(2) The most interesting relations (codimension 2) correspond to $(k-2)$-polytopes with $k+1$ vertices.

It would be extremely interesting to establish the connection between G_n^k with the Manin–Schechtmann *"higher braid groups"* [Manin and Schechtmann, 1990], where the authors study the fundamental group of complements to some configurations of *complex hyperplanes*.

It is also worth mentioning, that the relations in the group G_n^2 resemble the relations in *Kirillov–Fomin algebras*, see [Fomin and Kirillov, 1999]. For that reason it seems interesting to study the interconnections between those objects.

Finally, our invariants may not be just group-valued: some variations of G_n^k admit simplicial group structures, which is studied now in a joint work with S. Kim, F. Li and J. Wu.

Note that the present book is very much open-ended. On one hand, the invariants of manifolds are calculated in some explicit cases. On the other hand, one can vary "the G_n^k-principle" and "Γ_n^k-principle" together with the groups itself and try to find invariants of manifolds depending on complex structures, spin-structures, and other structures by looking at some "good codimension 1 properties" and creating interesting configuration spaces.

The present book has the following structure. In Part 1 we review basic notions of knot theory and combinatorial group theory: groups and their presentations, van Kampen diagrams, braid theory, knot theories and the theory of 2-dimensional knots. Part 2 is devoted to the parity theory and its applications to cobordisms of knots and free knots. In Part 3 we present the theory of G_n^k groups and their relations to invariants of dynamical systems. Part 4 deals with the notion of *manifold of triangulations*, higher dimensional braids, and investigates the groups Γ_n^k. In the final Part 5 we present a list of unsolved problems in the theories discussed in the present book.

Vassily Olegovich Manturov
2019

Acknowledgments

The authors would like to express their heartfelt gratitude to L. A Bokut', H. Boden, J. S. Carter, A. T. Fomenko, S. G. Gukov, Y. Han, D. P. Ilyutko, A. B. Karpov, R. M. Kashaev, L. H. Kauffman, M. G. Khovanov, A. A. Klyachko, P. S. Kolesnikov, I. G. Korepanov, S. V. Matveev, A. Yu. Olshanskii, W. Rushworth, G. I. Sharygin, V. A. Vassiliev, Zheyan Wan, Jun Wang, J. Wu and Zerui Zhang for their interest and various useful discussions on the present work. We are grateful to Efim I. Zelmanov for pointing out the resemblance between the groups G_n^2 and Kirillov–Fomin algebras.

The first named author would like to express his special thanks to $\boxed{\text{Patrick Dehornoy}}$ (11 Sep. 1952–4 Sep. 2019):

> *"I learnt a lot about word problems and conjugacy problems in the braid group from him. Meeting him many times during the last twenty years increased my knowledge in braid group theory. It is a great loss for the mathematical community that he passed away in 2019."*

—*Vassily Olegovich Manturov*

The first named author was supported by the Laboratory of Topology and Dynamics, Novosibirsk State University (grant No. 14.Y26.31.0025 of the government of the Russian Federation). The second named author was supported by the program "Leading Scientific Schools" (grant no. NSh-6399.2018.1, Agreement No. 075-02-2018-867) and by the Russian Foundation for Basic Research (grant No. 19-01-00775-a). The fourth named author was supported by the program "Leading Scientific Schools" (grant no. NSh-6399.2018.1, Agreement No. 075-02-2018-867) and by the Russian Foundation for Basic Research (grant No. 18-01-00398-a).

Acknowledgments

Contents

PART 1

Introduction

Chapter 1

Groups. Small Cancellations.
Greendlinger Theorem

In the present chapter we discuss certain basic notions of combinatorial group theory. In particular, we recall the notion of *small cancellation conditions* for groups and study the diagrammatic method of describing such groups.

In a sense this approach (made possible by the van Kampen lemma (1933) giving a geometric interpretation of group relations) is the first example of the main principle this book is devoted to: how to study different objects (in our case — groups) with geometric, "picture-valued" instruments; how to construct a picture which contains all the important information about the studied object.

1.1   Group diagrams language

In this section we discuss diagrammatic language of group description. This approach was first discovered and used by van Kampen [van Kampen, 1933]. The essence of his discovery was an interconnection between combinatorially-topological and combinatorially-group-theoretical notions. The gist of this approach is a presentation of groups by flat diagrams (that is, geometrical objects, — flat complexes, — on a plane or other surfaces, such as a sphere or a torus). We review this theory following [Olshanskii, 1989].

1.1.1   *Preliminary examples*

First, let us present several examples of the principle, which will be rigorously defined later in this section.

Example 1.1. Consider a group G with relations $a^3 = 1$ and $bab^{-1} = c$.

Clearly in such group we have $c^3 = 1$. This fact can be seen on the following diagram, see Fig. 1.1.

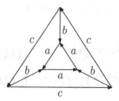

Fig. 1.1   Diagrammatic view of the $c^3 = 1$ relation

Indeed, if we go around the inner triangle of the diagram, we get the relation $a^3 = 1$ (to be precise, the boundary of this cell gives the left-hand side of the relation; if we encounter an edge whose orientation is compatible with the direction of movement, we read the letter which the edge is decorated with, otherwise we read the inverse letter; in this example we fix the counterclockwise direction of movement). Similarly, the quadrilaterals glued to the triangle lead the second given relation $c^{-1}bab^{-1} = 1$. Now if we look at the outer boundary of the diagram, we read $c^3 = 1$, and that is what we need to prove.

This simple example gives us a glimpse of the general strategy: we produce a diagram, composed of cells, along the boundary of which given relations can be read. Then the outer boundary of the diagram gives us a new relation which is a consequence of the given ones.

Let us consider a bit more complex example of the same principle.

Example 1.2. Consider a group G where the relation $x^3 = 1$ holds for every $x \in G$. It is a well-known theorem that in such a group every element a lies in some commutative normal subgroup $N \subset G$. Such situation arises, for example, in link-homotopy.

To prove this fact it is sufficient to prove that any element $y = bab^{-1}$ conjugate to a commutes with a. If that were the case, the subgroup N could be constructed as the one generated by all the conjugates of a.

So we need to prove that for every $b \in G$ the following holds:

$$a(bab^{-1}) = (bab^{-1})a,$$

or, equivalently

$$abab^{-1}a^{-1}ba^{-1}b^{-1} = 1.$$

This equality can be read walking clockwise around the outer boundary of the diagram in Fig. 1.2 composed of the relations $b^3 = 1, (ab)^3 = 1$, and $(a^{-1}b)^3 = 1$.

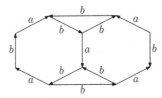

Fig. 1.2   Proof of the claim given in Example 1.2

1.1.2 *The notion of a diagram of a group*

Now we can move on to the explicit definitions of group diagrams and the overview of necessary results in that theory.

In accordance with [Olshanskii, 1989] in the present chapter a cell partitioning Δ of a surface S will be called a *map on S* for short. For some particular surfaces we will also use special names; for example, a map on a disc will be called a *disc map*, on an annulus an *annular map*, on a sphere or a torus — *spherical* or *toric*, respectively. Oriented sides of the partitioning are called *edges* of the map. Note that, if e is an edge of a map Δ, then e^{-1} is also its edge with the opposite orientation (consisting of the same points of the surface S as a side of the partitioning Δ).

Now consider an oriented surface S with a given map Δ and let us fix an orientation on its cells — e.g. let us walk around the boundary of each cell counterclockwise. In particular, the boundary of a disc map will be read clockwise and for an annular map, one boundary component ("exterior") will be read clockwise, and another ("interior") — counterclockwise.

Let a boundary component Y of a map or a cell consist of n sides. Walking around this component in accordance with the chosen orientation, we obtain a sequence of edges $e_1, \ldots, e_n$ forming a loop. This loop will be called a *contour* of the map or the cell. In particular, a disc map has one contour, and an annular map has two contours (exterior and interior). Contours are considered up to a cyclic permutation, that is, every loop $e_i \ldots e_n e_1 \ldots e_{i-1}$ gives the same contour. A contour of a cell Π will be denoted by $\partial \Pi$ and we will write $e \in \partial \Pi$ if an edge e is a part of the contour $\partial \Pi$ and we will call this situation "the edge e lies in the contour

$\partial\Pi$". Note that, even if an edge e lies in a contour $\partial\Pi$, its inverse e^{-1} does not necessarily lie in that contour. For example, in the situation depicted in Fig. 1.1 an edge a lies in the innermost triangular contour, but a^{-1} does not lie there.

Given a path p we may define a *subpath* in a natural way: a path q is a subpath of the path p if there exist two paths p_1, p_2 such that $p = p_1 q p_2$. In the same way a *subword* is defined.

Given an alphabet $\mathcal{A}$ we denote by $\mathcal{A}^1 = \mathcal{A} \cup \mathcal{A}^{-1} \cup \{1\}$, that is, the alphabet $\mathcal{A}^1$ consists of the letters from the alphabet $\mathcal{A}$, their inverses and the symbol "1". Let Δ be a map and for each edge e of the map Δ a letter $\varphi(e) \in \mathcal{A}^1$ is chosen (edges with $\varphi(e) \equiv 1$ are called *0-edges* of the map; other edges are called $\mathcal{A}-edges$).

Definition 1.1. If for each edge e of a map Δ the following relation holds:

$$\varphi(e^{-1}) \equiv (\varphi(e))^{-1},$$

then the map Δ is called a *diagram over* $\mathcal{A}$.

Here the symbol "$\equiv$" denotes the *graphical equality* of the words in the alphabet $\mathcal{A}$. In other words the notation $V \equiv W$ means that the words V and W are the same as sequences of letters of the alphabet. By definition we set $1^{-1} \equiv 1$.

When $p = e_1 \ldots e_n$ is a path in a diagram Δ over $\mathcal{A}$ let us define its *label* by the word $\varphi(p) = \varphi(e_1) \ldots \varphi(e_n)$. If the path is empty, that is $|p| = 0$, then we set $\varphi(p) \equiv 1$ by definition. As before, a label of a contour is defined up to a cyclic permutation (and thus forms a cyclic word).

Consider a group G with presentation

$$G = \langle \mathcal{A} \mid R = 1, R \in \mathcal{R} \rangle. \tag{1.1}$$

That means that $\mathcal{A}$ is a basis of a free group $F = F(\mathcal{A})$, $\mathcal{R}$ is a set of words in the alphabet $\mathcal{A} \cup \mathcal{A}^{-1}$ and there exists an epimorphism $\pi : F(\mathcal{A}) \rightarrow G$ such that its kernel is the normal closure of the subset $\{[r] \mid r \in \mathcal{R}\}$ of the set of words $F(\mathcal{A})$. Elements of $\mathcal{R}$ are called the *relations* of the presentation $\langle \mathcal{A}|\mathcal{R} \rangle$. We will always suppose that every element $r \in \mathcal{R}$ is a non-empty *cyclically-irreducible* word, that is, every element r of $\mathcal{R}$ or any of its cyclic permutations do not include subwords of the form ss^{-1} or $s^{-1}s$ for some $s \in F$.

Note that if a presentation of a group has a relation R, then it has all its cyclic permutations as relations as well.

Let Δ be a map over the alphabet $\mathcal{A}$.

Definition 1.2. A cell of the diagram Δ is called a $\mathcal{R}$-*cell* if the label of its contour is graphically equal (up to cyclic permutations) either to a word $R \in \mathcal{R}$, or its inverse R^{-1}, or to a word, obtained from R or from R^{-1} by inserting several symbols "1" between its letters.

This definition effectively means that choosing direction and the starting point of reading the label of the boundary of any cell of the map and ignoring all trivial labels (the ones with $\varphi(e) \equiv 1$) we can read exactly the words from the set of relations of the group G and nothing else.

Sometimes it proves useful to consider cell with effectively trivial labels. To be precise, we give the following definition.

Definition 1.3. A cell Π of a map Δ is called a *0-cell* if the label W of its contour $e_1 \ldots e_n$ graphically equals $\varphi(e_1) \ldots \varphi(e_n)$, where either $\varphi(e_i) \equiv 1$ for each $i = 1, \ldots, n$, or for some two indices $i \neq j$ the following holds:

$$\varphi(e_i) \equiv a \in \mathcal{A},$$

$$\varphi(e_j) \equiv a^{-1},$$

and

$$\varphi(e_k) \equiv 1 \ \forall k \neq i, j.$$

Finally, we can define a diagram of a group.

Definition 1.4. Let G be the group given by a presentation (1.1). A diagram Δ on a surface S over the alphabet $\mathcal{A}$ is called a *diagram on a surface S over the presentation (1.1)* (or a *diagram over the group G* for short) if every cell of this map is either an $\mathcal{R}$-cell or a 0-cell.

1.1.3   *The van Kampen lemma*

Earlier we gave two examples of diagrams used to show that a certain equality of the type $W = 1$ holds in a group given by its presentation. In fact this process is made possible by the following lemma due to van Kampen:

Lemma 1.1 (van Kampen [van Kampen, 1933]). *Let W be an arbitrary non-empty word in the alphabet $\mathcal{A}^1$. Then $W = 1$ in a group G given by its presentation (1.1) if and only if there exists a disc diagram over the presentation (1.1) such that the label of its contour graphically equals W.*

Proof. 1) First, let us prove that if Δ is a disc diagram over the presentation (1.1) with contour p, its label $\varphi(p) = 1$ in the group G.

If the diagram Δ contains exactly one cell Π, then in the free group F we have either $\varphi(p) = 1$ (if Π is a 0-cell) or $\varphi(p) = R^{\pm 1}$ for some $R \in \mathcal{R}$ (if Π is an $\mathcal{R}$-cell). In any case, $\varphi(p) = 1$ in the group G.

If Δ has more than one cell, then the diagram can be cut by a path t into two disc diagrams Δ_1, Δ_2 with fewer cells. We can assume that their contours are $p_1 t$ and $p_2 t^{-1}$ where $p_1 p_2 = p$. By induction it holds that $\varphi(p_1 t) = 1$ and $\varphi(p_2 t^{-1}) = 1$ in the group G. Therefore

$$\varphi(p) = \varphi(p_1 p_2) = \varphi(p_1 t t^{-1} p_2) = \varphi(p_1 t)\varphi(t^{-1} p_2) = 1$$

in the group G.

2) Now let us prove the inverse implication. To achieve it we need for a given word W such that $W = 1$ in the group G to construct a diagram Δ with contour p such that $\varphi(p) = W$.

It is well-known that in the free group F the word W equals a word $V = \prod_{i=1}^{n} X_i R_i^{\pm 1} X_i^{-1}$ for some $R_i \in \mathcal{R}$.

Construct a polygonal line t_1 on a plane and mark its segments with letters so that the line reads the word X_1. Connect a circle s_1 to the end of this line and mark it so that it reads $R_1^{\pm 1}$ if we walk around it clockwise. Now we glue 0-cells to t_1, s_1 and t_1^{-1} to obtain a set homeorphic to a disc. We obtain a diagram with contour of the form $e_1 \ldots e_k$ with $\varphi(e_1) \equiv 1 \equiv \varphi(e_k)$ and $\varphi(e_2 \ldots e_{k-1}) \equiv X_1 R_1^{\pm 1} X_1^{-1}$.

Construct the second diagram analogously for the word $X_2 R_2^{\pm 1} X_2^{-1}$ and glue it to the first diagram by the edge e_k.

Continue the process until we obtain a diagram Δ' such that $\varphi(\partial \Delta') \equiv V$, see Fig. 1.3.

Finally, gluing some 0-cells to the diagram Δ' we can transform the word V into the word W getting a diagram Δ such that $\varphi(\partial \Delta) \equiv W$. That completes the proof. $\square$

This lemma means that disc diagrams can be used to describe the words in a group which are equal to the neutral element of the group. It turns out that annular diagrams can be used in a similar manner.

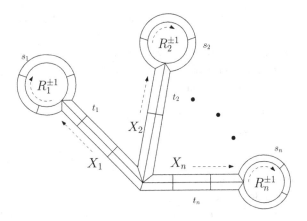

Fig. 1.3   The diagram Δ' with boundary label $\varphi(\partial\Delta') \equiv V$

Lemma 1.2 (Schupp [Schupp, 1968]). *Let V, W be two arbitrary non-empty words in the alphabet $\mathcal{A}^1$. Then they are conjugate in a group G given by its presentation (1.1) if and only if there exists an annular diagram over the presentation (1.1) such that it has two contours p and q with the labels $\varphi(p) \equiv V$ and $\varphi(q) \equiv W^{-1}$.*

Let p be a loop on a surface S such that its edges form a boundary of some subspace $\tilde{S} \subset S$ homeomorphic to a disc. Then the restriction of the cell partitioning Δ to the subspace $\tilde{S}$ is a cell partitioning on the space $\tilde{S}$ which is called a *submap* Γ of the map Δ. Note that by definition a submap is always a *disc* submap.

A *subdiagram* of a given diagram Δ is a submap Γ of the map Δ with edges endowed with the same labels as in the map Δ. Informally speaking, a subdiagram is a disc diagram cut out from a diagram Δ.

Let us state an additional important result about the group diagrams.

Lemma 1.3. *Let p and q be two (combinatorially) homotopic paths in a given diagram Δ over a presentation (1.1) of a group G. Then $\varphi(p) = \varphi(q)$ in the group G.*

In the next section the diagrammatic approach will be used to deal with *groups satisfying the small cancellation conditions*. In that theory a process of cancelling out pairs of cells of a diagram is useful (in addition to the usual process of cancelling out pairs of letters a and a^{-1} in a word). The problem is that two cells which are subject to cancellation do not always form a disc submap, so to define the cancellation process correctly we need to prepare the map prior to cancelling a suitable pair of cells. Let us define those notions in detail.

First, for a given cell partitioning Δ we define its *elementary transformations* (note that elementary transformations are defined for any cell partitioning, not necessarily diagram).

Definition 1.5. The following three procedures are called the *elementary transformations* of a cell partitioning Δ:

(1) If the degree of a vertex o of Δ equals 2 and this vertex is boundary for two different sides e_1, e_2, delete the vertex o and replace the sides e_1, e_2 by a single side $e = e_1 \cup e_2$;
(2) If the degree of a vertex o of a n-cell Π ($n \geq 3$) equals 1 and this vertex is boundary for a side e, delete the side e and the vertex o (the second boundary vertex of the side e persists);
(3) If two different cells Π_1 and Π_2 have a common side e, delete the side e (leaving its boundary vertices), naturally replacing the cells Π_1 and Π_2 by a new cell $\Pi = \Pi_1 \cup \Pi_2$.

Now we can define a *0-fragmentation* of a diagram Δ. First, consider a diagram Δ' obtained form the diagram Δ via a single elementary transformation. This transformation is called an *elementary 0-fragmentation* if one of the following holds:

(1) The elementary transformation is of type 1 and either $\varphi(e_1) \equiv \varphi(e)$, $\varphi(e_2) \equiv 1$ or $\varphi(e_2) \equiv \varphi(e)$, $\varphi(e_1) \equiv 1$ and all other labels are left unchanged;
(2) The elementary transformation is of type 2 and $\varphi(e) \equiv 1$;
(3) The elementary transformation is of type 3 and one of the cells Π_1, Π_2 became a 0-cell.

Definition 1.6. A diagram Δ' is a *0-fragmentation* of a diagram Δ if it is obtained from the diagram Δ by a sequence of elementary 0-fragmentations.

Note that 0-fragmentation does not change the number of $\mathcal{R}$-cells of a diagram.

Now consider an oriented diagram over a presentation (1.1). Let there be two $\mathcal{R}$-cells Π_1, Π_2 such that for some 0-fragmentation Δ' of the diagram Δ the copies Π_1', Π_2' of the cells Π_1, Π_2 have vertices O_1, O_2 with the following property: those vertices can be connected by a path ξ without selfcrossings such that $\varphi(\xi) = 1$ in the free group F and the labels of the contours of the cells Π_1', Π_2' beginning in O_1 and O_2 respectively are mutually inverse in the group F. In that case the pair $\{\Pi_1, \Pi_2\}$ is called *cancelable* in the diagram Δ.

Such pairs of cells are called cancelable because if a diagram Δ over a group G on a surface S has a pair of cancelable $\mathcal{R}$-cells, there exists a diagram Δ' over the group G with two fewer $\mathcal{R}$-cells on the same surface S. Moreover, if the surface S has boundary, then the cancellation of cells of the diagram Δ leaves the labels of its contours unchanged.

Given a diagram Δ and performing cell cancellation we get a diagram Δ' with no cancelable pairs of cells. Such diagrams are called *reduced*. Since this process of reduction does not change the boundary label of a diagram, we obtain the following enhancements of Lemma 1.1 and Lemma 1.2:

Theorem 1.1. *Let W be an arbitrary non-empty word in the alphabet $\mathcal{A}^1$. Then $W = 1$ in a group G given by its presentation (1.1) if and only if there exists a reduced disc diagram over the presentation (1.1) such that the label of its contour graphically equals W.*

Theorem 1.2. *Let V, W be two arbitrary non-empty words in the alphabet $\mathcal{A}^1$. Then they are conjugate in a group G given by its presentation (1.1) if and only if there exists a reduced annular diagram over the presentation (1.1) such that it has to contours p and q with the labels $\varphi(p) \equiv V$ and $\varphi(q) \equiv W^{-1}$.*

1.1.4 *Unoriented diagrams*

In the present section we introduce a notion of *unoriented diagrams* — a slight modification of van Kampen diagrams which is useful in the study of a certain class of groups.

Consider a diagram Δ over a group G with a presentation (1.1). With the alphabet $\mathcal{A}$ we associate an alphabet $\bar{\mathcal{A}}$ which is in a bijection with the alphabet $\mathcal{A}$ and is an image of the natural projection $\pi : \mathcal{A}^1 \to \bar{\mathcal{A}}$ defined as

$$\pi(a) = \pi(a^{-1}) = \bar{a}$$

for all $a \in \mathcal{A}$ with $\bar{a}$ being the corresponding element of $\bar{\mathcal{A}}$.

Now we take the diagram Δ and "forget" the orientation of its edges. The resulting 1-complex will be called an *unoriented diagram over the group* G and denoted by $\bar{\Delta}$. All definitions for diagrams (such as disc and annular diagrams, cells, contours, labels, etc.) are repeated verbatim for unoriented diagrams.

Since each edge of the diagram Δ was decorated with a letter from the alphabet $\mathcal{A}$ and we obtained $\bar{\Delta}$ from Δ just by forgetting the orientation of the edges, each edge of the diagram $\bar{\Delta}$ is decorated with an element of the alphabet $\mathcal{A}$ as well. Therefore, walking around the boundary of a cell in a chosen direction we obtain a sequence of letters but, unlike the oriented case, we do not have an orientation of the edges to determine the sign of each appearing letter. Therefore we shall say that the label $\varphi(\partial\Pi)$ of the contour of a cell Π of the diagram $\bar{\Delta}$ is a cyclic word in the alphabet $\bar{\mathcal{A}}$.

Given a word $\bar{w} = \bar{a}_1 \ldots \bar{a}_n$ in the alphabet $\bar{\mathcal{A}}$ we can produce 2^n words in the alphabet $\mathcal{A}^1$ of the form $a_1^{\varepsilon_1} \ldots a_n^{\varepsilon_n}$ with $\varepsilon_i \in \{+1, -1\}$. We shall call each of those words a *resolution* of the word $\bar{w}$.

Unoriented diagrams are very useful when describing group presentations such that the relation $a^2 = 1$ holds for all generators of the group, in other words, groups with the presentation

$$G = \langle \mathcal{A} | R = 1, R \in \mathcal{R}; a^2 = 1, a \in \mathcal{A} \rangle. \tag{1.2}$$

In fact, the following analog of the van Kampen lemma holds:

Lemma 1.4. *Let W be an arbitrary non-empty word in the alphabet $\mathcal{A}^1$. Then $W = 1$ in a group G given by its presentation (1.2) if and only if there exists a reduced unoriented disc diagram over the presentation (1.2) such that there is a resolution of the label of its contour which graphically equals W.*

Proof. First let W be a non-empty word in the alphabet $\mathcal{A}^1$ such that $W = 1$ in the group G. Let us show that there exists an unoriented diagram with the corresponding label of its contour.

Since $W = 1$, due to the strong van Kampen lemma (Theorem 1.1) there exists a reduced disc diagram over the presentation (1.2) such that the label of its contour graphically equals W. Denote this diagram by Δ. Now transform every edge e_i of this diagram into a bigon with the label $\varphi(e_i)^2$. Note that the result of this transformation is still a disc diagram. Indeed, we replaced every edge with a $\mathcal{R}$-cell (since $\varphi(e_i)^2 = 1$ in the group G) and reversed the orientation of some edges in the boundary contours of the cells of the diagram Δ but they remain $\mathcal{R}$-cells due to the relations $a^2 = 1$.

Now we "collapse" those 0-cells: replace every bigonal cell with the label of the form $\varphi(e)^2$ with an unoriented edge. Thus we obtain an unoriented diagram and the label of its contour by construction has a resolution graphically equal to W.

To prove the inverse implication, consider an unoriented diagram $\bar{\Delta}$ with the label of its contour $\bar{W} \equiv \varphi(\partial\bar{\Delta})$. We need to show that for each resolution W of the word $\bar{W}$ the relation $W = 1$ holds in the group G.

First, note that if this relation holds for one resolution of the word $\bar{W}$, it holds for every other resolution of this word. Indeed, due to the relations $a^2 = 1$ we may freely replace letters with their inverses: the relation $uav = ua^{-1}v$ holds in the group G for any subwords u, v and any letter a.

But by definition the diagram $\bar{\Delta}$ is obtained from some diagram Δ over the group G by forgetting the orientation of its edges. Therefore there exists a van Kampen diagram over the group G with the label of its contour graphically equal to some resolution R of the word $\bar{W}$. Therefore due to the van Kampen lemma $R = 1$ in the group G. And thus for every other resolution W of the word $\bar{W}$ the relation $W = 1$ holds in the group G. □

1.2 Small cancellation theory

1.2.1 *Small cancellation conditions*

We will introduce the notion of *small cancellation conditions* $C'(\lambda)$, $C(p)$ *and* $T(q)$. Roughly speaking, the conditions $C'(\lambda)$ and $C(p)$ mean that if one takes a free product of two relations, one gets "not too many" cancellations. To give exact definition of those objects, we need to define *symmetrisation* and *a piece*.

As before, let G be a group with a presentation (1.1). Given a set of relations $\mathcal{R}$ its *symmetrisation* $\mathcal{R}_*$ is a set of all cyclic permutations of the relations $r \in \mathcal{R}$ and their inverses. A word u is called *a piece* with respect to $\mathcal{R}$ if there are two distinct elements $w_1, w_2 \in \mathcal{R}_*$ with the common beginning u, that is $w_1 = uv', w_2 = uv''$. The length of a word w (the number of letters in it) will be denoted by $|w|$.

Definition 1.7. Let λ be a positive real number. A set of relations $\mathcal{R}$ is said to *satisfy the small cancellation condition* $C'(\lambda)$ if

$$|u| < \lambda|r|$$

for every $r \in \mathcal{R}_*$ and its any beginning u which is a piece with respect to $\mathcal{R}$.

Definition 1.8. Let p be a natural number. A set of relations $\mathcal{R}$ is said to *satisfy the small cancellation condition* $C(p)$ if every element of $\mathcal{R}_*$ is a product of at least p pieces.

The small cancellation conditions are given as conditions on the set of relations $\mathcal{R}$. If a group Γ admits a presentation $\langle S|\mathcal{R}\rangle$ with the set of relations satisfying the small cancellation condition, then the group Γ is said to satisfy this condition as well.

The condition $C'(\lambda)$ is sometimes called *metric* and the condition $C(p)$ — *non-metric*. Note that $C'(\frac{1}{n})$ always yields $C(n+1)$.

There exists one more small cancellation condition. Usually it is used together with either of the conditions $C'(\lambda)$ or $C(p)$.

Definition 1.9. Let q be a natural number, $q > 2$. A set of relations $\mathcal{R}$ is said to *satisfy the small cancellation condition* $T(q)$ if for every $l \in \{3, 4, \ldots, q-1\}$ and every sequence $\{r_1, r_2, \ldots, r_l\}$ of the elements of $\mathcal{R}_*$ the following holds: *if* $r_1 \neq r_2^{-1}, \ldots, r_{l-1} \neq r_l^{-1}, r_l \neq r_1^{-1}$, *then at least one of the products* $r_1 r_2, \ldots, r_{l-1} r_l, r_l r_1$ *is freely reduced.*

Remark 1.1. Those conditions have a very natural geometric interpretation in terms of van Kampen diagrams. Namely, the $C(p)$ condition means that every interior cell of the corresponding disc partitioning has at least p sides; the $T(q)$ condition means that every interior vertex of the partitioning has the degree of at least q.

Note that every set $\mathcal{R}$ satisfies the condition $T(3)$. Indeed, no interior vertex of a van Kampen diagram has degree 1 or 2.

1.2.2   *The Greendlinger theorem*

An important problem of combinatorial group theory is the *word problem*: the question whether for a given word W in a group G holds the equality $W = 1$ (or, more generally, whether two given words are equal in a given group). Usually the difficult question is to construct a group where a certain set of relations holds but a given word is nontrivial. In other words, to prove that a set of relations $\mathcal{R}$ does not yield $W = 1$ (note that there may exist groups where both the relations $\mathcal{R}$ and $W = 1$ hold due to the presence of *additional* relations). Small cancellation theory proves to be a very powerful and useful instrument in that situation. In particular, an important role in solution of that kind of problems plays the Greendlinger theorem which we will formulate in this section. We will always assume that the presentation (1.1) is symmetrised.

The Greendlinger theorem deals with the length of the common part of cells' boundary. Sometimes two cells are separated by 0-cells. Naturally, we should ignore those 0-cells. To formulate that accurately we need some preliminary definitions.

Let Δ be a reduced diagram over a symmetrised presentation (1.1). Two $\mathcal{A}$-edges e_1, e_2 are called *immediately close* in Δ if either $e_1 = e_2$ or e_1 and e_2^{-1} (or e_1^{-1} and e_2) belong to the contour of some 0-cell of the diagram Δ. Furthermore, two edges e and f are called *close* if there exists a sequence $e = e_1, e_2, \ldots, e_l = f$ such that for every $i = 1, \ldots, l - 1$ the edges e_i, e_{i+1} are immediately close.

Now for two cells Π_1, Π_2 a subpath p_1 of the contour of the cell Π_1 is a *boundary arc between* Π_1 *and* Π_2 if there exists a subpath p_2 of the contour of the cell Π_2 such that $p_1 = e_1 u_1 e_2 \ldots u_{n-1} e_n$, $p_2^{-1} = f_1 v_1 f_2 \ldots v_{n-1} f_n$, the paths u_i, v_i consist of 0-edges, e_i, f_i are $\mathcal{A}$-edges such that for each $i = 1, \ldots, n$ the edge f_i is close to the edge e_i. In the same way a boundary arc between a cell and the contour of the diagram Δ is defined.

Informally we can explain this notion in the following way. Intuitively, a boundary arc between two cells is the common part of the boundaries of those cells. The boundary arc defined above becomes exactly that if we collapse all 0-cells between the cells Π_1 and Π_2.

Now consider a *maximal* boundary arc, that is a boundary arc which does not lie in a longer boundary arc. It is called *interior* if it is a boundary arc between two cells, and *exterior* if it is a boundary arc between a cell and the contour $\partial \Delta$.

Remark 1.2. It is easy to see that the small cancellation conditions $C'(\lambda)$ and $C(p)$ have a natural geometric interpretation in terms of boundary arcs. Namely, an interior arc of a cell Π of a reduced diagram of a group satisfying the $C'(\lambda)$ condition has length smaller than $\lambda|\partial\Pi|$. Likewise, the $C(p)$ condition means that the boundary of every cell of the corresponding diagram consists of at least p arcs.

Now we can formulate the Greendlinger theorem.

Theorem 1.3. *Let Δ be a reduced disc diagram over a presentation of a group G satisfying the small cancellation condition $C'(\lambda)$ for some $\lambda \leq \frac{1}{6}$ and let Δ have at least one $\mathcal{R}$-cell. Further, let the label $\varphi(q)$ of the contour $q = \partial\Delta$ be cyclically irreducible and such that the cyclic word $\varphi(q)$ does not contain any proper subwords equal to 1 in the group G.*

Then there exists an exterior arc p of some $\mathcal{R}$-cell Π such that

$$|p| > \frac{1}{2}|\partial\Pi|.$$

Remark 1.3. If we formulate the Greendlinger theorem for unoriented diagrams, the theorem still holds.

Before proving this theorem, let us interpret it in terms of group presentation and the word problem. Consider a group G with a presentation (1.1) satisfying the small cancellation condition $C'(\lambda)$, $\lambda \le \frac{1}{6}$. Due to van Kampen lemma 1.1 (and its strengthening, Theorem 1.1) for a word $W = 1$ in the group G there exists a diagram with boundary label W. Boundary label of every $\mathcal{R}$-cell of the diagram by definition is some relation from the set $\mathcal{R}$ (or its cyclic permutation). Then, due to Greendlinger theorem 1.3 there is an $\mathcal{R}$-cell such that at least half of its boundary "can be found" in the boundary of the diagram. Thus we obtain the following corollary (sometimes it is also called the Greendlinger theorem):

Theorem 1.4 (Greendlinger [Greendlinger, 1961]).
Let G be a group with a presentation (1.1) satisfying the small cancellation condition $C'(\lambda)$, $\lambda \le \frac{1}{6}$. Let $W \in F$ be a nontrivial freely reduced word such that $W = 1$ in the group G. Then there exist a subword V of W and a relation $R \in \mathcal{R}$ such that V is also a subword of R and

$$|V| > \frac{1}{2}|R|.$$

This theorem is very useful in solving the word problem.

Example 1.3 (A.A. Klyachko). Consider the relation $R = [x, y]^2$ and the word

$$W = [x^{1000}, y^{1000}]^{1000}.$$

The question is, whether in every group with the relation R the equality $W = 1$ holds.

First, it is easy to see that the set of relations $\mathcal{R}_*$ obtained from R by symmetrisation satisfies the condition $C'(\frac{1}{6})$. Therefore, due to the Greendlinger theorem every word V such that $V = 1$ in the group G has a cyclic permutation such that both its irreducible form and some relation $\tilde{R} \in \mathcal{R}_*$ have a common subword p of length $|p| > \frac{1}{2}|\tilde{R}| = \frac{8}{2} = 4$.

On the other hand, the longest common subwords of the word $W = [x^{1000}, y^{1000}]^{1000}$ and any of the relations have length 2: those are

$$x^{-1}y^{-1}, xy, y^{-1}x, yx^{-1}.$$

Therefore we can state that there exists a group G where for some two elements $a, b \in G$ $[a, b]^2 = 1$ but $[a^{1000}, b^{1000}]^{1000} \neq 1$.

Now let us prove Theorem 1.3.

Proof. Let G be a group with a presentation (1.1) and Δ be a diagram as in the statement of the theorem. Since Δ is a disc diagram, we can place it on the sphere. We construct the dual graph Φ to the 1-skeleton of the diagram Δ in the following way.

Place a vertex of the new graph Φ into each $\mathcal{R}$-cell of the diagram Δ and one vertex O into the outer region on the sphere. To construct the edges of the graph Φ (note that they are not edges of a diagram) we perform the following procedure. For every interior boundary arc between cells Π_1 and Π_2 we chose an arbitrary $\mathcal{R}$-edge e_1 and an edge e_2 close to it such that e_2^{-1} lies in $\partial \Pi_2$. Now we connect the vertices of the graph Φ, lying inside the cells Π_1, Π_2 by a smooth curve γ transversally intersecting the interior of the edges e_1 and e_2 once and crossing all 0-cells lying between them. This curve γ is now considered as an edge of the graph Φ. This procedure is performed for every vertex of the graph Φ and every boundary arc (if the arc is exterior, we connect the vertex with the "exterior" vertex O).

The main idea of the proof is to study the dual graph Φ and to prove the necessary inequality by the reasoning of Euler characteristic of sphere.

First, note that among the faces of the graph Φ there are no 1-gons (because the boundary labels of the cells of the diagram Δ are cyclically irreducible) or 2-gons (because in the previous paragraphs we have constructed exactly one edge crossing every maximal boundary arc of Δ). Therefore every face of Φ is at least a triangle. Thus, denoting the number of vertices, edges and faces of the graph Φ by V, E and F respectively, and considering the usual Euler formula $V - E + F = 2$ we get the following inequality:

$$V > \frac{1}{3}E + 1. \tag{1.3}$$

Now, suppose that the statement of the theorem does not hold. For each edge of the graph Φ connecting the vertices o_i, o_j lying in the cells Π_i, Π_j of the diagram Δ we attribute this edge to each of those vertices with the coefficient $\frac{1}{2}$; if an edge connects a vertex o_k with the vertex O, we attribute it to the vertex o_k with the coefficient 1.

Fix an arbitrary vertex $o \neq O$. Denote the cell where the vertex o lies by Π and consider the following possibilities.

(1) Let the contour $\partial \Pi$ consist only of interior arcs. Due to the $C'(\frac{1}{6})$ condition the length of each of those arcs is smaller than $\frac{1}{6}|\partial \Pi|$ (see Remark 1.2). Therefore, their number is not smaller than seven and thus at least $\frac{7}{2}$ edges of the graph Φ is attributed to the vertex o.

(2) Let there be exactly one exterior arc p. Since we suppose that $|p| \leq \frac{1}{2}|\partial \Pi|$, the number of interior arcs of the contour of the cell Π is at least four. Therefore we attribute at least $1 + 4 \cdot \frac{1}{2} = 3$ edges to the vertex o.

(3) Let the cell Π have two exterior edges p_1, p_2. Note that the end of the arc p_1 cannot be the beginning of the arc p_2 (and vice versa) and they cannot be separated by 0-edges only because otherwise we could cut the diagram Δ with edges $f_1, \ldots, f_l$ such that $\varphi(f_i) = 1$ for all $i = 1, \ldots, l$ and due to van Kampen lemma obtain two proper subwords $\varphi(q_1), \varphi(q_2)$ of the word $\varphi(q)$ (where q is the contour of the diagram Δ) equal to 1 in the group G. That contradicts the condition of $\varphi(q)$ being cyclically irreducible.

Therefore, the cell Π has at least two distinct interior boundary arcs and there are at least $2 \cdot 1 + 2 \cdot \frac{1}{2} = 3$ edges attributed to the vertex o.

(4) If the cell Π has at least three exterior arcs, there are at least three edges of the graph Φ attributed to the vertex o.

Considering those possibilities for all vertices of the graph Φ except the "exterior" vertex O we see that

$$V - 1 \leq \frac{1}{3} E$$

and that contradicts the inequality (1.3). This contradiction completes the proof. □

1.3 Algorithmic problems and the Dehn algorithm

In this section we give a brief overview of two algorithmic problems in the combinatorial group theory which were already mentioned in the context of group diagrams — the *word problem* and the *conjugacy problem*.

Consider a group G given by its presentation (1.1) which is effective in the sense that for every word R we can algorithmically determine whether it lies in the set $\mathcal{R}$ or not. There are two fundamental problems in the combinatorial group theory:

(1) **The word problem:** is there an algorithm which for any pair of words U, V in the group alphabet determines whether those words are equal in the group G or not?

(2) **The conjugacy problem:** is there an algorithm which for any pair of words U, V in the group alphabet determines whether those words are conjugate in the group G or not?

Naturally the word problem can be reformulated in the form

"is there an algorithm which for any word W in the group alphabet determines whether $W = 1$ is in the group G or not?"

because $U = V$ is in the group G if and only if $UV^{-1} = 1$ is in the group G.

It turns out that the small cancellation theory is a very useful instrument in the study of the word and conjugacy problems. In particular, the following algorithm (originally due to Dehn, see [Schupp, 1968]) solves the word problem in the groups with the condition $C'(\lambda)$, $\lambda \leq \frac{1}{6}$.

The Dehn algorithm. The algorithm is inductive, so we may assume that for any word V of the length less than n we can algorithmically determine whether $V = 1$ is in the group G or not.

Now let W be a word of length n. We may assume that W is cyclically irreducible: for any proper subword of the cyclic word W we may determine by induction whether it equals 1 in the group G or not. If such a subword exists, replacing it with 1 we reduce the length of the word W and by induction solve the word problem for W.

So consider a cyclically irreducible word W, $|W| = n$. By the Greendlinger theorem if $W = 1$ in the group G, then the cyclic word W has a subword X such that there exists a relation $R = XY \in \mathcal{R}$ and $|X| > \frac{1}{2}|R|$.

If W does not have such subwords, then we may state that $W \neq 1$ in the group G. If such a subword X exists, then replace it with the word Y^{-1} in the word W. We obtain a word W' such that $W = W'$ in the group G and $|W'| < |W|$, since $|Y| < |X|$. Therefore by induction we can determine whether $W' = 1$ in the group G. Thus the problem for the word W is solved.

The key element of this algorithm is, of course, the Greendlinger theorem which holds for the groups with the condition $C'(\lambda)$ for $\lambda \leq \frac{1}{6}$.

It is worth mentioning that Goldberg [Goldberg, 1978] proved that for $\lambda > \frac{1}{5}$ the condition $C'(\lambda)$ (and the condition $C(5)$ as well) does not give us any relevant information about the group: any group allows a presentation with these conditions. In particular, for such groups the word problem may be algorithmically unsolvable, since it was shown by P.S. Novikov [Novikov, 1955] that there exist finitely defined groups with algorithmically unsolvable word problem.

Note that the Dehn and Goldberg results leave the "gap" $(\frac{1}{6}, \frac{1}{5})$ for the parameter λ. In fact, Lyndon [Lyndon, 1966] constructed an algorithm for the word problem solution for all groups with the condition $C'(\lambda)$, $\lambda \leq \frac{1}{5}$ and for groups with the condition $C(6)$. It is also worth mentioning that the original Dehn algorithm does not work for the groups with the condition $C(6)$. Those results were obtained by using geometric methods which may be regarded as discreet analogues to the Gauß–Bonnet theorem.

1.4   The Diamond lemma

In the present section we briefly consider a well-known Newman lemma (also known as the Diamond lemma in the literature) [Newman, 1942]. This lemma is a very powerful tool for constructing a minimal form, proving uniqueness theorems and solving word and other problems.

Let A be a set and let $\to$ denote a binary relation on the set A.

Definition 1.10. A relation $\to$ is called *well-founded* (or *Noetherian*) if every non-empty subset $X \subset A$ has a minimal element, that is, there exists $a \in X$ such that the relation $a \to b$ does not hold for any $b \in X$.

Alternatively, one can say that a relation is well-founded if the set A does not contain any infinite decreasing chains in the sense of that relation.

For a relation $\to$ let us denote by $\dashrightarrow$ its reflexive transitive closure.

Definition 1.11. A relation $\to$ is called *locally confluent* if for every three elements $a, b, c \in A$ such that

$$a \to b, \ a \to c,$$

there exists an element $d \in A$ such that

$$b \dashrightarrow d, \ c \dashrightarrow d.$$

This condition is sometimes formulated in the form "every covering is bounded below" in the system $(A, \rightarrow)$.

Definition 1.12. The relation $\rightarrow$ is called *confluent* if for every three elements $a, b, c \in A$ such that
$$a \dashrightarrow b, \ a \dashrightarrow c,$$
there exists an element $d \in A$ such that
$$b \dashrightarrow d, \ c \dashrightarrow d.$$

Lemma 1.5 (The Diamond lemma [Newman, 1942]). *If a relation $\rightarrow$ is well-founded and locally confluent, then the relation $\rightarrow$ is confluent.*

Proof. Let the relation $\rightarrow$ be well-founded and locally confluent and let $a, b, c \in A$ be elements such that $a \dashrightarrow b$, $a \dashrightarrow c$. We need to prove the existence of an element $d \in A$ such that $b \dashrightarrow d$, $c \dashrightarrow d$.

Let us call two elements $y, z \in A$ *joinable* if there exists an element $h \in A$ such that $y \dashrightarrow h$ and $z \dashrightarrow h$. The proof is by the following induction made possible by the well-foundedness of the relation $\rightarrow$. Let us say that *the property P is satisfied for an element $x \in A$* and write $P(x)$ if any elements $y, z \in A$, such that $x \dashrightarrow y$ and $x \dashrightarrow z$, are joinable. Obviously the statement of the lemma is equivalent to $P(x)$ for all $x \in A$.

If we denote by $\rightarrow^+$ the transitive closure of the relation $\rightarrow$, the induction is of the form "to show $P(x)$ under the assumption $P(t)$ for all t such that $x \rightarrow^+ t$". This induction is valid because the relation $\rightarrow$ is well-founded.

Now let us consider an element $x \in A$ and the divergence $x \dashrightarrow y$, $x \dashrightarrow z$. If $x = y$ or $x = z$, y and z are joinable immediately. Otherwise, we have $x \rightarrow y_1 \dashrightarrow y$, $x \rightarrow z_1 \dashrightarrow z$. Due to local confluency of the relation $\rightarrow$, there exists an element u such that $y_1 \dashrightarrow u$, $z_1 \dashrightarrow u$. By induction, there exists an element v such that $y \dashrightarrow v$, $u \dashrightarrow v$. Finally, by induction again, there exists an element w such that $v \dashrightarrow w$, $z \dashrightarrow w$. Therefore the elements y and z are joinable. This construction is depicted in Fig. 1.4. □

This lemma essentially means that, if the conditions of well-foundedness and local confluency hold for a relation $\rightarrow$, then every connected component of the graph of the relation $\rightarrow$ contains a *unique* minimal element a and for every element b of that connected component the relation $b \dashrightarrow a$ holds.

Example 1.4. A simple, yet structurally very important example of the use of the diamond lemma is the identity problem and the conjugacy problem

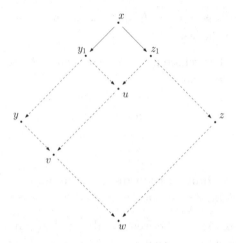

Fig. 1.4   Proof of the Diamond lemma

in the free group. Given a set of letters (alphabet) $a_1, \ldots, a_n$, we consider all possible words in this alphabet. Two words w, w' are in the relation $w \to w'$ if w' is obtained from w by contraction of two successive letters aa^{-1} or $a^{-1}a$. These transformations are called *elementary contractions*. The Noetherian property is evident.

How one can check the diamond lemma in this case? Let w be a word which admits two elementary contractions $w \to w'$ and $w \to w''$. If the contracted letters don't intersect, then we can contract them all in any order and get a word w''' which is a descendant for w' and w''. If the letters intersect then we have a subword $aa^{-1}a$ or $a^{-1}aa^{-1}$ in w which is contracted by two possible ways. In this case the words w' and w'' coincide.

Analogously, the conjugacy problem in the free group can be solved. For a word $w = w_1 \ldots w_k$ of length k, we consider words $w_{(j)} = w_{1+j}w_{2+j} \ldots w_{k+j}$ obtained from w by cyclic permutations $1 \to 1 + j, \ldots, k \to k + j$, where the sums are modulo k. It is clear that all these word are conjugate to w. The word w is called *cyclically contractible* if some of the words w_j are contractible. We introduce the following relation on the cyclic words: $\mathbf{w} \to \mathbf{w}'$ if there exist representatives w and w' of the words $\mathbf{w}, \mathbf{w}'$ such that $w \to w'$.

Evidently, two cyclic words $\mathbf{w}, \mathbf{w}'$ are conjugated if and only if there is a sequence of cyclic words $\mathbf{w} = \mathbf{w}_0 \to \mathbf{w}_1 \ldots \mathbf{w}_p = \mathbf{w}'$ such that for any successive words either $\mathbf{w}_j \to \mathbf{w}_{j+1}$ or $\mathbf{w}_{j+1} \to \mathbf{w}_j$. Thus, the conjugacy is the equivalence generated by the elementary relation $\to$.

For cyclic words of length ≥ 3 the conditions of the diamond lemma can be checked in the same manner as for ordinary words. The verification of the conditions for words of length 2 is obvious.

Remark 1.4. As we could see in the examples considered above, two word contractions $a \to b, b \to c$ can be *independent* if the contracted letters of one contraction are not involved in the other, and *dependent* if the both contractions use common letters. In the first case the commutative diagram of the diamond lemma can be constructed in one step: the word d is obtained from the word c "in the same way" as the word b is obtained from the word a, meanwhile the word d is obtained from the word b "in the same way" as the word c is obtained from the word a. Thus, we have $a \to b \to d$ and $a \to c \to d$.

Such cases of "independent contractions" are connected usually to the phenomenon of "far commutativity".

As for the case of *dependent contractions*, in the simplest situation dependence of the contractions $a \to b$ and $a \to c$ implies $b \equiv c$ where $\equiv$ is the identity relation. In the general case application of the diamond lemma to dependent contractions is much more difficult.

Chapter 2

Braid Theory

The present chapter is an introduction to the theory of braids, which has some nice intrinsically interesting properties. Here we highlight some ideas and definitions which will be important for our purposes later. In our overview of the basics of braid theory we will closely follow [Manturov, 2018].

2.1 Definitions of the braid group

Below, we are going to give some definitions of the braid groups and to discuss some of their properties.

First we consider a *geometrical definition* of the braid group. Consider the lines $\{y = 0, z = 1\}$ and $\{y = 0, z = 0\}$ in $\mathbb{R}^3$ and choose m points on each of these lines having $x-$coordinates $1, \ldots, m$.

Definition 2.1. An *m-strand braid* is a set of m non-intersecting smooth paths connecting the chosen points on the first line to the points on the second line (in arbitrary order), such that the projection of each of these paths to Oz represents a diffeomorphism.

These smooth paths are called *strands* of the braid.

An example of a braid is shown in Fig. 2.1. It is natural to consider braids up to isotopy in $\mathbb{R}^3$.

Definition 2.2. Two braids B_0 and B_1 are *equal* if they are *isotopic*; i.e., if there exists a continuous family $B_t, t \in [0, 1]$ of braids starting at B_0 and finishing at B_1.

Definition 2.3. The set of (isotopy classes of) m-strand braids generates

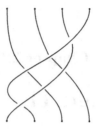

Fig. 2.1   A braid

a group. The operation in this group is just juxtaposing one braid under the other and rescaling the z-coordinate.

The *unit element* or the *unity* of this group is the braid represented by all vertical parallel strands. The inverse element for a given braid is just its mirror image; see Fig. 2.2.

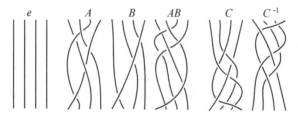

Fig. 2.2   Unity. Operations in the braid group

Definition 2.4. The *Artin m-strand braid group*[1] is the group of braids with the operation defined above. The braid group is denoted by $Br(m)$.

The second definition of the braid group considered in the present chapter is an *algebraic definition*.

Definition 2.5. The *m-strand braid group* is the group given by the presentation with $(m-1)$ generators $\sigma_1, \ldots, \sigma_{m-1}$ and the following relations

$$\sigma_i \sigma_j = \sigma_j \sigma_i$$

for $|i - j| \geq 2$ and

$$\sigma_i \sigma_{i+1} \sigma_i = \sigma_{i+1} \sigma_i \sigma_{i+1}$$

for $1 \leq i \leq m - 2$.

---

[1]In fact, there are other braid groups called *Brieskorn braid groups*. For more details see [Brieskorn, 1971; Brieskorn, 1973].

These relations are called *Artin's relations.*

Definition 2.6. Words in the alphabet of σ's and σ^{-1}'s will be referred to as *braid words.*

In fact, those two definitions give the same object:

Theorem 2.1. *The two definitions of the braid group $Br(m)$ given above are equivalent.*

Proof. In order to prove the equivalence of the definitions, let us introduce the notion of the *planar braid diagram.*

To see what it is, let us project a braid on the plane Oxz.

In the general case we obtain a diagram that can be described as follows.

Definition 2.7. A *braid diagram* (for the case of m strands) is a graph lying inside the rectangle $[1, m] \times [0, 1]$ endowed with the following structure and having the following properties:

(1) Points $(i, 0)$ and $(i, 1), i = 1, \ldots, m$, are vertices of valency one; there are no other graph vertices on the lines $\{z = 0\}$ and $\{z = 1\}$.
(2) All other graph vertices (crossings) have valency four.
(3) Unicursal curves; i.e., lines consisting of edges of the graph, passing from an edge to the opposite one, go from vertices with ordinate one and come to vertices with ordinate zero and descend monotonously.
(4) Each vertex of valency four is endowed with an over and undercrossing structure.

Obviously, all isotopy classes of geometrical braids can be represented by their planar diagrams. Moreover, after a small perturbation, all crossings of the braid can be set to have different z-coordinates.

It is easy to see that each element of the geometrical braid group can be decomposed into a product of the following generators σ_i's: the element σ_i for $i = 1, \ldots, m - 1$ consists of $m - 2$ segments connecting $(k, 1)$ and $(k, 0), k \neq i, k \neq i + 1$, and two segments $(i, 0) - (i + 1, 1), (i + 1, 0) - (i, 1)$, where the latter goes over the first one; see Fig. 2.3.

Different braid diagrams can generate the same braid. Thus we obtain some relations in $\sigma_1, \ldots, \sigma_{m-1}$.

Let us suppose that we have two equal geometrical braids B_1 and B_2. Let us represent the process of isotopy from B_1 to B_2 in terms of their planar diagrams. Each interval of this isotopy either does not change the

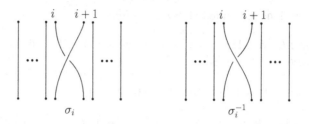

Fig. 2.3   Generators of the braid group

disposition of their vertex ordinates, or in this interval at least two crossings have (in a moment) the same ordinate; in the latter case the diagram becomes irregular.

We are interested in those moments where the algebraic description of our braid changes. We see that there are only three possible cases (all others can be reduced to these ones). The first case gives us the relation $\sigma_i \sigma_j = \sigma_j \sigma_i, |i - j| \geq 2$ (this relation is called *far commutativity*), or an equivalent relation $\sigma_i^{\pm 1} \sigma_j^{\pm 1} = \sigma_j^{\pm 1} \sigma_i^{\pm 1}, |i - j| \geq 2$, in the second case we get $\sigma_i \sigma_i^{-1} = 1$ (or $\sigma_i^{-1} \sigma_i = 1$), and in the third case we obtain one of the following three relations:

$$\sigma_i \sigma_{i+1} \sigma_i = \sigma_{i+1} \sigma_i \sigma_{i+1},$$
$$\sigma_i \sigma_{i+1} \sigma_i^{-1} = \sigma_{i+1}^{-1} \sigma_i \sigma_{i+1}, \quad \sigma_i^{-1} \sigma_{i+1} \sigma_i = \sigma_{i+1} \sigma_i \sigma_{i+1}^{-1}.$$

Obviously, each of the latter two relations can be obtained from the first one. This simple observation is left to the reader as an exercise. This completes the proof of the theorem.   □

2.2   The stable braid group and the pure braid group

For natural numbers $m < n$, there exists a natural embedding $Br(m) \subset Br(n)$: a braid from $Br(m)$ can be treated as a braid from $Br(n)$ where the last $(n - m)$ strands are vertical and unlinked (separated) with the others.

Definition 2.8. The *stable braid group Br* is the limit of groups $Br(n)$ as $n \to \infty$ with respect to these embeddings.

With each braid one can associate its permutation which takes an element k to l if the strand starting with the k-th upper point ends at the l-th lower point.

Definition 2.9. A braid is said to be *pure* if its permutation is identical. Obviously, pure braids generate a subgroup $PB_n \subset Br_n$.

An interesting problem is to find an explicit finite presentation of the pure braid group on n strands.

Here we shall present some concrete generators (according to [Artin, 1947]). A presentation of this group can be found in e.g. [Makanina, 1992].

There exists an algebraic *Reidemeister–Schreier method* that allows us to construct a presentation of a finite–index subgroup having a presentation of a finitely defined group, see e.g. [Crowell and Fox, 1963].

The following theorem holds.

Theorem 2.2. *The group $PB(m)$ is generated by braids*

$$b_{ij}, 1 \leq i < j \leq n \tag{2.1}$$

(see Fig. 2.4).

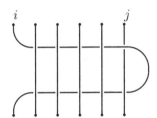

Fig. 2.4   Generator b_{ij} of the pure braid group

2.3   The curve algorithm for braids recognition

Below, we shall give a proof of the completeness of the invariant for the braid group elements invented by Artin, see [Gaifullin and Manturov, 2002].

The invariant to be constructed has a simple algebraic description as a map (non-homeomorphic) from the braid group $Br(n)$ to the n copies of the free group in n generators.

Several generalisations of this invariant, such as the spherical and cylindrical braid group invariants, are also complete. The key point of such a completeness is that these invariants originate from several curves, and the braid can be uniquely restored from these curves.

Moreover, this approach finally led to the algorithmic recognition of virtual braids due to Oleg Chterental [Chterental, 2015].

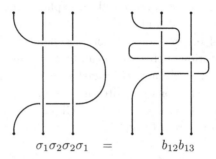

$$\sigma_1\sigma_2\sigma_2\sigma_1 \quad = \quad b_{12}b_{13}$$

Fig. 2.5   Decomposing a pure braid

2.3.1   *Construction of the invariant*

Let us begin with the definition of notions that we are going to use, and let us introduce the notation.

Definition 2.10. By an *admissible system of n curves* we mean a family of n non-intersecting non-self-intersecting curves in the upper half plane $\{y \geq 0\}$ of the plane Oxy such that each curve connects a point having ordinate zero with a point having ordinate one and the abscissas of all curve ends are integers from 1 to n. All points $(i,1)$, where $i = 1,\ldots,n$, are called *upper points*, and all points $(i,0), i = 1,\ldots,n$, are called *lower points*.

Definition 2.11. Two admissible systems of n curves A and A' are *equivalent* if there exists a homotopy between A and A' in the class of curves with fixed endpoints lying in the upper half plane such that no interior point of any curve can coincide with any upper or lower point during the homotopy.

Analogously, the equivalence is defined for one curve (possibly, self-intersecting) with fixed upper and lower points: during the homotopy in the upper half plane no interior point of the curve can coincide with an upper or lower point.

In the sequel, admissible systems will be considered up to equivalence.

Let β be a braid diagram on the plane connecting the set of lower points $\{(1,0),\ldots,(n,0)\}$ with the set of upper points $\{(1,1),\ldots,(n,1)\}$. Consider the topmost crossing C of the diagram β and push the lower branch along the upper branch to the upper point of it as shown in Fig. 2.6.

Naturally, this move spoils the braid diagram: the result, shown in Fig. 2.6 is not a braid diagram. The advantage of this "diagram" is that we have a smaller number of crossings.

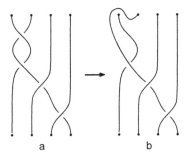

Fig. 2.6   Pushing the upper crossing

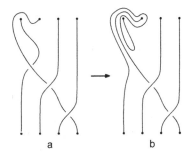

Fig. 2.7   Pushing the next crossing

Now, let us do the same with the next crossing. Namely, let us push the lower branch along the upper branch towards the end. If the upper branch is deformed during the first move, we push the lower branch along the deformed branch (see Fig. 2.7).

Reiterating this procedure for all crossings (until the lowest one), we get an admissible system of curves. Denote its equivalence class by $f(\beta)$.

Theorem 2.3. *The function f is a braid invariant; i.e., for two diagrams β, β' of the same braid we have $f(\beta) = f(\beta')$.*

Proof. Having two braid diagrams, we can write the corresponding braid-words, and denote them by the same letters β, β'. We must prove that the admissible system of curves is invariant under braid isotopies. As we shall see, this statement is very simple from the algebraic point of view, but here it is useful for our purposes to consider it using curves techniques.

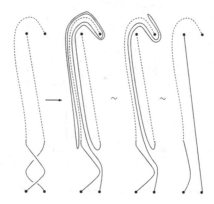

Fig. 2.8   Invariance of f under the second Reidemeister move

The invariance under the commutation relations $\sigma_i\sigma_j = \sigma_j\sigma_i, |i-j| \geq 2$, is obvious: the order of pushing two "far" branches does not change the result.

The invariance under $\sigma_i\sigma_i^{-1} = e$ can be readily checked; see Fig. 2.8.

In the leftmost part of Fig. 2.8, the dotted line indicates the arbitrary behaviour for the upper part of the braid diagram. The rightmost part of Fig. 2.8 corresponds to the system of curves without $\sigma_i\sigma_i^{-1}$.

Finally, the invariance under the transformation $\sigma_i\sigma_{i+1}\sigma_i \to \sigma_{i+1}\sigma_i\sigma_{i+1}$ is shown in Fig. 2.9. In the upper part (over the horizontal line) in Fig. 2.9 we demonstrate the behaviour of $f(A\sigma_i\sigma_{i+1}\sigma_i)$, and in the lower part in Fig. 2.9 we show that of $f(A\sigma_{i+1}\sigma_i\sigma_{i+1})$ for an arbitrary braid A. In the middle-upper part, a part of the curve is shown by a dotted line. By removing it, we get the upper-right picture which is just the same as the lower-right picture.

Note that the behaviour of the diagram in the upper part A of the braid diagram is arbitrary. For the sake of simplicity it is pictured by three straight lines.

Thus we have proved that $f(A\sigma_i\sigma_{i+1}\sigma_i) = f(A\sigma_{i+1}\sigma_i\sigma_{i+1})$.

This completes the proof of the theorem. $\square$

In fact, the following statement holds.

Theorem 2.4. *The function f is a complete invariant.*

In order to prove Theorem 2.4, we should be able to restore the braid from its admissible system of curves.

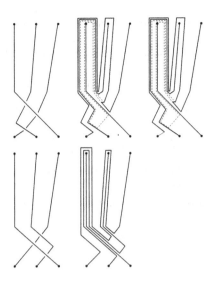

Fig. 2.9 Invariance of f under the third Reidemeister move

In the sequel, we shall deal with braids whose end points are $(i, 0, 0)$ and $(j, 1, 1)$ with all strands coming upwards with respect to the third projection coordinates. They obviously correspond to standard braids with upper points $(j, 0, 1)$. This correspondence is obtained by moving neighbourhoods of upper points along Oy.

Consider a braid B and consider the plane $P = \{y = z\}$ in $Oxyz$. Let us place B in a small neighbourhood of P in such a way that its strands connect points $(i, 0, 0)$ and $(j, 1, 1), i, j = 1, \ldots, n$. Both projections of this braid on Oxy and Oxz are braid diagrams. Denote the braid diagram on Oxy by β.

The next step now is to transform the projection on Oxy without changing the braid isotopy type; we shall just deform the braid in a small neighbourhood of a plane parallel to Oxy.

It turns out that one can change abscissas and ordinates of some intervals of strands of b in such a way that the projection of the transformed braid on Oxy constitutes an admissible system of curves for β.

Indeed, since the braid lies in a small neighbourhood of P, each crossing on Oxy corresponds to a crossing on Oxz. Thus, the procedure of pushing a branch along another branch in the plane parallel to Oxy deletes a crossing on Oxy, preserving that on Oxz.

Thus, we have described the geometric meaning of the invariant f.

Definition 2.12. By an *admissible parametrisation* (in the sequel, all para-metrisations are thought to be smooth) of an admissible system of curves we mean a set of parametrisations for all curves by parameters $t_1, \ldots, t_n$ such that at the upper points all t_i are equal to one, and at the lower points t_i are equal to zero.

Any admissible system A of n curves with an admissible parametrisation T generates a braid representative: each curve on the plane becomes a braid strand when we consider its parametrisation as the third coordinate. The corresponding braid has end points $(i, 0, 0)$ and $(j, 1, 1)$, where $i, j = 1, \ldots, n$. Denote it by $g(A, T)$.

Lemma 2.1. *The result $g(A, T)$ does not depend on T.*

Proof. Indeed, let us consider two admissible parametrisations T_1 and T_2 of the same system A of curves. Let $T_i, i \in [1, 2]$, be a continuous family of admissible parametrisations between T_1 and T_2, say, defined by the formula $T_i = (i - 1)T_1 + (2 - i)T_2$. For each $i \in [1, 2]$, the curves from T_i do not intersect each other, and for each $i \in [1, 2]$ the set of curves $g(A, T_i)$ is a braid, thus $g(A, T_i)$ generates the desired braid isotopy.   □

Thus, the function $g(A) \equiv g(A, T)$ is well defined.

Now we are ready to prove the main theorem. First, let us prove the following lemma.

Lemma 2.2. *Let A, A' be two equivalent admissible systems of n curves. Then $g(A) = g(A')$.*

Proof. Let $A_t, t \in [0, 1]$, be a homotopy from A to A'. For each $t \in [0, 1]$, A_t is a system of curves (possibly, not admissible). For each curve $\{a_{i,t}, i = 1, \ldots, n, t \in [0, 1]\}$ choose points $X_{i,t}$ and $Y_{i,t}$, such that the interval from the upper point (upper interval) of the curve to $X_{i,t}$ and the interval from $Y_{i,t}$ to the lower point (lower interval) do not contain intersection points. Denote the remaining part of the curve (the middle interval) between $X_{i,t}$ and $Y_{i,t}$ by $S_{i,t}$. Now, let us parametrise all curves for all t by parameters $\{s_{i,t} \in [0, 1], i = 1, \ldots, n\}$ in the following way: for each t, the upper point of each curve has parameter $s = 1$, and the lower point has parameter $s = 0$. Besides, we require that for $i < j$ and for each $x \in S_{i,t}, y \in S_{j,t}$ we have $s_{i,t}(x) < s_{j,t}(y)$. This is possible because we can vary parametrisations of upper and lower intervals on $[0, 1]$; for instance, we parametrise the middle interval of the j-th strand by a parameter in $[\frac{j}{n+2}, \frac{j+1}{n+2}]$.

It is obvious that for $t = 0$ and $t = 1$ these parametrisations are admissible for both A and A'. For each $t \in [0, 1]$ the parametrisation s generates a braid B_t in $\mathbb{R}^3$: we just take the parameter $s_{i,t}$ for the strand $a_{i,t}$ as the third coordinate. The strands do not intersect each other because parameters for different intervals cannot be equal to each other.

Thus the system of braids B_t induces a braid isotopy between $B_0 = g(A)$ and $B_1 = g(A')$. $\qquad \square$

So, the function g is well defined on the equivalence classes of admissible systems of curves.

Now, to complete the proof of the main theorem, we need only to prove the following lemma.

Lemma 2.3. *For any braid B, we have $g(f(B)) = B$.*

Proof. Indeed, let us place B in a small neighbourhood of the "inclined plane" P in such a way that the ends of B are $(i, 0, 0)$ and $(j, 1, 1)$, $i, j = 1, \ldots, n$.

Consider $f(B)$ that lies in Oxy. It is an admissible system of curves for B. So, there exists an admissible parametrisation that restores B from $f(B)$. By Lemma 2.1, each admissible parametrisation of $f(B)$ generates B. So, $g(f(B)) = B$. $\qquad \square$

2.3.2 Algebraic description of the invariant

The general situation in the construction of a complete invariant is the following: one constructs a new object that is in one-to-one correspondence with the described object. However, the new object might also be badly recognisable.

Now, we shall describe our invariant algebraically. It turns out that the final result is very easy to recognise. Namely, the problem is reduced to the recognition problem of elements in a free group. So, there exists an injective map from the braid group to the (n copies of) the free group with n generators that is not homomorphic.

Each braid B generates a permutation. This permutation can be uniquely restored from any admissible system of curves corresponding to B. Indeed, for an admissible system A of curves, the corresponding permutation maps i to j, where j is the ordinate of the strand with the upper point $(i, 1)$. Denote this permutation by $p(A)$. It is obvious that $p(A)$ is invariant under equivalence of A.

Let n be an integer. Consider the free product G of n copies of the group $\mathbb{Z}$ with generators $a_1, \ldots, a_n$. Denote by E_i the right residue classes in G by $\{a_i\}$; i.e., $g_1, g_2 \in G$ represent the same element of E_i if and only if $g_1 = a_i^k g_2$ for some k.

Definition 2.13. An *n-system* is a set of elements $e_1 \in E_1, \ldots, e_n \in E_n$;

An *ordered n-system* is an n-system together with a permutation from S_n.

Proposition 2.1. *There exists an injective map from equivalence classes of admissible systems of curves to ordered n-systems.*

Proof. Since the permutation for equivalent admissible systems of curves is the same, we can fix the permutation $s \in S_n$ and consider only equivalence classes of admissible systems of curves with permutation s (i.e., with all lower points fixed depending on the upper points in accordance with s). Thus we only have to show that there exists an injective map from the set of admissible systems of n curves with fixed lower points to n-systems.

To complete the proof of the proposition, it suffices to prove the following.

Lemma 2.4. *Equivalence classes of curves with fixed points $(i, 1)$ and $(j, 0)$ are in one-to-one correspondence with E_i.*

Proof. Denote $\mathbb{R}^2 \setminus \cup_{i=1,\ldots,n} \{(i, 1)\}$ by P_n. Obviously, $\pi_1(P_n) \cong G$. Consider a small circle C centred at $(i, 1)$ for some i with the lowest point X on it. Let ρ be a curve with endpoints $(i, 1)$ and $(j, 0)$. Without loss of generality, assume that ρ intersects C in a finite number of points. Let Q be the first such point that one meets while walking along ρ from $(i, 1)$ to $(j, 0)$. Thus we obtain a curve ρ' coming from C to $(j, 0)$. Now, let us construct an element of $\pi_1(P_n, X)$. First it comes from X to Q along C clockwise. Then it goes along ρ until $(j, 0)$. After this, it goes along Ox to the point $(i, 0)$. Then it goes vertically upwards till the intersection with C in X. Denote the constructed element by $W(\rho)$.

If we deform ρ outside C, then we obtain a continuous deformation of the loop, thus $W(\rho)$ stays the same as an element of the fundamental group. The deformations of ρ inside C might change $W(\rho)$ by multiplying it by a_i on the left side. So, we have constructed a map from equivalence classes of curves with fixed points $(i, 1)$ and $(j, 0)$ to E_i.

The inverse map can be easily constructed as follows. Let W be an element of $\pi_1(P_n, X)$. Consider a loop L representing W. Now consider the curve L' that first goes from $(i, 1)$ to X vertically, then goes along L', after this goes vertically downwards until $(i, 0)$ and finally, horizontally until $(j, 0)$. Obviously, $W(L') = W$. It is easy to see that for different representatives L of W we obtain the same L'. Besides, for $L_1 = a_i L_2$, the curves L'_1 and L'_2 are isotopic. This completes the proof of the lemma. $\square$

Thus, for a fixed permutation s, admissible systems of curves can be uniquely encoded by n-systems, which completes the proof of the proposition. $\square$

Now, we see that this invariant is a quite simple object: elements of E_i can be easily compared.

Let us describe the algebraic construction of the invariant f in more detail.

Let β be a braid word, written as a product of generators $\beta = \sigma_{i_1}^{\varepsilon_1} \ldots \sigma_{i_k}^{\varepsilon_k}$, where each ε_j is either $+1$ or -1; $1 \le i_j \le n-1$ and $\sigma_1, \ldots, \sigma_{n-1}$ are the standard generators of the braid group $Br(n)$.

We are going to construct the n-system step-by-step while writing the word β. First, let us write n empty words (in the alphabet $a_1, \ldots, a_n$). Let the first letter of β be σ_j. Then all words except for the word e_{j+1} should stay the same (i.e., empty), and the word e_{j+1} becomes a_j^{-1}. If the first crossing is negative; i.e., σ_j^{-1} then all words except e_j stay the same and e_j converts to a_{j+1}. While considering each next crossing, we do the following. Let the crossing be $\sigma_j^{\pm 1}$. Let p and q be the numbers of strands coming from the left side and from the right side respectively. If this crossing is positive; i.e., σ_j, then all words except e_q stay the same, and e_q becomes $e_q e_p^{-1} a_p^{-1} e_p$. If it is negative, then all crossings except e_p stay the same, and e_p becomes $e_p e_q^{-1} a_q e_q$. After processing all the crossings, we get the desired n-system.

Example 2.1. For the trivial braid written as $\sigma_1 \sigma_2 \sigma_1 \sigma_2^{-1} \sigma_1^{-1} \sigma_2^{-1}$ the construction operation works as follows:

$$(e, e, e) \to (e, a_1^{-1}, e) \to (e, a_1^{-1}, a_1^{-1}) \to (e, a_1^{-1}, a_2^{-1} a_1^{-1}) \to$$

$$(e, e, a_2^{-1} a_1^{-1}) \to (e, e, a_2^{-1}) \to (e, e, e).$$

A priori these words may be non-trivial; they must only represent trivial residue classes, say, (a_1, a_2^2, a_3^{-1}).

However, it is not the case.

Proposition 2.2. *For the trivial braid, the algebraic algorithm described above gives trivial words.*

Proof. Indeed, the algebraic number of occurrences of a_i in the word e_i equals zero. This can be easily proved by induction on the number of crossings. In the initial position all words are trivial. The induction step is obvious. Thus, the final word e_i equals a_i^p, where $p = 0$.   □

From this approach, one can easily obtain the well known invariant (action) as follows. Instead of a set of n words $e_1, \ldots, e_n$, one can consider the words $e_1 a_1 e_1^{-1}, \ldots, e_n a_n e_n^{-1}$. Since e_i's are defined up to a multiplication by a_i's on the left, the obtained elements are well defined in the free groups. Besides, these elements $\mathfrak{E}_i = e_i a_i e_i^{-1}$ are generators of the free group. This can be checked by a step-by-step confirmation. Thus, for each braid B we obtain a set $Q(B)$ of generators for the braid group. So, the braid B defines a transformation of the free group $\mathbb{Z}^{*n}$. It is easy to see that for two braids, the transformation corresponding to the product equals the composition of transformation. Thus, one can speak about the *action of the braid group on the free group*. Since f is a complete invariant, this action has an empty kernel.

Definition 2.14. This action is called *the Hurwitz action* of the braid group Br_n on the free group $\mathbb{Z}^{*n}$.

2.4   Virtual braids. Inclusion of classical braids into virtual braids

Just as classical knots can be obtained as closures of classical braids, virtual knots can be similarly obtained by closing *virtual braids*. Virtual braids were suggested by V. V. Vershinin, [Vershinin, 2001].

2.4.1   *Definitions of virtual braids*

Virtual braids have a purely combinatorial definition. Namely, one takes virtual braid diagrams and factorises them by virtual Reidemeister moves (all moves with the exception of the first classical and the virtual moves; the latter moves do not occur).

Definition 2.15. A *virtual braid diagram* on n strands is a graph lying in $[1, n] \times [0, 1] \subset \mathbb{R}^2$ with vertices of valency one (there should be exactly $2n$ such vertices with coordinates $(i, 0)$ and $(i, 1)$ for $i = 1, \ldots, n$) and a finite number of vertices of valency four. The graph is a union of n smooth curves without horizontal tangent lines connecting a point on the line $\{y = 1\}$ with those on the line $\{y = 0\}$; their intersection makes crossings (four-valent vertices). Each crossing should be either endowed with a structure of over- or undercrossing (as in the case of classical braids) or marked as a virtual one (by encircling it).

Definition 2.16. A *virtual braid* is an equivalence class of virtual braid diagrams by planar isotopies and all virtual Reidemeister moves (see Figs. 3.2 and 3.13) except the first classical move and the first virtual move.

A virtual braid diagram is called *regular* if any two different crossings have different ordinates.

Remark 2.1. We shall also treat braid words and braids familiarly, saying, e.g. "a strand of a braid word" and meaning "a strand of the corresponding braid".

Let us describe the construction of the word by a given regular virtual braid diagram as follows. Let us walk along the axis Oy from the point $(0, 1)$ to the point $(0, 0)$ and watch all those levels $z = t \in [0, 1]$ having crossings. Each such crossing permutes strands $\#i$ and $\#(i + 1)$ for some $i = 1, \ldots, n - 1$. If the crossing is virtual, then we write the letter ζ_i, if not, we write σ_i if overcrossing is the "northeast-southwest" strand, and σ_i^{-1} otherwise.

Thus, we have got a braid word by a given regular virtual braid diagram; see Fig. 2.10. Let us describe this presentation of virtual braids formally.

Like classical braids, virtual braids form a group (with respect to juxtaposition and rescaling the vertical coordinate). The generators of this group are:

$\sigma_1, \ldots, \sigma_{n-1}$ (for classical crossings) and $\zeta_1, \ldots, \zeta_{n-1}$ (for virtual crossings).

The inverse elements for the σ's are defined as in the classical case.

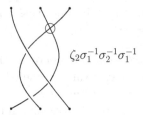

$$\zeta_2\sigma_1^{-1}\sigma_2^{-1}\sigma_1^{-1}$$

Fig. 2.10   A virtual braid diagram and the corresponding braid word

Obviously, for each $i = 1, \ldots, n-1$ we have $\zeta_i^2 = e$ (this follows from the second virtual Reidemeister move).

One can show that the following set of relations [Vershinin, 2001] is sufficient to generate this group:

(1) (Braid group relations):

$$\sigma_i\sigma_j = \sigma_j\sigma_i \quad \text{for } |i - j| \geq 2;$$

$$\sigma_i\sigma_{i+1}\sigma_i = \sigma_{i+1}\sigma_i\sigma_{i+1};$$

(2) (Permutation group relations):

$$\zeta_i\zeta_j = \zeta_j\zeta_i \quad \text{for } |i - j| \geq 2;$$

$$\zeta_i\zeta_{i+1}\zeta_i = \zeta_{i+1}\zeta_i\zeta_{i+1};$$

$$\zeta_i^2 = e;$$

(3) (Mixed relations):

$$\sigma_i\zeta_{i+1}\zeta_i = \zeta_{i+1}\zeta_i\sigma_{i+1};$$

$$\sigma_i\zeta_j = \zeta_j\sigma_i \quad \text{for } |i - j| \geq 2.$$

The proof of this fact is left to the reader.

2.4.2   *Invariants of virtual braids*

In this section, we are going to present an invariant of virtual braids proposed by the first-named author in [Manturov, 2008] and show that the classical braid group is a subgroup of the virtual one. For an elementary proof of this fact see [Manturov, 2016b]. More precisely, we give a generalisation of the complete braid invariant described before for the case of virtual braids. The new "virtual invariant" is quite strong. The question of whether the invariant is complete was answered negatively by O. Chterental [Chterental, 2015]. The completeness of the multi-variable extension of the invariant (see [Manturov, 2003]) is unknown.

Thus the main question is the word problem for the virtual braid group: how to recognise whether two different (regular) virtual braid diagrams β_1 and β_2 represent the same braid B.

Remark 2.2. The recognition problem for virtual braids was solved by O. Chterental [Chterental, 2017].

Given two braid diagrams one can apply the virtual braid group relations to one of those diagrams without getting the other diagram and one does not know whether he has to stop and say that they are not isomorphic or he has to continue.

A partial answer to this question is the construction of a virtual braid group invariant; i.e., a function on virtual braid diagrams (or braid words) that is invariant under all virtual braid group relations. In this case, if for an invariant f we have $f(\beta_1) \neq f(\beta_2)$, then β_1 and β_2 represent two different braids.

Here we give a generalisation of the complete classical braid group invariant for the case of virtual braids.

Let G be the free group in generators $a_1 \ldots, a_n, t$. Let E_i be the quotient set of right residue classes $\{a_i\}\backslash G$ for $i = 1, \ldots, n$.

Definition 2.17. A *virtual n-system* is a set of elements $\langle e_1 \in E_1, e_2 \in E_2, \ldots, e_n \in E_n \rangle$.

The aim of this subsection is to construct an invariant map f (non-homomorphic) from the set of all virtual n-strand braids to the set of virtual n-systems.

Let β be a braid word. Let us construct the corresponding virtual n-system $f(\beta)$ step-by-step. Namely, we shall reconstruct the function $f(\beta\psi)$ from the function $f(\beta)$, where ψ is σ_i or σ_i^{-1} or ζ_i.

First, let us take n residue classes of the unit element of G: $\langle e, e, \ldots, e \rangle$. This means that we have defined

$$f(e) = \langle e, e, \ldots, e \rangle.$$

Now, let us read the word β. If the first letter is ζ_i, then all words but e_i, e_{i+1} in the n-systems stay the same, e_i becomes equal to t and e_{i+1} becomes t^{-1} (here and in the sequel, we mean, of course, residue classes, e.g. $[t]$ and $[t^{-1}]$. But we write just t and t^{-1} for the sake of simplicity).

Now, if the first letter of our braid word is σ_i, then all classes but e_{i+1} stay the same, and e_{i+1} becomes a_i^{-1}. Finally, if the first letter is σ_i^{-1}, then the only changing element is e_i: it becomes a_{i+1}.

The procedure for each next letter (generator) is the following. Denote the index of this letter (the generator or its inverse) by i. Assume that the left strand of this crossing originates from the point $(p, 1)$, and the right one originates from the point $(q, 1)$. Let $e_p = P, e_q = Q$, where P, Q are some words representing the corresponding residue classes. After the crossing all residue classes but e_p, e_q should stay the same.

Then if the letter is ζ_i then e_p becomes $P \cdot t$, and e_q becomes $Q \cdot t^{-1}$. If the letter is σ_i, then e_p stays the same, and e_q becomes $QP^{-1}a_p^{-1}P$. Finally, if the letter is σ_i^{-1}, then e_q stays the same, e_p becomes $PQ^{-1}a_qQ$. Note that this operation is well defined.

Actually, if we take the words $a_p^l P, a_q^m Q$ instead of the words P, Q, then we get: in the first case

$$a_p^l Pt \sim Pt, a_q^m Qt^{-1} \sim Qt^{-1},$$

and in the second case we obtain

$$a_p^l P \sim P, \quad a_q^m QP^{-1}a_p^{-l}a_p^{-1}a_p^l P = a_q^m QP^{-1}a_p^{-1}P \sim QP^{-1}a_p^{-1}P.$$

In the third case we obtain

$$a_p^l PQ^{-1}a_q^{-m}a_q^{-1}a_q^m P = a_p^l PQ^{-1}a_q^{-1}Q \sim PQ^{-1}a_q^{-1}Q, \quad a_q^m Q \sim Q.$$

Thus, we have defined the map f from the set of all virtual braid diagrams to the set of virtual n-systems.

Theorem 2.5. *The function f, defined above, is a braid invariant. Namely, if β_1 and β_2 represent the same braid β, then $f(\beta_1) = f(\beta_2)$.*

Proof. We have to demonstrate that the function f defined on virtual braid diagrams is invariant under all virtual braid group relations. It suffices to prove that, for the words $\beta_1 = \beta\gamma_1$ and $\beta_2 = \beta\gamma_2$ where $\gamma_1 = \gamma_2$ is a relation we have proved, we can also prove $f(\beta_1) = f(\beta_2)$. During the proof of the theorem, we shall call it the A-statement.

Indeed, having proved this claim, we also have $f(\beta_1\delta) = f(\beta_2\delta)$ for arbitrary δ because the invariant $f(\beta_1\delta)$ (as well as $f(\beta_2\delta)$) is constructed step-by-step; i.e., knowing the value $f(\beta_1)$ and the braid word δ, we easily obtain the value of $f(\beta_1\delta)$. Hence, for braid words β, δ and for each braid group relation $\gamma_1 = \gamma_2$ we prove that $f(\beta\gamma_1\delta) = f(\beta\gamma_2\delta)$. This completes the proof of the theorem.

Let us return to the A-statement.

To prove the A-statement, we must consider all virtual braid group relations. The commutation relation $\sigma_i\sigma_j = \sigma_j\sigma_i$ for "far" i, j is obvious: all four strands involved in this relation are different, so the order of applying the operation does not affect on the final result. The same can be said about the other commutation relations, involving one σ and one ζ or two ζ's.

Now let us consider the relation $\zeta_i^2 = e$ which is pretty simple too.

Actually, let us consider a braid word β, and let the word β_1 be defined as $\beta\zeta_i^2$ for some i. Let $f(\beta) = (P_1, \ldots, P_n), f(\beta_1) = (P'_1, \ldots, P'_n)$. Let p and q be the numbers of strands coming to the crossing from the left side and from the right side. Obviously, for $j \neq p, q$ we have $P_j = P'_j$. Besides, $P'_p = (P_p \cdot t) \cdot t^{-1} = P_p, P'_q = (P_q \cdot t^{-1}) \cdot t = P_q$.

Now let us consider the case $\beta_1 = \beta \cdot \sigma_i \cdot \sigma_i^{-1}$ (obviously, the case $\beta_1 = \beta \cdot \sigma_i^{-1} \cdot \sigma_i$ is quite analogous to this one).

As before, denote $f(\beta)$ by $(\ldots P_i \ldots)$, and $f(\beta_1)$ by $(\ldots P'_i \ldots)$, and the corresponding strand numbers by p and q. Again, we have: for $j \neq p, q$: $P'_j = P_j$. Moreover, $P_p = P'_p$ by definition of f (since the p-th strand makes an overcrossing twice), and $P'_q = (P_q P_p^{-1} a_p^{-1} P_p) P_p^{-1} a_p P_p = P_q$.

Now let us check the invariance under the third Reidemeister move. Let β be a braid word, $\beta_1 = \beta\zeta_i\zeta_{i+1}\zeta_i$, and $\beta_2 = \beta\zeta_{i+1}\zeta_i\zeta_{i+1}$. Let p, q, r be the global numbers of strands occupying positions $n, n + 1, n + 2$ at the bottom of β.

Denote $f(\beta)$ by $(P_1, \ldots, P_n)$, $f(\beta_1)$ by $(P_1^1, \ldots, P_n^1)$, and $f(\beta_2)$ by $(P_1^2, \ldots, P_n^2)$. Obviously, $\forall i \neq p, q, r$ we have $P_i = P_i^1 = P_i^2$. Direct calculations show that $P_p^1 = P_p^2 = P_p \cdot t^2, P_q^1 = P_q^2 = P_q$ and $P_r^1 = P_r^2 = P_r \cdot t^{-2}$.

Now, let us consider the mixed move by using the same notation: $\beta_1 = \beta\zeta_i\zeta_{i+1}\sigma_i$, $\beta_2 = \sigma_{i+1}\zeta_i\zeta_{i+1}$. As before, $P_j^1 = P_j^2 = P_j$ for all $j \neq p, q, r$. Now, direct calculation shows that

$$P_p^1 = P_p t^2, P_q^1 = P_q t^{-1}, P_r^1 = P_r t^{-1}(P_q t^{-1})^{-1}a_q^{-1}(P_q t^{-1}) = P_r P_q a_q^{-1} P_q t^{-1}$$

and

$$P_p^2 = P_p t^2, P_q^2 = P_q t^{-1}, P_r^2 = P_r P_q^{-1} a_q^{-1} P_q t^{-1}.$$

Finally, consider the "classical" case $\beta_1 = \beta\sigma_i\sigma_{i+1}\sigma_i, \beta_2 = \beta\sigma_{i+1}\sigma_i\sigma_{i+1}$; the notation is the same. Again $\forall j \neq p, q, r: P_j^1 = P_j^2 = P_j$. Besides this, since the p-th strand forms two overcrossings in both cases then $P_p^1 = P_p^2 = P_p$. Then,

$$P_q^1 = P_q P_p^{-1} a_p^{-1} P_p, P_r^1 = (P_r P_p^{-1} a_p^{-1} P_p) \cdot (P_q P_p^{-1} a_p^{-1} P_p)^{-1} a_q^{-1}.$$

$$(P_q P_p^{-1} a_p^{-1} P_p) = P_r P_q^{-1} a_q^{-1} P_q P_p^{-1} a_p^{-1} P_p$$

and

$$P_q^2 = P_q P_p^{-1} a_p^{-1} P_p, \quad P_r^2 = P_r P_q^{-1} a_q^{-1} P_q P_p^{-1} a_p^{-1} P_p.$$

As we see, the final results coincide and this completes the proof of the theorem.   □

Thus, we have proved that f is a virtual braid invariant; i.e., for a given braid B the value of f does not depend on the diagram representing B. So, we can write simply $f(B)$.

Remark 2.3. In fact, we can think of f as a function valued not in $(E_1, \ldots, E_n)$, but in n copies of G: all these invariants were proved for the general case of $(G, \ldots, G)$. The present construction of $(E_1, \ldots, E_n)$ is considered for the sake of simplicity.

Classical braids (i.e., braids without virtual crossings) can be considered up to two equivalences: classical (modulo only classical moves) and virtual (modulo all moves). Now, we prove that they are the same. This fact is not new. It follows from [Fenn,Rimanyi and Rourke, 1997]. An elementary proof was given in [Manturov, 2016b].

Theorem 2.6. *Two virtually equal classical braids B_1 and B_2 are classically equal.*

Proof. Since B_1 is virtually equal to B_2, we have $f(B_1) = f(B_2)$. Now, taking into account that f is a complete invariant on the set of classical braids, we have $B_1 = B_2$ (in the classical sense).   □

As in the case of virtual knots, in the case of virtual braids there exists a forbidden move, namely, $X = \sigma_i \sigma_{i+1} \zeta_i = \zeta_{i+1} \sigma_i \sigma_{i+1} = Y$. Now, we are going to show that it cannot be represented by a finite sequence of the virtual braid group relations.

Theorem 2.7. *A forbidden move (relation) cannot be represented by a finite sequence of legal moves (relations).*

Proof. Actually, let us calculate the values $f(\sigma_1\sigma_2\zeta_1)$ and $f(\zeta_2\sigma_1\sigma_2)$. In the first case we have:

$$(e,e,e) \to (e, a_1^{-1}, e) \to (e, a_1^{-1}, a_1^{-1}) \to (e, a_1^{-1}t, a_1^{-1}t^{-1}).$$

In the second case we have:

$$(e,e,e) \to (e, t, t^{-1}) \to (e, t, t^{-1}a_1^{-1}) \to (e, ta_1^{-1}, t^{-1}a_1^{-1}).$$

As we see, the final results are not the same (i.e., they represent different virtual n-systems); thus, the forbidden move changes the virtual braid. $\square$

Remark 2.4. If we put $t = 1$, the results $f(X)$ and $f(Y)$ become the same. Thus that is the variable t that "feels" the forbidden move.

2.4.2.1 *A 2n-variable generalisation of the invariant*

In the present section we define a stronger version of the f invariant described in the previous section. This generalised invariant was invented by V. O. Manturov soon after the paper [Chterental, 2015] was published; however, the definition remained unpublished since the invariant of $(n + 1)$ variables itself was conjecturally complete.

We begin with the group $\tilde{G}$ — the free group in generators $a_1, \ldots, a_n$, $t_1, \ldots, t_n$ and denote by $\tilde{E}_i$ the quotient sets of right residual classes $\{a_i\}\backslash G$ for $i = 1, \ldots, n$.

Definition 2.18. An *extended virtual n-system* is a set of elements $\langle e_1 \in \tilde{E}_1, \ldots, e_n \in \tilde{E}_n \rangle$.

Now we construct an invariant $\tilde{f}$ which takes values in extended virtual n-systems. The construction follows the same pattern as in the case of the f invariant, but with a different approach to virtual crossings. We begin with a system $\langle e, \ldots, e \rangle$ and process the crossings one by one.

To be precise, consider a crossing corresponding to an i-th generator or its inverse: either σ_i, ζ_i or σ_i^{-1}. Assume that the left strand of this crossing originates from the point $(p, 1)$, and the right one originates from the point $(q, 1)$. Let $e_p = P, e_q = Q$, where P, Q are some words representing the corresponding residue classes. After the crossing of all residue classes but e_p, e_q should stay the same.

Then if the letter is σ_i, then e_p stays the same, and e_q becomes $QP^{-1}a_p^{-1}P$. If the letter is σ_i^{-1}, then e_q stays the same, and e_p becomes

$PQ^{-1}a_qQ$. Finally, if the letter is ζ_i, then e_p becomes $P \cdot t_q$, and e_q becomes $Q \cdot t_p^{-1}$. This operation is well defined.

Obviously, the function $\tilde{f}$ collapses to the function f defined in the previous section if we "forget" the distinction between the variables $t_1, \ldots, t_n$.

Along the same lines as in the previous section the following theorem is verified:

Theorem 2.8. *The function $\tilde{f}$ is an invariant of virtual braids.*

Unlike the case of invariant f, though, the following conjecture remains open.

Conjecture 2.1. *The invariant $\tilde{f}$ is complete.*

Chapter 3

Curves on Surfaces. Knots and Virtual Knots

3.1 Basic notions of knot theory

We start with basic definitions of knot theory [Manturov, 2018].

A *classical knot* (a *classical link*) is a smooth embedding of the circle S^1 (a disjoint union of circles $\bigsqcup_{i=1}^{n} S_i^1$) to three dimensional space $\mathbb{R}^3$ (or three dimensional sphere S^3). Knots and links are usually considered up to isotopies in $\mathbb{R}^3$.

The natural orientation of the circle S^1 induces an orientation of the knot (link).

A conventional way to present knots and links is based on their plane generic projections — link diagrams.

A *classical link diagram* is a 4-valent plane graph, each vertex of which is endowed with undercrossing-overcrossing structure, see Fig. 3.1. The graph can have also circle components without crossings.

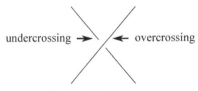

Fig. 3.1   Knot crossing

Isotopic links may give different diagrams after projection, but this freedom is controlled by Reidemeister's theorem [Reidemeister, 1948].

Theorem 3.1. *Two link diagrams D_1 and D_2 correspond to the same link isotopy class if and only if the diagram D_2 can be obtained from D_1 by a sequence of diagram transformations, called* Reidemeister moves, *see Fig. 3.2, and diagram isotopies.*

Invariants and Pictures

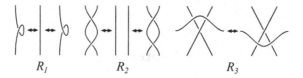

Fig. 3.2 Reidemeister moves

Thus, one can define links (and knots) as equivalence classes of link diagrams modulo Reidemeister moves and diagram isotopies.

Link diagrams do not carry natural orientations so their equivalence classes determine nonoriented links. In order to define an oriented link, one should orient all the edges of a link diagram so that opposite edges of any crossing of the diagram have the same orientation. Such an orientation is compatible with Reidemeister moves and the corresponding equivalence class of oriented diagrams determines an oriented link.

Diagrams of the simplest knots and links are given in Fig. 3.3

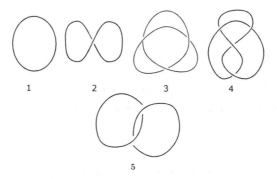

Fig. 3.3 Simplest knots and links

The main question of knot theory is *the knot recognition problem*: which two knots are (*isotopic*) and which are not? A partial case of the knot recognition problem is the *trivial knot recognition problem*. Here, *trivial knot* (or *unknot*) means the simplest knot that can be represented as the boundary of a 2-disc embedded in $\mathbb{R}^3$. Both questions are very difficult.

As usual, in order to prove that two knot diagrams correspond to the same knot, one should present a sequence of Reidemeister moves which transforms the first diagram to the second one. The difficulty is that the intermediate diagrams can be much more complicated than the initial ones. For example, the diagram of the unknot in Fig. 3.4 cannot be reduced by Reidemeister moves to the trivial diagram (Fig. 3.3, 1) without adding new crossings to the diagram.

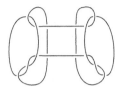

Fig. 3.4   A diagram of the unknot

In order to prove two knot diagrams are not equivalent, one should construct a knot invariant that distinguish these diagrams. A *knot (link) invariant* is a function on the representatives of knots and links (embeddings, diagrams etc.) whose value does not change if one replaces a representative of a knot (link) with another representative of the same knot (link). So if an invariant has different values on two diagrams, then the corresponding knots (links) are different.

One of the most famous and useful knot invariants is Jones polynomial, see for example [Manturov, 2018].

Given a (nonoriented) link diagram D with the set of crossings $\chi(D)$, consider the set of *states* $S(D) = \{0,1\}^{\chi(D)}$. For each state $s \in S(D)$ we can define a diagram D_s which appears from D by smoothing of the diagram according the state s. The rule for smoothing is shown in Fig. 3.5.

Fig. 3.5   Types of smoothing

Let $\alpha(s)$ be the number of 0 in s and $\beta(s)$ be the number of 1 in s. The diagram D_s has no crossings, i.e. it is a union of circles. Let $\gamma(s)$ be the number of circles in the diagram D_s. The polynomial

$$\langle D \rangle = \sum_{s \in S(D)} a^{\alpha(s)-\beta(s)}(-a^2 - a^{-2})^{\gamma(s)-1} \in \mathbb{Z}[a, a^{-1}] \qquad (3.1)$$

is called the *Kauffman bracket* of the diagram D.

The Kauffman bracket is invariant under second and third Reidemeister moves, but the first Reidemeister move multiplies the Kauffman bracket by $-a^{\pm 3}$. A genuine invariant appears after normalising the bracket by an appropriate factor. The normalisation uses a knot orientation. Let

$$w(D) = \sum_{c \in \xi(D)} \text{sign}(c)$$

be the *writhe number* of an oriented link diagram D where the sign of a crossing c is calculated according to Fig. 3.6.

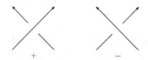

Fig. 3.6   Sign of a crossing

The polynomial $X(D) = (-a)^{-3w(D)}\langle D \rangle \in \mathbb{Z}[a, a^{-1}]$ is called the *Jones polynomial* of the link diagram D with given orientation. The properties of Jones polynomial can be summarized as follows, see for example [Manturov, 2018].

Theorem 3.2.

(1) Jones polynomial $X(D)$ is an invariant of oriented links;
(2) Jones polynomial obeys the skein relation

$$a^4 X(\text{⤬}) - a^{-4} X(\text{⤬}) = (a^{-2} - a^2)X(\text{)(}).$$

The arguments here are any oriented link diagrams which coincide everywhere except a small neighbourhood inside which they look like the corresponding icons.
(3) For any oriented links L_1 and L_2 we have $X(L_1 \# L_2) = X(L_1)X(L_2)$ where $L_1 \# L_2$ is a connected sum of the links, see Fig. 3.7.

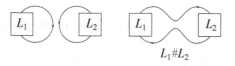

Fig. 3.7   Connected sum of links

It is yet unknown, whether the Jones polynomial recognises the trivial knot.

Another way to present an oriented knot (not link) is its *Gauß diagram* (also called a *chord diagram*). Given a knot diagram D, it can be treated as an immersion $S^1 \to \mathbb{R}^2$. Consider the preimages of the double points and connect any two preimages, corresponding to the same crossing, by an edge, see Fig. 3.8.

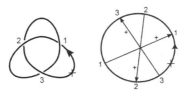

Fig. 3.8 A Gauß diagram

The edges of the resulting chord diagrams have orientation and signs. The orientation is induced by the undercrossing-overrossing structure; any edge is oriented from the overcrossing to the undercrossing. The edge signs come from the orientation of the immersion and coincides with the signs of the corresponding crossings, see Fig. 3.6.

The knot diagram can be restored from its Gauß diagram up to isotopy (and pass of arcs through the infinite point of $\mathbb{R}^2$).

Given a link diagram, one can apply the same construction and obtain a Gauß diagram which will have several oriented circles and chord (with orientations and signs) between them, see Fig. 3.9.

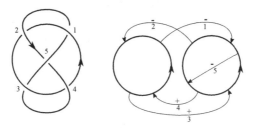

Fig. 3.9 Whitehead link and its Gauß diagram

Reidemeister moves on knot diagrams induce moves on Gauß diagrams, see Fig. 3.10.

On the other hand, there are Gauß diagrams which do not correspond to any classical knot diagram, for example see in Fig. 3.11. Any attempt to draw a diagram of the knot in the plane leads to an additional crossing

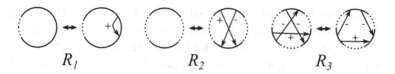

Fig. 3.10   Reidemeister moves on Gauß diagrams

(marked with a circle in the figure). This fact can be proved by the parity argument: any chord in the Gauß diagram of a classical knot can intersect only even number of the other chords, but in the given Gauß diagram the both chords are odd in this sense.

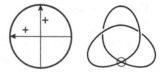

Fig. 3.11   Nonclassical Gauß diagram

This disparity was one of the motivations to enhance the notion of knots and to introduce virtual knots and links.

One can define a *virtual knot* as an equivalence class of a Gauß diagram modulo Reidemeister moves on Gauß diagrams.

Another way to define virtual knots (and links) is to consider virtual link diagrams. A *virtual link diagram* is a 4-valent plane graph, whose vertices are split into two types: classical vertices with undercrossing-overcrossing structure (see Fig. 3.1) and *virtual vertices* marked with circles (see Fig. 3.12).

Fig. 3.12   Virtual crossing

The admissible transformation of virtual diagrams include classical Reidemeister moves (see Fig. 3.2) and *virtual Reidemeister moves*, see Fig. 3.13.

Local virtual Reidemeister moves can be replaced with one *detour move*, see Fig. 3.14. A detour move replaces an arc, containing only virtual crossing, with another arc with the same ends that contains only virtual crossings as well.

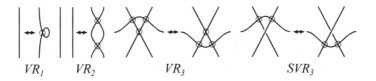

Fig. 3.13   Virtual Reidemeister moves

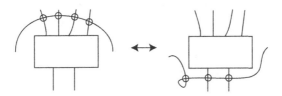

Fig. 3.14   Detour move

Now, we can define a *virtual link* as an equivalence class of virtual link diagrams modulo classical and virtual Reidemeister moves (or classical Reidemeister moves and the detour move).

The third approach to virtual links employs considering knots and links in thickenings of two dimensional surfaces. Let Σ be an oriented closed two dimensional surface. A *link in the thickening of the surface* Σ is an embedding a disjoint union of circles $\bigsqcup_{i=1}^{n} S^1$ into $\Sigma \times [0,1]$ considered up to isotopies.

As in the classical case, links in the thickening of the surface Σ can be presented by their diagrams — 4-valent graphs embedded into Σ whose vertices have undercrossing-overcrossing structure. The equivalence of the diagrams is generated by diagram isotopies and Reidemeister moves (Fig. 3.2).

Given a link L in $\Sigma \times [0,1]$, a *stabilisation* operation is defined by the attaching a thickened handle along a pair of annuli $C_1 \times [0,1]$ and $C_2 \times [0,1]$ which do not intersect the link, see Fig. 3.15. The result is a link in the thickening of a surface of higher genus. The inverse operation is called *destabilisation*.

Fig. 3.15   Stabilisation

N. Kamada and S. Kamada [Kamada and Kamada, 2000] showed that virtual links can be defined as equivalence classes of pairs (Σ, L), where Σ is an oriented closed surface and L is a link in the thickening of Σ, modulo isotopies of L, natural isomorphisms of these pairs and stabilisations/destabilisation.

Classical knots and links can be considered as links in the thickening of the sphere S^2.

Many invariants of classical knots can be straightforwardly extended to virtual knots. For example, the Kaufman bracket (3.1) is invariant under virtual moves on virtual diagram, so Jones polynomial is an invariant of virtual knots. In some cases the Jones polynomial shows that a virtual diagram defines a non classical link.

Remark 3.1. Historically, virtual knots and links were defined first by L. H. Kauffman in [Kauffman, 1997] as equivalence classes of plane diagrams with virtual crossings. Later M. N. Goussarov, M. B. Polyak and O. Ya. Viro in [Goussarov, Polyak and Viro, 2000] introduced moves on Gauß diagrams and showed their theory was equivalent to Kauffman's virtual knots. And finally, N. Kamada and S. Kamada [Kamada and Kamada, 2000] proposed an approach to virtual knots that uses thickenings of surfaces and the stabilisation.

The fact that virtual knots extend classical knots (more precisely, that the natural map from classical knots to virtual ones is injective) was established by G. Kuperberg [Kuperberg, 2002]. Further the fact was reproved many times. A proof based on the parity theory is given in Section 5.6.1.

Given a virtual link diagram, forgetting undercrossing-overcrossing structure at the vertices of the diagram yields a *flat link diagram*. In other words, flat diagram is an equivalence class of link diagrams modulo crossing switches, see Fig. 3.16.

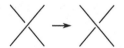

Fig. 3.16   Crossing switch

Equivalence classes of flat link diagrams modulo classical and virtual Reidemeister moves are called *flat links*.

Given a closed surface Σ, the flat knots on it can be identified with the homotopy classes of free loops in the surface Σ. This identification implies all flat classical knots (i.e. free loops in the sphere) are trivial.

Further simplification of knot structure leads to the notion of free knots and links. A *free link* is an equivalence class of a virtual link diagram modulo Reidemeister moves, crossing switchs and *the virtualisation*, see Fig. 3.17.

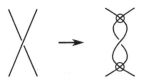

Fig. 3.17   Virtualization

On the other hand, free knots can be defined as equivalence classes of chord diagrams modulo Reidemeister moves given in Fig. 3.18. Comparing with Gauß diagrams, chords here do not have orientation nor marks.

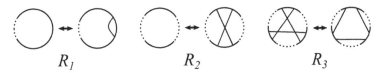

Fig. 3.18   Reidemeister moves of chord diagrams

3.2   Curve reduction on surfaces

Let us fix a two-dimensional manifold Σ. For simplicity we assume Σ to be oriented and closed, i.e. to be the sphere with g handles, $g \geq 0$.

We are interested in homotopy classes of curves in Σ, that are also called *free loops*.

From the topological point of view these curves correspond to conjugacy classes of the group $\pi_1(S_g)$. We denote the set of homotopy classes of curves in Σ as $\mathcal{H}(\Sigma)$.

We shall consider *immersions $S^1 \to \Sigma$ in general position*, i.e. immersions such that all the points having more than one preimage are transverse double points, i.e. points of transverse intersections of the images of two arcs in S^1.

Note that any curve can be approximated by a curve in general position. Since curves, which are close to each other, represent the same homotopy class, we can consider immersions in general positions as representatives of the free loops.

A *diagram* D in a closed surface Σ is a 4-graph Γ embedded in Σ, so that the complement $\Sigma \backslash \Gamma$ is a union of two dimensional cells. The graph Γ is called the *diagram graph* of D.

Analogously, one can define a diagram in a non compact surface or surface with boundary.

We shall not distinguish immersions which are combinatorially equivalent. We shall call such immersions *diagrams*.

The general position argument implies

Proposition 3.1. *Two immersions are homotopy equivalent if and only if one immersion can be obtained from the other by a sequence of Reidemeister moves, see Fig. 3.19.*

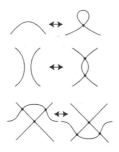

Fig. 3.19   Reidemeister moves for circle immersions in a two dimensional surface

Proposition 3.1 can be thought of as *definition* of the set $\mathcal{H}(\Sigma)$.

In the present section we consider a combinatorial algorithm of free loop recognition, following mainly the article [Hass and Scott, 1994] of J. Hass and P. Scott. Their paper gives a simple answer to the question due to V. G. Turaev [Turaev, 1989]:

Let s_0 and s_1 be homotopical curves in a surface and each of them have k double points. Does there exist a homotopy s_t between s_0 and s_1 such that any curve s_t has no more than k double points?

The theorems given below are proved in [Hass and Scott, 1994]:

Theorem 3.3. *Let s_0 and s_1 be homotopic curves in general position in a surface and each of them minimise the number of self-crossings in its homotopy class. Then there exists a homotopy s_t from s_0 to s_1 such that s_t is self transversal for every t and the number of double points (counted with multiplicities) of s_t does not depend on t.*

Theorem 3.4. *Let s_0 be a curve in general position in the surface Σ. Then there is a homotopy s_t from s_0 to a curve s_1 which has the minimal number of double points in the homotopy class, and the number of double points in s_t does not increase with t.*

A solution to Turaev's problem follows immediately from these theorems: given two curves s_0 and s_1, one can homotop them to minimal curves s_0', s_1' thanks to Theorem 3.4, and take a homotopy between s_0' and s_1' from Theorem 3.3.

Remark 3.2. Theorem 3.3 is not valid for unions of curves. For example, one cannot permute the components in Fig. 3.20 without adding new intersection points. On the other side, Theorem 3.4 remains true.

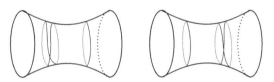

Fig. 3.20   Non-equivalent minimal pairs of curves

3.2.1   *The disc flow*

In paper [Hass and Scott, 1994] a simple construction of flow of curves is given which homotops a finite set of curves in a surface in such a way that:

(1) The number of self-intersections does not increase for each curve,
(2) The number of intersections does not increase for each pair of curves,
(3) Each curve either disappears in finite time or becomes close to a geodesic,
(4) This flow can be extended continuously to k-parameter families of curves.

This algorithm can be programmed easily and admits a generalisation to higher dimensions.

Let γ be a piecewise smooth immersed curve in a Riemannian surface F. Let us cover F by convex discs $D_1, \ldots, D_n$ whose radii are smaller than the injectivity radius. We choose discs to be in general position so that any point in F belongs to the boundary of no more than two discs, the boundaries of the discs intersect transversely the curve γ, and the discs of halved radii with the same centers cover still the surface F. Such a cover will be called *well situated* with respect to γ. Let us number the discs $D_1, D_2, \ldots$ cyclically so that $D_{n+i} = D_i$ for any $i \geq 1$.

Roughly speaking, a disc flow will be defined as homotopy of each arc in $\gamma \cap D_1$ to the unique geodesics with the same ends, then the process repeats for each disc $D_2, D_3, \ldots$.

We investigate properties and convergence of the flow, then we show that the number of intersection points does not increase.

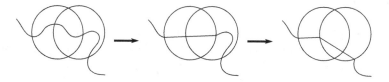

Fig. 3.21   Disc flow

We start with several combinatorial lemmas.

We call an *embedded loop* of the curve γ any embedded subarc of γ with coinciding ends which bounds an embedded disc, see Fig. 3.22 left. An *embedded bigon* is a pair of subarcs of γ with common ends that bound an embedded disc, see Fig. 3.22 right. An embedded loop (bigon) is called *innermost* if the correspondent embedded disc does not contain any other arcs of γ.

Fig. 3.22   An embedded loop and an embedded bigon

Lemma 3.1. *1. Given a triangle ABC and several embedded curves* *crossing it, such that any two curves have at most one intersection point in the triangle and neither curve intersects the side BC, then on each side AB and AC there is an innermost triangle adjacent to it.*

2. Given an innermost bigon with embedded curves crossing it, for each edge of the bigon there is an innermost triangle adjacent to it.

Proof. 1. The proof is by induction on the number k of the intersecting curves. For $k = 1$ the statement is evident. Assume that the statement holds for $k = n$. Consider the case of $k = n + 1$ intersecting curves. Let D be the intersection point on the side AB closest to A, and E be the intersection point on the side AC which belongs to the same intersecting curve as D. Then the triangle EAD contains at most n intersecting curves and none of them intersects the side AD. By induction assumption there is an innermost triangle adjacent to the side EA, thus, adjacent to AC. The reasonings for the side AB are analogous.

2. Since the bigon is innermost, any two curves inside intersect in at most one point. Draw a curve intersecting the bigon near to one of its vertices and not intersecting other curves. This curve splits the bigon into two triangles one of which is innermost. By the first statement, there are innermost triangles adjacent to edges of the triangle of the splitting. These innermost triangles cannot be adjacent to the splitting curve, thus, we can remove the supplementary splitting curve. $\square$

Remark 3.3.

(1) The lemma does not require the intersecting curves to be in general position. There can be multiple intersection points.
(2) The second statement of the lemma remains valid if one takes an embedded loop, which does not contains any bigons, instead of the bigon. That is, there is an innermost triangle adjacent to the edge of such loop.

Lemma 3.2. *A finite set of piecewise smooth transversal curves in a convex disc can be homotoped (with respect to the boundary) to a set of geodesics such that the number of self-intersection and intersection points of the curves do not increase during the homotopy.*

Proof. The proof is based on induction on double point number (with multiplicity). We count here an intersection point of multiplicity k as $\frac{k(k-1)}{2}$ double points.

Assume that the number of double points is minimal. Then the curves do not have self-intersections and any two curves intersect in at most one point.

If there are closed curves, then take an innermost closed curve. This curve can be contracted to a point without intersecting other curves. Repeating this operation we homotop our set of curves to a set without closed curves. So we can suppose there are no closed curves in the set.

Let $\{a_i\}, i = 1, \ldots, n$, be the set of curves and let d_i be the geodesics connecting the ends of the curve a_i in the disc for each $i = 1, \ldots, n$. We can isotop slightly (with respect to the boundary) the curves $\{a_i\}$ so that they lie in general position with $\{d_i\}$. So we can suppose that all intersections of $\{a_i\}$ and $\{d_i\}$ are transversal.

Assume that there are extra intersections of d_1 with the curves $\{a_i\}$. Then there is a bigon formed by a subarc of d_1 and a subarc of some curve a_j.

Take an innermost bigon among all bigons adjacent to d_1. Its edges belong to d_1 and a_j for some j. Since each two curves in $\{a_i\}$ intersect in at most one point, this bigon is an innermost bigon among all the bigons in the disc. If some curves $\{a_i\}$ intersect the bigon, then by Lemma 3.1 there is an innermost triangle in the bigon adjacent to a_j. So, by moving a_j, we can either decrease the number of intersection points in the bigon (if the triangle is adjacent to one edge of the bigon, see Fig. 3.23 left) or decrease the number of arcs intersecting the bigon (if the triangle is adjacent to two edges of the bigon, see Fig. 3.23 right).

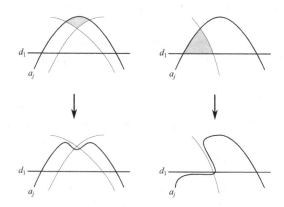

Fig. 3.23   Removing intersecting curves from the bigon

Repeating this operation, we obtain an empty bigon. Then we eliminate the bigon by moving the curve a_j.

In this manner we can homotop our curves to a position where the curves $a_j, j > 1$, intersect d_1 in at most one point and the curves d_1 and a_1 form a bigon. This bigon is innermost so a_1 can be homotop to d_1 without increasing the number of intersection points.

Applying these reasonings consequently to a_2 and d_2, a_3 and d_3 etc., we homotop the set of curves to the geodesics.

Assume now that the number of double points is not minimal. Then there is either a self-intersection of some curve or two curves have two or more intersection points. Then there is an embedded loop or a bigon in the disc. Take an innermost loop or a bigon among all embedded loops and bigons.

An innermost loop is empty and can be contracted, so the number of intersection points will reduce, see Fig. 3.24.

Fig. 3.24   Removing intersecting curves from the bigon

An innermost bigon may contain only arcs without self-intersections, and any two of these arcs can have at most one intersection point. Then there is an innermost triangle adjacent to an edge of the bigon. Applying the triangle move, we can reduce either the number of intersection points inside the bigon or the number of arcs intersecting the bigon. After all, we shall have an empty bigon which can be removed with eliminating two intersection points in the disc.

Thus, the configuration can be reduced to a case with the minimal number of intersection points and lemma is proved. $\qquad\square$

Note that the constructed homotopy will be regular everywhere except the moment when a loop is contracted to a point.

We can define the disc flow as follows. Consider a curve $s = s_0$. It can have several components and self-intersections. We define a family of curves $s_t, t \geq 0$ inductively. Given a curve $s_{i-1}, i \in \mathbb{N}$, we reduce the radius of the disc D_i by a factor $\lambda_i, \frac{1}{2} < \lambda_i \leq 1$ so that s_{i-1} is in general position with the boundary ∂D_i. It means that s_{i-1} and ∂D_i intersect transversely and ∂D_i contains no self-intersection points of s_{i-1}. Then we define $s_t, t \in [i-1, i]$

as the image of s_{i-1} under the straightening homotopy of Lemma 3.2 in the (shrinked) disc D_i. Note that the length of curves s_t is not monotonous but the length sequence of the curves $s_i, i \in \mathbb{N}$, is non increasing. The family of curves $s_t, t \geq 0$, is called the *disk flow of the curve* $s = s_0$.

The disc flow is not canonical since the choice of homotopy in Lemma 3.2 is not unique. Nonetheless, it possesses several useful properties whose formulation requires some additional notation. Let Δ be the map defined on the set of curves F which transforms a curve $s = s_0$ to the curve s_n obtained from s by consecutive straightening in the discs $D_1, \ldots, D_n$. When we talk about convergence of curves, we use the topology in the curve space such that the ε-neighbourhood of a curve γ consists of curves γ' that admits a parametrisation for which the curves γ and γ' are ε-close in the Frechet topology. This topology is induced by the Frechet metric in the space of curves.

Theorem 3.5. *Let γ, γ' be transversally intersecting curve in a surface F. Let $D_1, \ldots, D_n$ be a disc covering of the surface F which is good with respect to $\gamma \cup \gamma'$. Let γ_t, γ'_t be the images of the curves by the disc flow.*

(1) The number of self-intersection points of the curve γ_t does not increase with the grow of $t \in [0, +\infty)$.

(2) The number of intersection points between the curves γ_t and γ'_t does not increase with the grow of $t, t \in [0, +\infty)$.

(3) Either γ_t disappears in a finite time or a subsequence of the curves $\{\gamma_t\}$ converges to a geodesic with $t \to \infty$. In the second case, if U is an open neighbourhood of the set of geodesics which are homotopic to γ_t, then there exists $T > 0$ such that γ_t belongs to U with $t > T$.

(4) If the sequence $\{\gamma_i\}$ converges to a geodesic γ_∞ with $i \to \infty$, then $length(\gamma_i)$ converges to $length(\gamma_\infty)$.

(5) $Length(\Delta(\gamma)) \leq length(\gamma)$, and the equality takes place only if γ is a geodesic or a point.

Proof. The first two properties follows from Lemma 3.2.

Assume γ_i do not disappear, then the sequence $length(\gamma_i)$ is bounded from below. The sequence γ_i is a sequence of curves of bounded lengths in a compact manifold. Then by Ascoli theorem there is a subsequence $\{\gamma_j\}$ which converges uniformly to a curve δ with $length(\delta) \leq \lim_{j \to \infty} length(\gamma_j)$. Let us show δ is a geodesic.

If δ is not a geodesic, then there is a small subarc $\delta' \in \delta$ such that δ' is not a geodesic and δ' lies in some disc D_i with the halved radius.

Let $\varepsilon > 0$ be the value by which the straightening process decreases the length of δ. Since γ_j converges to δ we have $length(\Delta(\gamma_j)) < length(\gamma_j) - \varepsilon/2$ for sufficiently large j. On the other hand, $length(\gamma_j) < length(\delta) + \varepsilon/2$. Then $length(\Delta(\gamma_j)) < length(\delta)$ that gives a contradiction. Thus, any convergent subsequence of γ_i must converge to a geodesic.

Assume that we can form a subsequence $\{\gamma_k\}$ with all γ_k lying outside the neighbourhood U of the set of geodesics. The reasonings above show that there is a subsequence in $\{\gamma_k\}$ converging to a geodesic, a contradiction. Thus, the third statement of the theorem is true. The statements four and five can be proved analogously. □

Note that although there is a convergent subsequence in γ_i, hypothetically the entire sequence can oscillate between different geodesics. Nevertheless, in many cases we can show convergence of the sequence. We call two geodesics *parallel* if they cobound a flat annulus.

Theorem 3.6. *Let F be a closed surface of negative of zero curvature. The disc flow applied to a curve γ_0 gives a homotopy γ_t under which γ_t either disappears in finite time or γ_t converges to a unique geodesic γ_∞ as $t \to \infty$.*

Proof. If γ_0 is contractible, then it disappears under the disc flow since there is no homotopically trivial geodesic in F.

Consider a cover F_γ of the surface F corresponding the curve γ_0. Then F_γ is a surface of negative or zero curvature and it is a topological cylinder. We denote the lift of γ_t with the same letter.

If F has negative curvature, then there is a unique closed geodesic g in F_γ homotopic to γ_0. There is a subsequence γ_j which converges to g. But once a curve is sufficiently close to g under disc flow, it remains to stay near g and the disc flow converges to g as $t \to \infty$.

If F is flat, then F_γ is a flat cylinder. Define the *width* of a curve γ in F_γ as minimal distance between two parallel geodesics containing γ. The disc flow does not increase the width of $\gamma_i, i \in \mathbb{N}$, and the width converges to zero, since there is a subsequence converging to a geodesic. Moreover, if γ_i lies between geodesics γ' and γ'', then the disc flow applied to $\gamma_i \cup \gamma' \cup \gamma''$ shows that $\gamma_t, t > i$, lies between γ' and γ''. It shows that the disc flow cannot shift the curves family γ_t. Thus, γ_t converges to a unique geodesic. □

3.2.2 *Minimal curves in an annulus*

Let A be the annulus $S^1 \times I$ with a flat product metric. Let us describe the curves in A minimizing the number of self-intersections.

Denote the coordinates in A by $(\varphi, x), \varphi \in [0, 2\pi], x \in [0, 1]$. A tangent vector $\xi \in TA$ is called *horizontal* if its x-component is zero, and ξ is *vertical* if its φ-component is zero. A curve $\gamma \in A$ is called *almost horizontal* if neither of the tangent vectors to γ is vertical, and γ is *horizontal* if all of its tangent vectors are horizontal.

Proposition 3.2. *Let γ be a curve of degree n in general position in A. Then there is a homotopy g_t from $\gamma = g_0$ to the curve $\gamma_k = g_1$ (Fig. 3.25), which has minimal self-intersections, such that the number of double points of the curve g_t is non increasing with t.*

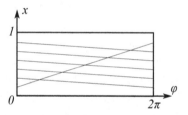

Fig. 3.25   The curve γ_k (case $k = 6$)

Thus, the minimal number of self-intersections for a curve of degree k is equal to $k - 1$.

Proof. Let g_t be a disc flow to the curve γ. By Theorem 3.6 the curves g_t converge to a geodesic $\tilde{\gamma}$ in the annulus A. Since $\tilde{\gamma}$ is horizontal, for a large T the curve $\gamma' = g_T$ will be almost horizontal. Thus, it suffice to show that there is a homotopy from γ' to γ_k which does not increase the number of self-intersections. We shall give an algebraic proof to this statement.

The annulus A can be considered as the rectangle $\Pi = [0, 2\pi] \times [0, 1]$ with two sides identified. Any almost horizontal curve γ' of degree n can be viewed as the closure of a flat braid diagram with n strands in Π (flat here means we don't distinguish over- and undercrossings at intersection points). We can code this braid by a word with letters $\sigma_1, \ldots, \sigma_{n-1}$. Braid isotopies are generated by relations

$$\sigma_i^2 = 1, \ 1 \leq i \leq n-1, \tag{3.2}$$

$$\sigma_i \sigma_j = \sigma_j \sigma_i, 1 \leq i, j \leq n-1, \ |i-j| \geq 2, \tag{3.3}$$

$$\sigma_i \sigma_{i+1} \sigma_i = \sigma_{i+1} \sigma_i \sigma_{i+1}. \tag{3.4}$$

The relation (3.2) corresponds to removing an innermost bigon, the relation (3.3) is realised by an ambient isotopy of Π and the relation (3.4) corresponds the third Reidemeister move which is a homotopy that does not change the number of double point in the diagram, see Fig. 3.26.

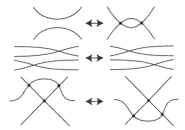

Fig. 3.26   Flat braid relations

The relations (3.2)–(3.4) yield Artin's presentation for the permutation group S_n. With some abuse of notation, about a word w with letters $\sigma_1, \ldots, \sigma_k, k < n$, we shall say w lies in the permutation group S_{k+1}. We say a word w in S_{k+1} is *reduced* to a word w' in S_{k+1} if w' can be obtained from w by a sequence of letter contractions (3.2) and relations (3.3), (3.4). The reduction sequence does not increase the word length (and the corresponding homotopy does not increase the number of double points).

Let $w = uv$ be a word in S_n. The word $w' = vu$ is called a *cyclic shift* of the word w. Cyclic shift can be realised by an ambient isotopy (in fact, an isometry) of the annulus A.

A word w_1 in S_n is *cyclically reduced* to a word w_2 in S_n if there is a sequence of letter contractions (3.2), relations (3.3), (3.4) and cyclic shifts which transforms w_1 to w_2. In this situation we denote $w_1 \sim w_2$. Cyclic reduction describes a homotopy of the closure of the corresponding flat braid diagram in the annulus A, and this homotopy does not increase the number of double points.

Lemma 3.3.

(1) Any word w in S_n can be reduced to a word w' which contains at most one generator σ_{n-1}.

(2) *Any word w in S_n can be cyclically reduced to a word $\prod_{i=1}^{n-1} \sigma_i^{k_i}$, $k_i = 0, 1$.*

Proof. 1. Prove the first statement by induction on n. The case $n = 1$ is trivial.

Assume that the statement is proved for n. Let w be a word in S_{n+1} and let w' be a word with minimal number l of the generator σ_n which the word w can be reduced to. If $l \geq 2$, then w' can be written as $a\sigma_n b\sigma_n c$ where $a, b \in S_n$, $c \in S_{n+1}$. By assumption of induction, the word b contains at most one generator σ_{n-1}, i.e. $b \in S_{n-1}$ or $b = b_1 \sigma_{n-1} b_2$ with $b_1, b_2 \in S_{n-1}$. In the first case,

$$w' = a\sigma_n b\sigma_n c = a\sigma_n^2 bc = abc;$$

in the second case,

$$w' = a\sigma_n b_1 \sigma_{n-1} b_2 \sigma_n c = ab_1 \sigma_n \sigma_{n-1} \sigma_n b_2 c = ab_1 \sigma_{n-1} \sigma_n \sigma_{n-1} b_2 c.$$

So, we can reduce the number of σ_n that contradicts to the minimality of l. Hence, for w' we must have $l \leq 1$.

2. Let us show by induction that w can be cyclically reduced to a word $w_l \prod_{i=l}^{n-1} \sigma_i^{k_i}$, $k_i = 0, 1$, $w_l \in S_l$, for any $1 \leq l \leq n-1$. The case $l = 1$ gives us the second statement of the lemma.

Let $l = n - 1$. Then by the first statement of the lemma $w = a\sigma_{n-1}^{k_{n-1}} b$ with $a, b \in S_{n-1}, k_{n-1} \in \{0, 1\}$. Hence,

$$w = a\sigma_{n-1}^{k_{n-1}} b \sim ba\sigma_{n-1}^{k_{n-1}}.$$

Assume $w \sim w' = w_l \prod_{i=l}^{n-1} \sigma_i^{k_i}$, $l > 1$. Denote $z = \prod_{i=l}^{n-1} \sigma_i^{k_i}$, then $w' = w_l z$. By the first statement of the lemma $w_l = a\sigma_{l-1}^{k_{l-1}} b$ with $a, b \in S_{l-1}$, $k_{l-1} \in \{0, 1\}$. Then

$$w \sim w' = a\sigma_{l-1}^{k_{l-1}} bz = a\sigma_{l-1}^{k_{l-1}} zb \sim ba\sigma_{l-1}^{k_{l-1}} z = w_{l-1} z'$$

where $w_{l-1} = ba \in S_{l-1}$ and $z' = \sigma_{l-1}^{k_{l-1}} z = \prod_{i=l-1}^{n-1} \sigma_i^{k_i}$.   □

The lemma implies that for any almost horizontal curve γ' there is a homotopy to the closure of a braid $\prod_{i=1}^{n-1} \sigma_i^{k_i}$ which does not increase the number of double points. Since the closure is a connected curve, all k_i are equal to 1. Thus the braid closure coincides with the curve γ_k.   □

3.2.3 Proof of Theorems 3.3 and 3.4

Let Σ be a closed oriented surface. We choose a constant curvature metric on Σ. For a curve γ in general position in Σ we denote the number of self-intersections of γ by $d(\gamma)$.

Proof of Theorem 3.3. Let s_0 and s_1 be homotopic curves in general position in the surface Σ which minimize the number of self-crossings. Then the homotopy class of s_0 contains a unique (up to a parallel shift in the genus 1 case) geodesic s_2.

1. Let s_0 and s_1 be contractible curves. Then they are embedded and isotopic.

2. Let the homotopy class of s_0 be primitive in $\pi_1(\Sigma)$. Disk flows for s_0 and s_1 yield homotopies to s_2 which do not increase the number of self-intersections. The composition of these homotopies is a homotopy from s_0 to s_1 which does not change the number of self-intersections.

3. Let the homotopy class $\alpha \in \pi_1(\Sigma)$ be not primitive. Then $\alpha = \beta^k$ for a primitive $\beta \in \pi_1(\Sigma)$ and s_2 is a k-fold cover over a simple geodesic r.

Let g'_t and g''_t are disk flows for s_0 and s_1 with $s_0 = g'_0$ and $s_1 = g''_0$. The disk flows don't change the number of self-intersections, so it suffices to show that for some t there is a good homotopy between g'_t and g''_t. Since g'_t and g''_t converge to s_2, for T large enough the curves g'_T and g''_T can be considered as compositions $g'_T = q \circ p'$, $g''_T = q \circ p''$ where $p' \colon S^1 \to A$ and $p'' \colon S^1 \to A$ are maps into the annulus of degree k and $q \colon A \to \Sigma$ maps the annulus to a thin regular neighbourhood of the geodesic r. Then $d(g'_T) = d(p') + kd(r)$ and $d(g''_T) = d(p'') + kd(r)$. Since $d(g'_T)$ and $d(g''_T)$ are minimal, the numbers $d(p')$ and $d(p'')$ are minimal and by Proposition 3.2 are equal to $k - 1$. Moreover, there is a homotopy p_t between p' and p'' which does not change the number of self-crossings. Hence, $q \circ p_t$ is a required homotopy between g'_T and g''_T. □

Proof of Theorem 3.4. Let s_0 be a curve in general position in the surface Σ.

1. Let s_0 be contractible. If Σ is a sphere, then take a disk D in Σ which covers s_0 except a small arc. After applying the reduction process of Lemma 3.2, we get an arc without self-intersections inside the disk. Hence, the homotopy transforms s_0 to an embedded curve in Σ.

If Σ is not a sphere, then apply a disk flow to s_0. Since there is no contracting geodesics in Σ, the disk flow s_t must disappear after a finite time. Thus for some t the curve s_t will lie inside some disk D_i. Then we can repeat the reasonings for the sphere to obtain a homotopy of s_t to an embedded curve.

2. Let the homotopy class of s_0 be primitive in $\pi_1(\Sigma)$. Then a disk flow s_t for s_0 gives a homotopy to a unique (up to a parallel shift for the torus) geodesic $\tilde{s}$ that minimizes the number of self-intersections.

3. Let the homotopy class $\alpha \in \pi_1(\Sigma)$ be not primitive. Then $\alpha = \beta^k$ for a primitive $\beta \in \pi_1(\Sigma)$ and any disk flow s_t for s_0 converges to a geodesic $\tilde{s}$ which is a k-fold cover over a simple geodesic r. For T large enough the curve s_T is a composition $s_T = q \circ p$ where $p \colon S^1 \to A$ is a map into the annulus of degree k and $q \colon A \to \Sigma$ maps the annulus to a thin regular neighbourhood of the geodesic r. By Proposition 3.2 there is a homotopy p_t from p to γ_k (Fig. 3.25) which does not increase the number of self-intersections. Then $q \circ p_t$ is a homotopy from s_T to the curve $q \circ \gamma_k$ whose number of self-intersections $k - 1 + kd(r)$ is minimal. □

3.2.4 *Operations on curves on a surface*

It turns out that on the set of pairs [a closed oriented 2-surface, a curve on this surface] there are interesting operations which are invariant under Reidemeister moves. Namely let Σ be a fixed closed surface (we assume this surface to be oriented; otherwise, all arguments below work for the case of coefficients from $\mathbb{Z}_2$).

Let Γ_Σ be the set of all linear combinations of homotopy classes of oriented curves on the surface Σ with coefficients from a ground field k (in the non-orientable case it should be of characteristic 2).

Furthermore, let $\Gamma_\Sigma^2 = \Gamma_\Sigma \otimes \Gamma_\Sigma$ be the set of k-linear combinations of homotopy classes for ordered pairs of oriented curves on Σ, and let $\Gamma_{\Sigma,0}^2$ be the quotient space of the space Γ_Σ^2 modulo the following relation: $K \sqcup \bigcirc = 0$, i.e., we take to zero all those links having a diagram on Σ with two connected components, one of which is homotopy trivial. The linear space $\Gamma_{\Sigma,0}^2$ can be considered as a subspace in Γ_Σ^2.

Let γ_1, γ_2 be a pair of oriented curves generically immersed in Σ, and let $\gamma_1 \cap \gamma_2 = \{X_1, \ldots, X_N\}$ be the set of intersection points (it is finite becouse the curves are in general position). Let $m(\gamma_1, \gamma_2)_i$ be the oriented curve obtained by smoothing $\gamma_1 \cup \gamma_2$ at X_i in the way compatible with the orientation. We set

$$[\gamma_1, \gamma_2] = \sum_{i=1}^{N} \text{sign}_i(1,2) m(\gamma_1, \gamma_2)_k \in \Gamma_\Sigma, \qquad (3.5)$$

where $\text{sign}_i(1,2)$ denotes the sign of the crossing X_i, i.e., it is equal to one if the basis formed by tangent vectors $(\dot{\gamma}_1, \dot{\gamma}_2)$, is positive, and to -1, otherwise.

This sum is considered as an element from Γ_Σ. A direct check shows that the following theorem holds [Goldman, 1986]

Theorem 3.7.

(1) The map $[\cdot, \cdot] \colon \Gamma_\Sigma^2 \to \Gamma_\Sigma$ *is well defined.*

(2) For any γ_1, γ_2 *we have* $[\gamma_1, \gamma_2] = -[\gamma_2, \gamma_1]$.

(3) For any $\gamma_1, \gamma_2, \gamma_3$ *we have*

$$[[\gamma_1, \gamma_2], \gamma_3] + [[\gamma_2, \gamma_3], \gamma_1] + [[\gamma_3, \gamma_1], \gamma_2] = 0.$$

Thus, Γ_Σ has a Lie algebra structure with the commutator $[\cdot, \cdot]$ called *Goldman's bracket*.

Remark 3.4. An analogue of Goldman's bracket can be defined for flat and free knots. The formula (3.5) determines a map from the set of linear combinations of two-component flat (free) links (with coefficients in $\mathbb{Z}_2$ for the free links) to the set of linear combinations of flat (free) knots.

Let γ be an oriented curve in general position on the oriented surface Σ, and let $X_1, \ldots, X_N$ be the self-intersection points of γ. Then, as a result of smoothing γ at X_i in the way compatible with the orientation, we get two curves, one of which, $\gamma_{i,L}$, can be naturally called the *left* one, and the other one $(\gamma_{i,R})$ is called the *right* one. Thus, we can define a map

$$\Delta \colon \gamma \mapsto \sum_{i=1}^{N} (\gamma_{i,L} \otimes \gamma_{i,R} - \gamma_{i,R} \otimes \gamma_{i,L}). \tag{3.6}$$

This map is skew-symmetrical, however, it is not quite well defined: if we apply the first increasing Reidemeister move to the curve γ, then in the right part of the equality (3.6) we shall have additional summands of the form $\pm\gamma_0 \otimes \gamma' \mp \gamma' \otimes \gamma_0$, where γ_0 is a contractible curve, and γ' is a curve homotopic to γ.

Thus, in order to have a well-defined cobracket, we have to take the quotient of the set of curves by the relation taking the contractible curve to zero (in the tensor product, we set $0 \otimes a = a \otimes 0 = 0$). Thus, Δ maps correctly Γ_Σ to $\Gamma_{\Sigma,0}^2$. After composition with the natural inclusion of $\Gamma_{\Sigma,0}^2$

into Γ_{Σ}^2, it gives a map $\Delta\colon \Gamma_{\Sigma} \to \Gamma_{\Sigma}^2$ called *Turaev's delta* [Turaev, 1991]. It defines a coalgebra structure on Γ_{Σ}.

Remark 3.5. For free knots an analogue of Turaev's delta (3.6)

$$\Delta\colon \gamma \mapsto \sum_{i=1}^{N} \gamma_i,$$

where γ_i is a free two-component link obtained by the smoothing of γ at the selfcrossing X_i in the way compatible with the orientation, defines a map from linear combinations of free knots Γ_{Fr} to the set $\Gamma_{Fr,0}^2$ of linear combinations of two-component free links (with coefficients in $\mathbb{Z}_2$) with relations $K \sqcup \bigcirc = 0$.

3.3   Links as braid closures

3.3.1   *Classical case*

With each braid diagram, one can associate a planar knot (or link) diagram as follows.

Definition 3.1. The *closure* of a braid B is the link $Cl(B)$ obtained from B by connecting the lower ends of the braid with the upper ends; see Fig. 3.27.

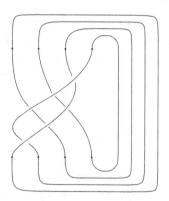

Fig. 3.27   A braid closure

Obviously, isotopic braids generate isotopic links.

Remark 3.6. Closures of braids are usually taken to be oriented: all strands of the braid are oriented from the top to the bottom.

Some braids generate knots; the others generate links. In order to calculate the number of components of the corresponding link, one should take into account the following simple observation. In fact, there exists a simple natural epimorphism from the braid group onto the permutation group $\Sigma : Br(n) \to S_n$, defined by $\sigma_i \mapsto s_i$, where s_i are natural generators of the permutation group, s_i transposes elements i and $i + 1$.

Consider a braid B. Obviously, for all numbers p belonging to the same orbit of the natural permutation action (of $\Sigma(B)$) on the set $1, \ldots, n$, all upper vertices with abscissas $(p, 0)$ belong to the same link component.

Consequently, we obtain the following proposition.

Proposition 3.3. *The number of components of the link of the closure $Cl(B)$ equals the number of orbits of action for $\Sigma(B)$.*

Obviously, non-isotopic braids might generate isotopic links. We will touch on this question later.

An interesting question is to define the minimal number of strands of a braid whose closure represents the given link isotopy class L. Denote this number by $Braid(L)$.

An interesting theorem on this theme belongs to Birman and Menasco [Birman and Menasco, 1991]:

Theorem 3.8. *For any knots K_1 and K_2, the following equation holds:*

$$Braid(K_1 \# K_2) = Braid(K_1) + Braid(K_2) - 1.$$

In Fig. 3.28 we show that if the knot K_1 can be represented by an n-strand braid, and K_2 can be represented by an m-strand braid, then $K_1 \# K_2$ can be represented by an $(n + m - 1)$-strand braid. This proves the inequality "$\leq$".

Let us also formulate the celebrated Alexander and Markov theorems, which are the most fundamental and important results on the interconnections between braids and links.

Theorem 3.9 (Alexander's theorem). *For each link L, there exists a braid B such that $Cl(B) = L$.*

Theorem 3.10 (Markov's theorem). *The closures of two braids β_1 and β_2 represent isotopic links if and only if β_1 can be transformed to β_2 by using a sequence of two transformations (Markov's moves), shown in Fig. 3.29 (on the right, both types of the additional crossing are admissible).*

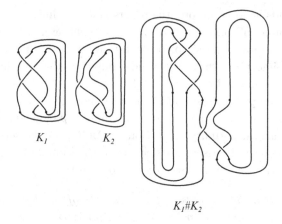

Fig. 3.28   Representing the connected sum of braids

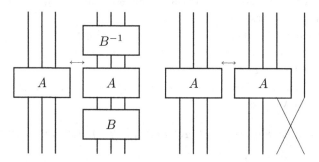

Fig. 3.29   Markov's moves

A systematic study of links via braid closures was done in the series of works by Birman [Birman, 1994] and Birman and Menasco [Birman and Menasco, 1991; Birman and Menasco, 1993; Birman and Menasco, 1990; Birman and Menasco, 1992a; Birman and Menasco, 1992b].

3.3.2   Virtual case

Analogously to classical braids, virtual braids admit *closures* as well; see Fig. 3.30. The obtained virtual link diagram will be *braided* with respect to some point A, see the right hand side of the figure.

The closure of a virtual braid is a virtual link diagram. Obviously, isotopic virtual braids generate isotopic virtual links. Furthermore, all virtual link isotopy classes can be represented by closures of virtual braids.

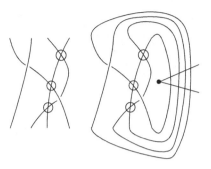

Fig. 3.30   Closure of a virtual braid

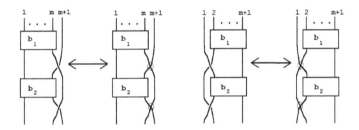

Fig. 3.31   The virtual exchange move

3.3.3   *An analogue of Markov's theorem in the virtual case*

In [Kamada, 2000], Seiichi Kamada proved an analogue of Markov's theorem for the case of virtual braids. Namely, he proved the following.

Theorem 3.11. *Two virtual braid diagrams have equivalent (isotopic) closures as virtual links if and only if they are related by a finite sequence of the following moves (VM0)–(VM3).*

(VM0) *braid equivalence;*
(VM1) *a conjugation (in the virtual braid group);*
(VM2) *a right stabilisation (adding a strand with additional positive, negative or* **virtual** *crossing) and its inverse operation;*
(VM3) *a right/left virtual exchange move; see Fig. 3.31.*

The moves (VM0)–(VM2) are analogous to those in the classical case.

The "new" move has two variants: the right one and the left one.

The necessity of the moves listed above is obvious; it is left for the reader as a simple exercise, see also [Manturov, 2015c]. For sufficiency, we refer the reader to the original work [Kamada, 2000].

Chapter 4

Two-dimensional Knots and Links

Similar to how classical 1-knots are embeddings of a circle into a 3-sphere, classical 2-knots are embeddings of a 2-sphere into $\mathbb{R}^4$ (or a sphere S^4), see Definition 4.1 later. The standard way of describing 2-knots is to consider their diagrams — two-dimensional complexes of a particular kind. A *diagram* is a generic projection of a knot in 4-space into a 3-subspace where for each double line it is said which sheet is "over" and which one is "under" and for every triple point the same is said for all three sheets incident to that point.

Two diagrams are *equivalent* if they correspond to the same knot; in other words their preimages are ambiently isotopic. In the one-dimensional case two diagrams represent the same knot if and only if they can be connected by a sequence of Reidemeister moves and trivial isotopies (that is, isotopies preserving the combinatorial type of the object). In the two-dimensional case it was shown by Roseman (see [Roseman, 1998]) that two diagrams correspond to the same knot if and only if they can be connected by a sequence of moves from a certain finite set (*Roseman moves*) and trivial isotopies. In the classical case it is imposed that all the complexes can be embedded into a 3-space and the moves to be compatible with the embedding.

Relaxing the embedding condition one obtains *abstract 2-knots* (see for example [Kamada and Kamada, 2000; Winter, 2015]). Forgetting the "over/under" structure one gets *free 2-knots*. They turn out not only to be important as a tool for the study of 2-knots in $\mathbb{R}^4$ but also are interesting all by themselves.

We begin with some basic definitions and constructions.

4.1   2-knots and links

Definition 4.1. A *2-knot* (resp. *an n-component 2-link*) is a generic smooth embedding of a 2-sphere S^2 (resp. disjoint union of n spheres) into $\mathbb{R}^4$ or S^4 up to isotopy.

There are several approaches to knot diagrams. We will use the "2-diagrams in 3-space" approach. Let us define it following Roseman [Roseman, 1998; Roseman, 2004] and Winter [Winter, 2015]. Note that in the same way one can define knot diagrams for any dimension.

Consider a 2-link $L \subset F \times [0,1]$ (in the classical case the 3-manifold F equals either $\mathbb{R}^3$ or S^3) and the restriction $\pi \colon L \to F$ of the natural projection $F \times [0,1] \to F$.

The *crossing set* D^* of a link L is the closure of the set $\{x \in \pi(L) \mid |\pi^{-1}(x)| \geq 2\}$. The *double point set* D is defined as $\pi^{-1}(D^*)$. The *branch point set* B is a subset of the link L where π fails to be an embedding. The images of branch points are called *cusps*. The set of *actual double points* D_0 is a subset of D consisting of $x \in L$ such that $|\pi^{-1}(\pi(x))| = 2$. We also define the *overcrossing set* D_+ as the set of $x \in D_0$ such that for the other point $y \in \pi^{-1}(\pi(x))$ holds $\pi_I(y) < \pi_I(x)$ with π_I denoting the natural projection $F \times [0,1] \to [0,1]$. The *undercrossing set* is defined as $D_- = D_0 \setminus D_+$.

We will say that a link L is in a *general position with respect to the projection* π if

(1) D is a union of embedded closed 1-submanifolds of the link L with normal crossings;

(2) B is a finite subset of the set D and for every $b \in B$ there exists an open neighborhood $U \subseteq D$ such that $U \setminus B$ consists of two components U_0, U_1 which are emebdded into F via π so that $\pi(U_0) = \pi(U_1)$.

For every link there is an isotopy sending it into a link in a general position with respect to π.

Definition 4.2. A diagram d of a link L in a general position with respect to the projection π is the image $\pi(L)$ of the link L under the projection π; every point of the crossing set with exactly two preimages is marked with information which preimage has greater coordinate in $[0,1]$.

Consider a 2-link diagram. It contains singularities of three types: double points, triple points and branch points (cusps), see Fig. 4.1. Triple points and cusps lie in the closure of the double point set. It is easy to see that this closure is an image of a one-dimensional manifold embedded into a 3-space. It is neat to include cusps into the set of double points. In that case the boundary of the double point manifold consists of cusps. Every triple point has exactly three preimages in the double points manifold.

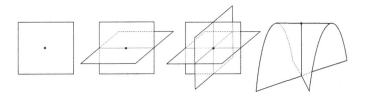

Fig. 4.1   Types of points of a 2-knot diagram

Roseman [Roseman, 1998] proved that two diagrams represent the same 2-link (so that preimages of the diagrams differ by an ambient isotopy) if and only if the diagrams can be connected by a finite sequence of isotopies of $\pi(L) \subset F$ and the so called *Roseman moves* shown in Fig. 4.2 ($\mathcal{R}_i$ stands for i-th Roseman move). Figures in Fig. 4.2 do not show which sheet is upper and which one is lower. The corresponding moves take place for every "over/under" structure. In other words, the following theorem holds:

Theorem 4.1 (Roseman [Roseman, 1998]). *Two diagrams represent equal 2-knots if and only if they can be connected by a finite sequence of* Roseman moves *shown in Fig. 4.2 and isotopies of the diagram in* $\mathbb{R}^3$.

Consider a 2-knot K. By definition it is an image of an embedding p of a 2-sphere S into $F \times [0,1]$. Consider its diagram D. It can be regarded as an image of the sphere S under a map $\psi = \pi \circ p$. Consider the preimages of double points, triple points and cusps on the sphere under this map. The resulting complex will be called a *spherical diagram* of the 2-knot. Formally one could say that a spherical diagram is obtained from a sphere by factorisation by the following relation: two points are equivalent if they correspond to the same point on a double line. Spherical diagram is a natural generalisation of the notion of a Gauß diagram of 1-knot and can be defined regardless of the 2-knot in 4-space. More precisely:

Definition 4.3. A *spherical diagram* is a 2-complex consisting of a sphere S and a set Γ of marked curves on it such that:

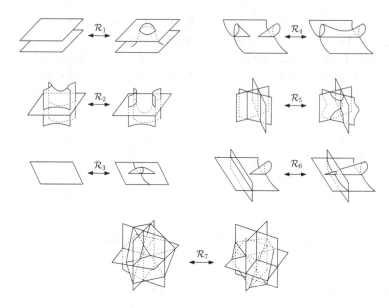

Fig. 4.2   Roseman moves

(1) Every curve is either closed or ends with two cusps, the total number of cusps is finite;

(2) Every curve of the set Γ is continuously pointwise paired with exactly one curve of that set, one of the paired curves is marked as *upper*, both curves are oriented (marked with arrows) up to simultaneous orientation change, the pairing respects the orientation given by the arrows;

(3) Two curves ending in the same cusp are paired and both arrows either look towards the cusp or away from it;

(4) If a point x of a curve $\alpha \in \Gamma$ is a crossing point, the corresponding point x' of the curve α' paired with the curve α is a crossing point as well; a triple point appears on the sphere S three times.

Remark 4.1. The pairing, the cusp, and the triple point conditions of Definition 4.3 essentially mean the following. First, consider all curves of the set Γ as parametrised curves. Consider a curve $\alpha \in \Gamma$ and its parametrisation $\alpha = \alpha(t), t \in I$. There is exactly one curve paired with it. Let us denote it by α' (this convention will be standard from now on). The curve α' is parametrised as well: $\alpha' = \alpha'(\tau), \tau \in I'$. Those curves are oriented. There is an orientation-respecting homeomorphism $h \colon I \to I'$, which establishes a pointwise pairing of those curves. Note that as a consequence a closed curve may only be paired with another closed curve.

The orientation is necessary for the following reason. Like every two points (ends of a chord) of a Gauß diagram of a 1-knot correspond to a crossing, every pair of paired curves corresponds to a double line of a 2-knot. But unlike the one-dimensional case, two curves can be glued in two ways differing by the curves' orientation. The arrows fix this ambiguity: the curves are glued with respect to the arrows.

The third condition of the definition (the cusp condition) means that the two curves ending in the same cusp yield one double line and the cusp is glued with itself, not the opposite end of one of the curves.

The fourth condition of the definition (the triple point condition) means the following. Consider a point $x \in \alpha$ at which the curve α intersects with a curve β (technically, that may be a selfcrossing point, in which case $\beta = \alpha$). If x is not a selfcrossing point, there is exactly one moment t_0, such that $\alpha(t_0) = x$. Otherwise there are two such moments: $\alpha(t_0) = \alpha(t_1) = x$. We shall call the moments t_0, t_1 the *crossing moments*.

Consider a crossing moment t. There is a point $x' = \alpha'(h(t)) \in \alpha'$. It must be a crossing point as well. Let us denote the curve intersected by α' at this point by γ. Thus we obtain a point $y = \gamma(\tau) \in \gamma$ coinciding with the point $x' = \alpha'(h(t)) \in \alpha'$. Therefore there is a point $y' \in \gamma'$ paired with it. It must be a crossing point as well. Continuing this procedure we obtain a set of crossing points naturally paired between themselves. The cardinality of this set must be equal to 3.

This last condition essentially means that there could be exactly two possibilities: (1) a curve α crosses a curve β, the curve α' crosses a curve γ and the curve β' crosses the curve γ'; or (2) a curve α has a selfcrossing point and the curve α' crosses the curves β and β'. The latter case appears, for example, on the spherical diagram of the right-hand side of the sixth Roseman move, see Fig. 4.2 and Fig. 4.3 below.

It is clear from the definition that every 2-knot has a spherical diagram. On the other hand, not every spherical diagram corresponds to a "real" 2-knot just like not every chord diagram corresponds to a classical 1-knot. It is worth noting that due to the decoration of the curves on a spherical diagrams with arrows and "over/under" information there is a bijection between space diagrams and spherical diagrams of any 2-knot K.

Equivalent diagrams of classical 2-knots are connected by Roseman moves $\mathcal{R}_1$–$\mathcal{R}_7$. Those moves have spherical counterparts $\mathcal{S}_1$–$\mathcal{S}_7$. Let us describe them in more detail.

Let us call a *local link* a proper embedding in a general position of a disjoint union of n discs into a ball D.

Remark 4.2. In the 1-dimensional case local links are well known under the name of *tangles*. Theory of tangles is well developed and useful in many contexts.

Images of the discs will be called the *sheets* of the local link. Diagrams of local links are defined exactly as those of (global) links. *Local moves* — moves on local links — remove a set of discs and replace it with a new one. Roseman moves can be regarded as local moves since no Roseman move changes diagram away from a small sphere. Local links appearing in Roseman moves can have one sheet (moves $\mathcal{R}_3$, $\mathcal{R}_4$), two (moves $\mathcal{R}_1$, $\mathcal{R}_2$, $\mathcal{R}_6$), three (move $\mathcal{R}_5$) or four sheets (move $\mathcal{R}_7$).

Let us introduce the notion of *spherical diagram of a local n-component link* (also called a *local diagram*).

Definition 4.4. A *spherical diagram of a local n-component link* is a 2-complex consisting of a sphere S, a set of disjoint discs U_i, $i = 1, \ldots, n$, on it and a set Γ of marked curves on it such that:

(1) Every curve lies on one of the discs U_i;
(2) Every curve is either closed or ends with a cusp or ends on the boundary of the corresponding disc, the number of cusps is finite;
(3) Every curve of the set Γ is continuously pointwise paired with exactly one curve of that set, one of the paired curves is marked as *upper*, both curves are oriented (marked with arrows) up to simultaneous orientation change, the pairing respects the orientation given by the arrows;
(4) Two curves ending in the same cusp are paired and both arrows either look towards the cusp or away from it;
(5) If a point x of a curve $\alpha \in \Gamma$ is a crossing point, the corresponding point x' of the curve α' paired with the curve α is a crossing point as well; a triple point appears on the sphere S three times.

A (Roseman) move on a local link induces a move on its spherical diagram. On the other hand, consider a given spherical diagram and a move on it changing the diagram only inside a finite set of disjoint discs $U_1, \ldots, U_n$ (though the set of curves Γ and their topology can change). Such move will be called *local*. More formally:

Local move consists of a removal of local link from a spherical diagram and glueing in another local link with the same boundary respecting the curves meeting the boundary.

It is naturally identified with a move on a local link diagram. Every Roseman move (or its spherical analog) corresponds to two local links: its left-hand side and right-hand side.

Thus, since every Roseman move $\mathcal{R}_i$ is a local move, it corresponds to a (local) move $\mathcal{S}_i$ on a spherical diagram. Those moves are depicted in Fig. 4.3. The same letters denote paired curves; if the same letters appear in the left-hand and right-hand sides of the moves, then they denote the curves in a natural one-to-one correspondence.

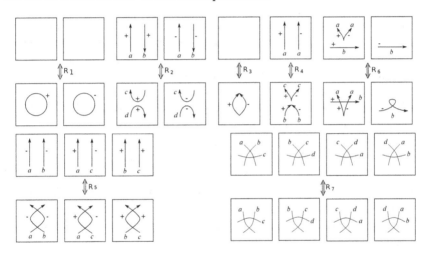

Fig. 4.3 Spherical Roseman moves; the upper curves are marked with plus

Note that the definition of the moves $\mathcal{S}_i$ does not require the spherical diagram to be a diagram of a "real" 2-knot. We introduce the following definitions:

Definition 4.5. An *abstract 2-knot* is an equivalence class of spherical diagrams modulo moves $\mathcal{S}_1 - \mathcal{S}_7$.

Definition 4.6. A *free 2-knot* is an abstract 2-knot without the "over/under" structure in double and triple points.

Additionally, note the following fact. Every Roseman move is performed as follows: one cuts a local link out and glues a different local link in.

Therefore, if a spherical diagram were a diagram of a real 2-knot, then the resulting diagram would also correspond to a real 2-knot.

4.2 Surface knots

In the previous section we considered 2-knots: embeddings of a sphere into a 4-space. This notion can be further generalised to the notion of *surface knots* and *links*.

Definition 4.7. A *surface knot* (resp. *an n-component surface link*) is a smooth embedding in general position of an oriented 2-surface S_g (resp. disjoint union of n surfaces $S_{g_1}, \ldots, S_{g_n}$) into $\mathbb{R}^4$ or S^4 up to isotopy.

Those objects are studied exactly in the same manner as 2-knots and 2-links: one may consider their diagrams as general position projection to the 3-subspace and then construct surface diagrams in the same way as spherical diagrams were constructed above:

Definition 4.8. A *surface diagram* is a 2-complex consisting of a orientable 2-surface S and a set Γ of marked curves on it such that:

(1) Every curve is either closed or ends with two cusps, the total number of cusps is finite;
(2) Every curve of the set Γ is continuously pointwise paired with exactly one curve of that set, one of the paired curves is marked as *upper*, both curves are oriented (marked with arrows) up to simultaneous orientation change, the pairing respects the orientation given by the arrows;
(3) Two curves ending in the same cusp are paired and both arrows either look towards the cusp or away from it;
(4) If a point x of a curve $\alpha \in \Gamma$ is a crossing point, the corresponding point x' of the curve α' paired with the curve α is a crossing point as well; a triple point appears on the sphere S three times.

Roseman moves are local and do not care for the global geometry of the object. Thus, they remain the same.

The notions of abstract and free surface knots are thus defined exactly as in the case of 2-knots substituting a sphere with an oriented 2-surface:

Definition 4.9. An *abstract surface knot* is an equivalence class of surface diagrams modulo moves $\mathcal{S}_1 - \mathcal{S}_7$.

Definition 4.10. A *free surface knot* is an abstract surface knot without the "over/under" structure in double and triple points.

4.3 Other types of 2-dimensional knotted surfaces

If we consider the definition of surface knots (Definition 4.7) and allow the surface S_g to have boundary, then the object appearing is known as a *surface knot with boundary*. When considering diagrams of such knots we impose an additional natural condition. Consider a boundary point x of the knot K and its image $y = \pi(x)$ under the projection to a 3-subspace. Then the full preimage $\pi^{-1}(y)$ must consist only of boundary points of the knot K. In other words, a boundary point and an interior point should not be projected to the same point in 3-space. To reformulate this condition in one more way: let the knot K be a knotted surface S with boundary $\partial S = \gamma_1 \sqcup \cdots \sqcup \gamma_k$; then the image of each boundary component γ_i should be a closed curve not intersecting the image of the interior of the knot $S \setminus \partial S$.

The definition of surface diagrams is also naturally extended to the case of knots with boundary:

Definition 4.11. A *surface diagram* of a surface knot with boundary is a 2-complex consisting of a orientable 2-surface S with boundary ∂S and a set Γ of marked curves on it such that:

(1) Every curve is either closed, ends with two cusps, the total number of cusps is finite, or ends on the boundary of the surface S;
(2) Every curve of the set Γ is continuously pointwise paired with exactly one curve of that set, one of the paired curves is marked as *upper*, both curves are oriented (marked with arrows) up to simultaneous orientation change, the pairing respects the orientation given by the arrows;
(3) Two curves ending in the same cusp are paired and both arrows either look towards the cusp or away from it;
(4) If a point x of a curve $\alpha \in \Gamma$ is a crossing point, the corresponding point x' of the curve α' paired with the curve α is a crossing point as well; a triple point appears on the sphere S three times.
(5) A curve with its endpoints on the boundary of the surface is paired with another curve with endpoints on the boundary of the surface S.

In the case of surface knots with boundary there are two types of double lines: the ones with endpoints on the boundary of the diagram (those double lines are called *boundary*) and the ones lying entirely in the interior of the diagram (they can be either closed, or end with two cusps; such double lines are called *interior*).

Moves on surface knots with boundary are of two types as well: either the ones performed away from the boundary, or the ones involving the boundary.

The first type of moves coincides with Roseman moves for surface knots without boundary. The second type can be obtained by cutting the local diagrams of Roseman moves by a plane so that the plane intersects all double lines transversally, does not go through triple points or cusps and and the cutting plane is naturally "the same" on the left-hand and on the right-hand side of the move.

Note that local diagrams (see Definition 4.4) provide a simple example of a diagram of a knot with boundary.

In the same way one can define *surface links with boundary*.

Another object, closely related to the theory of surface knots and links is *non-orientable surface knots and links*. Their definition is obtained directly from Definition 4.7 by relaxing the condition of orientability of the surface S.

Definitions of space and surface diagrams are repeated in this case verbatim with the condition of orientability of the surface S relaxed as well. Naturally, since Roseman moves are local and do not take into consideration the global topology of the surface, they remain the same in case of non-orientable knots or links as in case of oriented ones.

Allowing the surface S to have boundary, one obtains the notion of *non-orientable surface knots (links) with boundary*.

Remark 4.3. There exists one more type of objects closely connected with surface knots: *immersed 2-dimensional knotted surfaces*. This theory lies beyond the scope of the present book. A reader interested in that and other generalisations of the surface knot theory may be directed to the works [Kamada and Kamada, 2000; Carter, Kamada and Saito, 2004; Carter, 2015; Winter, 2015].

4.4   Smoothing on 2-dimensional knots

In the one-dimensional knot theory an important place belongs to the so-called *smoothing*: the process that deletes a crossing of a knot diagram connecting the endpoints of the four semi-arcs in one of the two possible non-intersecting ways (see Fig. 4.4).

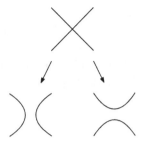

Fig. 4.4   Smoothing of a crossing of a 1-knot diagram

The notion of smoothing is important for the construction of many deep and well-known classical, virtual and free knot invariants (such as Kauffman bracket and parity bracket, see Section 5.1).

In this section we introduce the similar notion of smoothing on surface knots. In that case double lines are deleted and incident half-regions are reglued together to form a new knot or link.

4.4.1   *The notion of smoothing*

A *double line* of a surface knot in terms of surface diagrams is a pair of paired curves from the set Γ, see Definition 4.8.

A surface diagram S is stratified into 0-, 1-, and 2-dimensional components. 0-dimensional components of the diagram are *curve crossings* and *cusp* points, we denote their set by S_0; the set of 1-dimensional elements is denoted by S_1 and consists of the segments of curves $\gamma \in \Gamma$ between the elements of S_0 as well as the curves with no crossings and cusps on them; the set of 2-dimensional elements is denoted by S_2 and equals $S \setminus (S_0 \cup S_1)$.

We will call the elements of S_0 *vertices*, the elements of S_1 *edges* and the elements of S_2 *faces*.

The pairing of curves of the set Γ together with their orientation induces a natural pairing of the elements of the set S_1. Hence we can say that S_1 is broken into a disjoint union of two-element subsets: $S_1 = \sqcup S_1^i$. Likewise, S_0 is broken into a disjoint union of 1- and 3-element subsets: $S_0 = \sqcup S_0^i$. Here every 1-element subset consists of a single cusp point and every 3-element subset consists of three points from the 4th section of the surface diagram definition. By slightly abusing the notation we will call the former sets "cusps" and the latter sets "triple points"; we will also say that an edge α (or a curve $\gamma \in \Gamma$) is incident to a cusp (or a triple point) if it is incident to an element of the corresponding set S_0^i.

Note that the elements of S_2 are manifolds but may not be discs; we shall call them *pseudo-cells*. Note further that the Euler characteristic of the surface S is completely determined by the sets S_i by the usual formula if one takes into account that each pseudo-cell of the set S_2 and each edge of the set S_1 has its own Euler characteristic. Thus, if we denote $|S_0| := k_0, |S_1| := k_1, |S_2| := k_2$ with $\gamma_i \in S_1, i = 1, \ldots, k_1; C_i \in S_2, i = 1, \ldots, k_2$, then we have

$$\chi(S) = k_0 - \sum_1^{k_1} \chi(\gamma_i) + \sum_1^{k_2} \chi(C_i). \tag{4.1}$$

The set of edges S_1 is framed in the following sense. Consider a pair of paired edges $\alpha, \alpha' \in S_1^i$ for some i. There are two faces A_1, A_2 incident to α and two faces A_1', A_2' incident to α'. Those four faces are split into two pairs of formally opposite.

Moreover, the set of vertices S_0 is framed as well: for every element x of S_0 the elements of S_1 incident to x are split into pairs of formally opposite.

Now let us define *smoothing* of double lines.

Consider a surface knot K and a double line ξ. On the surface diagram S_K there is a pair of paired curves $\gamma, \gamma' \in \Gamma$ corresponding to the double line ξ. If we cut the surface S along the curves γ and γ', we obtain a disjoint union of pseudo-cells, the boundary of which consists of four curves: γ_1, γ_2 (obtained by cutting along γ) and γ_1', γ_2' (obtained by cutting along γ').

To smooth the double line ξ now we have to reglue those pseudo-cells along their boundary in one of the following ways: either γ_1 is glued to γ_1', γ_2 is glued to γ_2', or γ_1 is glued to γ_2', γ_2 is glued to γ_1'. The regluing is performed with respect to orientation of the curves and their splitting into the elements of the set S_1.

Consider a triple point $S_0^i = \{x_1, x_2, x_3\}$ incident to either of the curves γ, γ' (in other words, a triple point on the double line ξ). If there is exactly one $x_j \in S_0^i$ not incident to either of the curves γ, γ', the regluing induces a natural 1-dimensional smoothing of the that crossing (see Fig. 4.5).

After that regluing is performed, one may remove the curves γ, γ' from the diagram. That completes the smoothing process.

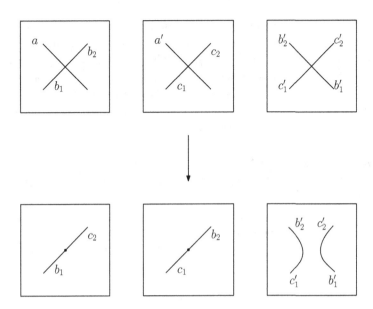

Fig. 4.5   Induced 1-dimensional smoothing

4.4.2   *The smoothing process in terms of the framing change*

The same procedure can be regarded from a different point of view.

Regluing of the pseudo-cells, defined in Section 4.4.1, essentially means a change of framing of the edges, which lie on the smoothed double line (with framing of vertices changed accordingly). The following algorithm describes the process of smoothing of a double line whose preimages on the surface diagram $\mathcal{S}$ will be denoted by γ, γ'.

We will need a preliminary definition.

Consider a surface diagram $\mathcal{S}$ and the curves $\gamma, \gamma' \in \Gamma$. Let them have a common point. In that case a small neighbourhood U of $\gamma \cup \gamma'$ is a connected 2-manifold. Consider a smoothing of the double line, corresponding to those curves. Under the smoothing the neighbourhood U transforms into a new 2-manifold U'.

Definition 4.12. The smoothing is called *locally connected* if the manifold U' is connected.

Now we have all the necessary tools to describe the smoothing process in the form of an algorithm. We adopt the following convention. Consider a double line γ of a surface knot K. On the surface diagram it corresponds to two curves — let us call them γ, γ'. Those curves consist of arcs (elements of the set S_1) and vertices (elements of the set S_0). From now on saying "remove a vertex v from a curve" we mean that v is removed from the set S_0 and the two arcs with the common endpoint v are replaced by one arc — the union of their closures.

The smoothing algorithm

0. Begin with a surface diagram $\mathcal{S} = (S_0, S_1, S_2)$ with the appropriate framing.

1. Change the framing of the edges corresponding to the curves γ, γ'.

2. For every triple point $p = (x_1, x_2, x_3) \subset S_0$ perform the following.

If the three vertices x_1, x_2, x_3 corresponding to this triple point are of the form

$$x_1 = \gamma \cap \alpha, \ x_2 = \gamma' \cap \beta, \ x_3 = \alpha' \cap \beta'$$

and both γ and γ' appear only in two crossings of those three (in other words, neither of the curves $\alpha, \alpha', \beta, \beta'$ is the same as γ or γ'), then transform the crossing $\alpha' \cap \beta'$ into a segment as shown in Fig. 4.6 choosing the transformation with respect to the framing from the step 1.

Clarification: in Fig. 4.6 we have segments $\alpha_1, \alpha_2, \alpha'_1, \alpha'_2, \beta_1, \beta_2, \beta'_1, \beta'_2 \in S_1$; every segment is an arc of a curve with the same name without an index, e.g. α'_1 is an arc of the curve α'.

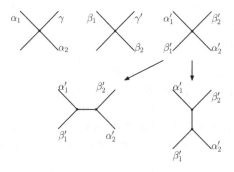

Fig. 4.6   Crossing transformation at the step 2

Transform the sets S_0, S_1 as follows: $S_0 \rightarrow (S_0 \setminus \{x_3\}) \cup \{x'_3, x''_3\}$, $S_1 \rightarrow S_1 \cup \{s\}$.

Denote the set of all segments added by $S^+ \subset S_1$ and the set of all vertices added by $V^+ \subset S_0$. Note that at this point the diagram is no longer a surface diagram of a free knot. The next steps restore it to the proper state.

3. For every point incident both to the curves γ and γ' endow this point (which can be either a crossing or a cusp) with index 1 if the chosen smoothing is locally connected and with index 2 in other case.

4. Reglue the pseudo-cells according to the chosen framing of the edges.

5. Transform the diagram in the following way: $S_0 \to S_0 \setminus V^+, S_1 \to S_1 \setminus (S^+ \cup \gamma \cup \gamma')$. S_2 is changed automatically. The resulting triple (S_0, S_1, S_2) is a diagram of a surface link.

Those 5 steps transform a surface knot into a surface link. To be more precise, they transform a knotted 2-manifold into a knotted 2-manifold (not necessarily oriented or connected).

4.4.3 Generalised F-lemma

The generalised F-lemma follows from the algorithm described in the previous section:

Lemma 4.1 (Generalised F-lemma). *Let K be a surface knot of genus g and K' be a complex obtained from K by smoothing a double line. Then the following holds:*

$$\chi(K) \leq \chi(K').$$

Remark 4.4. Note that in this lemma we don't require (and don't state) that the resulting complex represents a connected or oriented manifold. In other words, smoothing may transform a knot into a link and may preserve orientability or lose it. In any case, the Euler characteristic (understood as the sum of the Euler characteristics of all components of a link) can only increase.

Proof. To smooth a double line we perform the algorithm given above. Let us stop after the step 3. At that point the complex obtained is not a surface diagram but it is easy to calculate the Euler characteristic of the surface obtained.

Note that the pseudo-cells are unchanged by the algorithm. Only their framing was changed but it is irrelevant for the purposes of Euler characteristic.

The number of vertices was increased by the number of crossings, satisfying the conditions of the step 2 (that number will be denoted by n_2) and the number of crossings from the step 3 with index 2 (that number will be denoted by n_3).

The set of edges was expanded by n_2 segments since every transformation described produces exactly one new edge homeomorphic to a segment. That means that due to the formula (4.1) we obtain

$$\chi(K') = k_0 + n_2 + n_3 - \left(\sum_1^{k_1} \chi(\gamma_i) + n_2 \right) + \sum_1^{k_2} \chi(C_i) = \chi(K) + n_3.$$

In other words, the Euler characteristic was increased by the number of crossings satisfying the conditions of the step 3. Steps 4 and 5 do not change the Euler characteristic of the complex. That concludes the proof. $\quad\square$

That lemma has an immediate corollary:

Lemma 4.2. *Let K be a surface knot of genus g and K' be a complex obtained from K by smoothing a double line. Let K' be a connected oriented manifold. In that case, K' is a surface knot of genus $g' \leq g$.*

Considering a standard complex of genus 0, that is a diagram of a 2-knot, we obtain the following particular case of the generalised F-lemma:

Lemma 4.3 (F-Lemma, [Fedoseev and Manturov, 2019b]).
For each standard complex K of genus 0 with boundary $\partial K = \Gamma$, there exists a smoothing K' of K of genus 0.

Moreover, those statements have obvious analogs for surface links and the smoothing of their double lines. In particular, the following holds:

Lemma 4.4. *Let L be an n-component surface link and L' be an n'-component surface link obtained from L by smoothing of a double line. Let us denote the total genus of L by g and the total genus of L' by g'. Then we have*

$$g' \leq g + (n' - n). \tag{4.2}$$

Proof. The proof of that fact is trivial. Since all the link components are spheres with handles, we have

$$\chi(L) = \sum_{i=1}^{n}(2 - 2g_i),$$

$$\chi(L') = \sum_{i=1}^{n'}(2 - 2g_i'),$$

where g_i, g_i' denote the genera of the components of the links L, L', respectively. Since $\chi(L) \leq \chi(L')$, we obtain the necessary estimate. $\square$

PART 2
Parity Theory

Chapter 5

Parity in Knot Theories. The Parity Bracket

The foundation for *parity theory* was laid in the paper [Manturov, 2010] for one-dimensional (virtual) knots. The main idea behind it is:

All crossings of a diagram of a (virtual) knot can be naturally split into even *ones and* odd *ones so that this parity behaves nicely under Reidemeister moves.*

Many knot invariants may be refined via parity and it is possible to create new invariants of knots valued in *knot diagrams* using parity techniques, see [Ilyutko, Manturov and Nikonov, 2011]. Such invariants are studied in the present book. Due to existence of "picture-valued" (that is, diagram-valued) invariants the following principle holds for virtual knots:

If a diagram is complicated enough, then it can be found as a subdiagram in any equivalent diagram.

An example of the application of this principle can be seen in Fig. 0.3.

Virtual knots admit many different parities. The simplest of them, the Gaußian one, see Definition 5.1 below, comes from Gauß (chord) diagram of a knot: every crossing corresponds to a chord of the Gauß diagram; a crossing is called *even* if the corresponding chord is linked with an even number of chords, and *odd* in the other case. If the given knot is classical, then all chords of its Gauß diagram are even. This fact was known to Gauß himself. Therefore, the Gaußian parity on *all* classical knots is trivial. Moreover it is proved that all parities on classical knots are trivial (this fact is technical and we will not prove it here; for details see, for example, [Ilyutko, Manturov and Nikonov, 2014]). For virtual knots the Gaußian parity is defined exactly the same way and is non-trivial.

There exists a general approach to the parity theory. It is possible to define a set of *parity axioms* so that the theorems and invariants obtained by using parity hold for any parity (that is a map from the set of crossings to the set {even, odd}) satisfying the axioms.

Consider for example free knots. In the classical knot theory every crossing is decorated with "over/under" structure telling which arc of the crossing is over (its coordinate along the projection direction greater) and which is under. Thanks to the parity even without this structure one can create new strong invariants of coarse objects. Free knots are a rough factorisation of virtual knots obtained by "forgetting" the "over/under" structure in the crossings. Initially Turaev [Turaev, 2007] conjectured that free knots were trivial. By using parity arguments, the first named author disproved this conjecture in [Ilyutko, Manturov and Nikonov, 2011] (this conjecture was also disproved independently roughly at the same time by Gibson [Gibson, 2011] using different techniques).

There are many different approaches leading to different parities. One of the natural sources of parities is (co-)homology of the ambient space. Informally one can say that in the classical case there are no non-trivial parities since plane has trivial homology groups. Nevertheless, many parity theoretical methods can be applied to the classical case due to a number of topological constructions (see [Chrisman and Manturov, 2012; Krasnov and Manturov, 2013; Manturov, 2015a]).

5.1 The Gaußian parity and the parity bracket

In the present section we define the most basic but important example of parity: the Gaußian parity, and use the parity techniques to define the *parity bracket*, a mapping from knot diagrams into knot diagrams which is an invariant of free knots.

This construction is extremely important for it delivers a prominent example of the main principle studied in the present book:

if an object is complicated enough, then it serves as its own invariant.

In this particular case, as we will show in the present section, if a diagram of a free knot is *odd* and *irreducible*, then it can be found as a subdiagram inside any equivalent diagram, that is, it can be obtained from it as a result of smoothing.

That essentially means that we work with a *picture-valued* invariant: the invariant object is not a group or a number, but a linear combination of pictures (diagrams).

5.1.1 *The Gaußian parity*

Consider a chord (Gauß) diagram D of a knot K. Note that the notion of chord diagram is defined (see Section 3.1) for the elements of different knot theories (classical knots, virtual knots, free knots, etc.). Consider a chord c of the diagram D. We say that a chord c' is *linked* with the chord c if their endpoints alternate when walking around the base circle of the diagram D in a given direction. In particular, a chord is not considered linked with itself.

For each chord c we can calculate the number of chords linked with the chord c and take the parity of this number. That leads to the following definition:

Definition 5.1. A mapping p_G from the set of chords of a chord diagram to $\mathbb{Z}_2$ such that

$$p_G(c) \equiv |\{\text{chords linked with the chord } c\}| \mod 2$$

is called the *Gaußian parity*, the value $p_G(c)$ is called the Gaußian parity of the chord c.

Since every chord of a chord diagram of a classical knot is linked with evenly many other chords of the diagram, every chord of a classical knot chord diagram is *even* (mapped into 0) under the Gaußian parity. But if one considers virtual or free knots, then it is not necessarily true. Indeed, since a free knot is an equivalence class of chord diagrams modulo Reidemeister moves, one can easily construct chord diagrams with chords linked with oddly many other chords of the diagram (see, for example, Fig. 5.1: there are four chords, two of which (the "long" ones) are even and two (the "short" ones) are odd).

Let us point out some basic properties of the Gaußian parity.

1. Consider the first Reidemeister move. In terms of chord diagram (see Fig. 3.18) it means either addition or removal of a chord unlinked with all other chords of the diagram. Evidently it means that this chord is even in the Gaußian sense.

2. Consider the second Reidemeister move. In that move two adjacent chords are either added or deleted. Those chords are either linked with

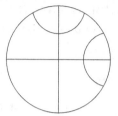

Fig. 5.1   An example of a chord diagram with *odd* chords

each other or not, but they are "close" to each other in the following sense. Let us denote the chords with letters a and b. Then walking around the base circle of the diagram starting at one of the endpoints of the chord a and writing down the letters corresponding to the encountered vertices we obtain a word W which for some choice of the orientation of the base circle is either of the form $W = abv_1abv_2$ or $W = abv_1bav_2$ where v_1, v_2 are some words without the letters a and b (possibly empty) see Fig. 5.2. That means that the Gaußian parity of the two chords participating in the second Reidemeister move is the same.

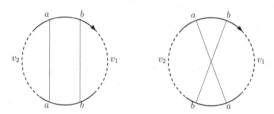

Fig. 5.2   The chords appearing in Ω_2 have the same Gaußian parity

3. Consider the third Reidemeister move. Three chords participate in this move. In all situations the total number of odd chords among them is even. Thus either two or zero chords, participating in the third Reidemeister move, are odd.

4. Note that Reidemeister moves are local in the sense that they do not interact with other chords of the diagram: the chords participating in the moves do not become linked or unlinked with other chords of the diagram. Therefore the parity of the chords not participating in a Reidemeister move is preserved by the move.

Later we will see that those properties are defining for the general notion of parity.

5.1.2 *Smoothings of knot diagrams*

Now we will proceed to the construction of the *parity bracket* for knots. But we still need some preliminary considerations on the subject of smoothings of a knot diagram. Since any knot diagram may be considered as a framed 4-graph (see Definition 5.5) with, possibly, additional information in the vertices, we will work in the most broad category: free knots, which can be interpreted as equivalence classes of framed 4-graphs modulo Reidemeister moves of their chord diagrams.

Let a framed 4-graph K be a diagram of a free knot. In Section 3.1 we discussed the notion of smoothing of a vertex of a diagram. Now we extend this notion: we allow a smoothing to be performed at *several* vertices. Later on "smoothing" is used both in the sense of the smoothing operation and the resulting graph. Given a smoothing s we denote the resulting diagram by K_s. Here s (also called a *state*) is a choice of smoothing for each smoothed vertex (recall that there are two possible smoothings for each vertex). By definition the result of the smoothing of a diagram in the empty set of vertices is the diagram itself.

Let K be a diagram and let $v_1, \ldots, v_n$ be all its even crossings under the Gaußian parity p_G. The smoothing at all even crossings of the diagram is called an *even smoothing* of K. There exist 2^n even smoothings. Moreover, if an even smoothing produces a graph with one unicursal component, then such smoothing is called a *1-even smoothing*.

For a given graph we single out the following sets of smoothings:

- the set S of all smoothings;
- the set S_{even} of all even smoothings;
- the set S_1 of all smoothings having one unicursal component;
- the set $S_{even,1}$ of all 1-even smoothings.

The elements of those sets will be denoted by s, s_{even}, s_1 and $s_{even,1}$, respectively.

Now consider framed 4-graphs with one unicursal component modulo the equivalence relation generated by the second Reidemeister move. Let us define the linear space $\mathfrak{G}$ as the set of $\mathbb{Z}_2$-linear combinations of such equivalence classes.

Definition 5.2. The linear space $\tilde{\mathfrak{G}}$ is the set of $\mathbb{Z}_2$-linear combinations of framed 4-graphs modulo the following equivalence relations:

(1) the second Reidemeister move Ω_2;
(2) the relation $K \sqcup \bigcirc = 0$, i.e. if a framed 4-graph has more than one component and at least one component is trivial, such graph is assumed to be zero in the space $\tilde{\mathfrak{G}}$.

We have a natural map $g : \tilde{\mathfrak{G}} \to \mathfrak{G}$ which act by taking to zero all equivalence classes of framed graphs with more than one unicursal component. The map g is an epimorphism of groups.

Now we have the necessary tools to define two bracket invariants.

5.1.3 *The parity bracket invariant*

Let K be a framed 4-graph. We define the bracket $\{\cdot\}$ by the formula

$$\{K\} = \sum_{s_{even}} K_{s_{even}}.$$

That is, we take all even smoothings of the diagram K (considered as elements of $\tilde{\mathfrak{G}}$) and take the formal sum of them. Naturally, $\{K\} \in \tilde{\mathfrak{G}}$.

Theorem 5.1. *The bracket $\{\cdot\}$ is an invariant of free links.*

Proof. To prove this theorem we need to consider two diagrams K, K' which differ by a Reidemeister move and prove that $\{K\} = \{K'\}$.

First, let K' be obtained from K by an increasing first Reidemeister move Ω_1. That means that in the diagram K' we have one extra vertex v, which is even as follows from the parity properties. One of the smoothings of the diagram K' at v leads us to a split component in such a way that every even state of the diagram K' where the vertex v is smoothed in the "wrong" way, will lead to a split trivial component. This will yield a trivial element from $\tilde{\mathfrak{G}}$.

The smoothing of K' at v performed in the "right" way will lead us to the diagram K, which yields a one-to-one correspondence for the even smoothings for K and K' with no unknots.

Thus we have proved the invariance of the bracket $\{\cdot\}$ under the first Reidemeister move.

For the second and third Reidemeister moves Ω_2 and Ω_3 we shall show that $\{K\} + \{K'\} \equiv 0 \mod \mathbb{Z}_2$ by means of partitioning of all diagrams from $\{K\} + \{K'\}$ representing non-trivial elements from $\tilde{\mathfrak{G}}$ into pairs. Since $\tilde{\mathfrak{G}}$ is a linear space over $\mathbb{Z}_2$, this will mean that $\{K\} = \{K'\}$.

Let K' be obtained from K by a second Reidemeister move adding two crossings v_1 and v_2. If both crossings are odd, then there is an obvious one-to-one correspondence between the set of even smoothings of the diagram K and the set of even smoothings of the diagram K'. The corresponding smoothings are obtained from each other by applying "the same" Reidemeister move to the vertices v_1 and v_2. If both v_1 and v_2 are even, then there exist four smoothings of K' at these vertices, see Fig. 5.3.

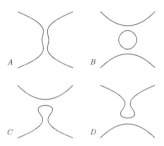

Fig. 5.3 Possible smoothings of the crossings appearing in the second Reidemeister move

First we note that the smoothing B has a trivial component and thus does not contribute to $\{K'\}$.

Next, even smoothings of type A are naturally in one-to-one correspondence with the even smoothings of the diagram K and give rise to framed 4-graphs.

Finally, the smoothings C and D are essentially identical and thus cancel each other in $\tilde{\mathfrak{G}}$ since it is a linear space over $\mathbb{Z}_2$.

Therefore, invariance under the second Reidemeister move is verified.

Now assume that the diagram K is taken to a diagram K' by a third Reidemeister move Ω_3. Among the three crossings of K taking part in the Reidemeister move, either all three ones are even, or one crossing is even, and the other two ones are odd.

First consider the case when all three vertices are even. In that case we have 8 possible smoothings for the crossings of the diagram K and 8 possible smoothings for the crossings of the diagram K' depicted in Fig. 5.4.

Studying those cases we deduce that the following equalities hold: $A = A' = 0$, $B = C'_1$, $C_1 = C_3$, $C'_2 = C'_3$, $B' = C_2$, $D_1 = D'_2$, $D_2 = D'_1$, $D_3 = D'_3$.

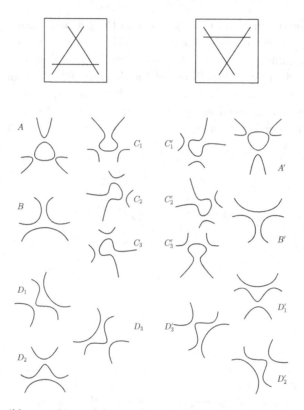

Fig. 5.4 Possible smoothings of the crossings appearing in the third Reidemeister move

We see that the cases C_1 and C_3 cancel each other as well as the cases C_2' and C_3'; also the cases A and A' do not contribute to the bracket. Therefore only five summands are left on both the left-hand side and the right-hand side. In those five cases we have a one-to-one correspondence and thus the equality $\{K\} = \{K'\}$ holds.

Similarly, in case of one even crossing we have two possible smoothings for the diagram K and two possible smoothings for the diagram K'. It is easy to see that they are in one-to-one correspondence as well: "vertical" smoothings of both diagrams yield the same summand and the summands obtained by the "horizontal" smoothings are related by the application of the second Reidemeister move twice.

That proves that $\{K\} = \{K'\}$ in $\tilde{\mathfrak{G}}$. $\qquad\square$

Definition 5.3. Let K be a diagram (a framed 4-graph). Then its parity bracket $[K]$ is defined by the formula

$$[K] = \sum_{s_{even,1}} K_{s_{even,1}}.$$

Obviously, $[K] = g(\{K\})$ and thus is an invariant of free links.

Now we will use this invariant to prove an important result.

We say that a framed 4-graph K is *irreducible* if a decreasing second Reidemeister move can't be applied to K. Now fix the Gaußian parity and consider an irreducibly odd graph with one unicursal component. An example of such graph is depicted in Fig. 5.5.

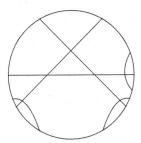

Fig. 5.5   An irreducibly odd framed 4-graph

Since the graph is odd, its bracket $[K]$ consists of only one summand — graph K itself. This graph is its own minimal representative. Moreover, since the graph K is irreducible, it equals itself in the space $\mathfrak{G}$. Therefore we obtain the following statement:

Lemma 5.1. *For an irreducibly odd framed 4-graph K with one unicursal component, the following equality holds:*

$$[K] = K.$$

Thus we can say that a *local* minimality yields *global* minimality: the fact that no single decreasing Reidemeister move can be applied to the diagram (the second due to irreducibility, the first and the third for the reason of the graph being odd) yields the minimality of the graph. Note that on the left-hand side of this equality, the graph K is considered as a representative of a free knot, and on the right-hand side it is considered as an element from $\mathfrak{G}$, that is, a single graph. That means that an irreducibly odd graph is its own invariant.

Lemma 5.1 yields the following important theorem:

Theorem 5.2. *Let K be a framed 4-graph, whose Gauß diagram is irreducibly odd. Then for every framed 4-graph K' representing the same free knot as K there exists a smoothing which is isomorphic to the framed 4-graph K (as a framed 4-graph).*

This theorem exactly means that if the graph is complicated enough (in this case it has to be irreducibly odd) it can be found as a subdiagram (with respect to smoothings) of any equivalent graph. Thus we obtain an example of the stated main principle.

Lemma 5.1 leads to another important result. Consider a graph Γ which is a representative of the trivial free knot. For obvious reasons we have $[\Gamma] = 0$. On the other hand, consider the irreducibly odd graph K depicted on Fig. 5.5. Due to Lemma 5.1 we have $[K] = K \neq 0$ in the space $\mathfrak{G}$. Therefore the graph K is a representative of a non-trivial free knot. Thus we get the following:

Corollary 5.1. *The theory of free knots is non-trivial.*

Remark 5.1. This corollary delivers an answer to the question originally stated by Turaev [Turaev, 2007]: whether there exist non-trivial free knots. This result was obtained by the first named author [Ilyutko, Manturov and Nikonov, 2011] and, almost at the same time, independently, by Gibson [Gibson, 2011] using completely different techniques.

5.1.4 *The bracket invariant with integer coefficients*

In the previous sections we defined a bracket invariant valued in a space of $\mathbb{Z}_2$-linear combinations of graphs. In the present section we develop this idea further, constructing a bracket invariant of knots valued in $\mathbb{Z}$-linear combinations of graphs. Note that unlike the previous section, here we work only with knots, not with links.

To be precise, consider the following linear space:

Definition 5.4. The linear space $\tilde{\mathfrak{G}}'$ is the set of $\mathbb{Z}$-linear combinations of framed 4-graphs modulo the following equivalence relations:

(1) the second Reidemeister move Ω_2;
(2) the relation $K \sqcup \bigcirc = (-2) \cdot K$, i.e. each trivial connected component of a framed 4-graph can be removed multiplying the rest by (-2) in the space $\tilde{\mathfrak{G}}'$.

Now we define the bracket invariant $\{\cdot\}_{\mathbb{Z}}$ sending free knot diagrams into the space $\tilde{\mathfrak{G}}'$ by the following formula:

$$\{K\}_{\mathbb{Z}} = (-1)^{n_K^e} \sum_{s_{even}} K_{s_{even}},$$

where n_K^e stands for the number of even crossings of the diagram.

A simple check alongside the lines of the proof of Theorem 5.1 gives the following

Theorem 5.3. *The bracket $\{\cdot\}_{\mathbb{Z}}$ is an invariant of free knots.*

Remark 5.2. The bracket $\{\cdot\}_{\mathbb{Z}}$ coincides with the value of the *even Jones polynomial* $X_{even}(a)$ defined in [Ilyutko and Manturov, 2015] at $a = -1$.

5.2 The parity axioms

Earlier we defined the Gaußian parity and discussed some of its properties. In this section we develop the general parity theory: a *parity* is understood as a mapping from crossings of a knot diagram (or, equivalently, from the set of chords of a chord diagram) to $\mathbb{Z}_2$ satisfying certain *parity axioms*. Those axioms essentially stem from the discussed properties of the basic parity example.

Consider a knot theory. As before saying "knot theory" we mean that we are given a class of diagrams and equivalence relation generated by some *moves*, that is local transformations of some sort. In each particular case we shall indicate the exact list of those transformations.

Each diagram represents some class of graphs (maybe with an additional structure at vertices), and some 4-valent vertices may have a framing. Here by *framed 4-graph* we mean the following:

Definition 5.5. A *framed 4-graph* is a finite 1-complex such that its every connected component is either a circle (with no vertices) or a 4-graph such that at every vertex four half-edges incident to it are organised into two pairs; half-edges in each pair are called *formally opposite*.

In some cases (for example, in classical knot theory) we impose some restrictions on the moves.

Examples of such theories are the theory of classical and virtual knots, the theory of flat knots, the theory of braids (classical or virtual), the theory of free knots. Recall that in the case of virtual knots we use the notion of "crossing" only for classical crossings unless otherwise indicated. We ignore the detour move since this move does not change a mutual position of crossings.

Now we take all diagrams of knots of this theory and denote it by $\mathcal{D}$. The set of all crossings of the diagrams of the set $\mathcal{D}$ will be denoted by $\mathfrak{C} = \mathfrak{C}(\mathcal{D})$. To define a *parity* for this knot theory, we decorate every crossing of the set $\mathfrak{C}$ with either 0 (such crossings are called *even*) or 1 (such crossings are called *odd*). In other words, a parity is a mapping from the set $\mathfrak{C}$ to the set $\mathbb{Z}_2$. But for such mapping to be a parity, certain axioms must be satisfied. To be precise:

(1) **Correspondence axiom:** for a move transforming a diagram D into a diagram D' and for every crossing v of the diagram D which has a naturally corresponding to it crossing v' of the diagram D' we have $p(v) = p(v')$;

(2) **Ω_1-axiom:** for a vertex v of a loop (which is the subject of a decreasing first Reidemeister move) we have $p(v) = 0$;

(3) **Ω_2-axiom:** for two vertices v_1, v_2 of a bigon (which are the subject of a decreasing second Reidemeister move) we have $p(v_1) + p(v_2) = 0$;

(4) **Ω_3-axiom:** for three vertices v_1, v_2, v_3 of a triangle (which are the subject of a third Reidemeister move) we have $p(v_1) + p(v_2) + p(v_3) = 0$.

Definition 5.6. A *parity* for a knot theory is a mapping $p \colon \mathfrak{C} \to \mathbb{Z}_2$ satisfying the axioms 1–4.

Let us now consider the Gaußian parity defined before from the standpoint of parity axioms.

Example 5.1. Consider the theory of classical, virtual, flat or free knots. Then define the following mapping p_G: for every crossing v $p_G(v) = 0$ if the corresponding chord of the knot chord diagram is linked with evenly many other chords and $p_G(v) = 1$ otherwise, in other words, consider the Gaußian parity as defined in Definition 5.1.

As was observed before, this parity possesses the properties 1–4 (see Section 5.1) and thus the axioms 1–4 are satisfied. Therefore, the Gaußian parity is indeed a parity.

5.3   Parity in terms of category theory

In this section we define the notion of parity even more generally using category language following the book [Ilyutko, Manturov and Nikonov, 2011].

Consider a knot $\mathcal{K}$. Exactly as in the previous section we consider an arbitrary knot theory and a knot is an equivalence class of diagrams modulo moves of this theory.

Let us define the category $\mathfrak{K}$ of diagrams of the knot $\mathcal{K}$. The objects of $\mathfrak{K}$ are diagrams of $\mathcal{K}$ and morphisms of the category $\mathfrak{K}$ are (formal) compositions of elementary morphisms. Here by an *elementary morphism* we mean either an isotopy or a Reidemeister move.

Reidemeister moves can change the number of vertices of a diagram (for example, under the first Reidemeister move the number of vertices is changed by one and under the second — by two), therefore, there may be no bijection between the sets of vertices of two diagrams connected by a sequence of the Reidemeister moves (we have the natural bijection only for the vertices of diagrams connected by the third Reidemeister move). But we still want to define some "correspondence" between the vertices after applying moves. To construct any connection between two sets of vertices for all cases of the moves we shall introduce the notion of a partial bijection which means just a bijection between the subsets of vertices, corresponding to each other in the two diagrams (see [Ilyutko and Manturov, 2013]).

Definition 5.7. A partial bijection of sets X and Y is a triple $(\tilde{X}, \tilde{Y}, \varphi)$, where $\tilde{X} \subset X$, $\tilde{Y} \subset Y$ and $\varphi \colon \tilde{X} \to \tilde{Y}$ is a bijection.

Now let us denote by $\mathcal{V}$ the vertex functor on $\mathfrak{K}$, i.e. a functor from $\mathfrak{K}$ to the category, objects of which are finite sets and morphisms are partial bijections. For each diagram K we define $\mathcal{V}(K)$ to be the set of classical crossings of K, i.e. the vertices of the underlying framed 4-graph. Any elementary morphism $f \colon K \to K'$ naturally induces a partial bijection $f_* \colon \mathcal{V}(K) \to \mathcal{V}(K')$.

Now we shall define a parity with coefficients in an arbitrary abelian group. Note that one can define a parity valued in a non-abelian group as well. Let A be an abelian group.

Definition 5.8. A parity p on diagrams of a knot $\mathcal{K}$ with coefficients in an abelian group A is a family of maps $p_K \colon \mathcal{V}(K) \to A, K \in ob(\mathfrak{K})$ is an object of the category, such that for any elementary morphism $f \colon K \to K'$

the following holds:

(1) $p_{K'}(f_*(v)) = p_K(v)$ in case if $v \in \mathcal{V}(K)$ and there exists $f_*(v) \in \mathcal{V}(K')$;
(2) $p_K(v_1) + p_K(v_2) = 0$ if f is a decreasing second Reidemeister move and v_1, v_2 are the disappearing crossings;
(3) $p_K(v_1) + p_K(v_2) + p_K(v_3) = 0$ if f is a third Reidemeister move and v_1, v_2, v_3 are the crossings participating in this move.

Note that the first condition in Definition 5.8 means that a parity respects the partial bijection, i.e. the corresponding crossings of two diagrams are of the same parity.

Note also that each knot class can have its own group A. So, one can speak of a parity for a given knot.

Originally, parity axioms were described in a slightly different language, as shown in the previous section. In particular the following axiom was required:

$$p_K(v) = 0 \text{ if } f \text{ is a decreasing first Reidemeister move and } v \text{ is the}$$
$$\text{disappearing crossing.}$$

It turns out that this axiom is redundant. More precisely, the following holds:

Lemma 5.2. *Let p be a parity and K be a knot diagram. Then $p_K(v) = 0$ if v is the disappearing crossing for some decreasing first Reidemeister move.*

Proof. Consider the second Reidemeister move depicted in Fig. 5.6.

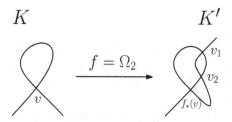

Fig. 5.6   Computation of the parity of the crossing appearing in the first Reidemeister move

From the definition of parity one gets $p_K(v) = p_{K'}(f_*(v))$. Vertices v_1 and v_2 appear via the second Reidemeister moves and thus

$$p_{K'}(v_1) + p_{K'}(v_2) = 0.$$

Finally, the vertices $f_*(v), v_1, v_2$ are the vertices of a triangle and thus are subject to the third Reidemeister move. Therefore, due to the third axiom we get

$$p_{K'}(f_*(v)) + p_{K'}(v_1) + p_{K'}(v_2) = 0.$$

Therefore, $p_K(v) = 0$. □

Note that, vice versa, by the similar construction (applying the first Reidemeister move to a bigon) one can deduce the axiom corresponding to the second Reidemeister move from the axiom, corresponding to the first move.

5.4 The L-invariant

In the present section we introduce the notion of L-*invariant* of free knots. Originally it was defined by the first named author in [Manturov, 2012a] using group techniques. Remarkably, although an invariant element of a certain group was constructed for a knot diagram (to be more precise, a conjugacy class of the element of a certain group), the invariant was essentially integer-valued. Later it was discovered that the same construction may be performed using a different approach. Here we follow the latter ideas, not resorting to group techniques.

Consider a chord diagram D of a free knot K. Fix a point x on the base circle of the diagram, not coinciding with an endpoint of a chord, and orient the diagram arbitrarily. A diagram with a fixed point and orientation will be called a *marked chord diagram*. A marked chord diagram is evidently a diagram of a long free knot obtained from the knot K by cutting at the fixed point.

Consider odd chords of the diagram in the sense of the Gaußian parity. Now we will define *signs* of their endpoints. First, let c be an odd chord of a marked diagram D and e_c its endpoint. We will denote by $n_e \in \mathbb{N}$ the position of e_c when walking around the base circle of the diagram according to the orientation starting from the fixed point. Then the sign of e_c is defined as follows:

Definition 5.9. The *sign* $s(e_c)$ of the endpoint e_c is defined by the formula $s(e_c) = (-1)^{n_e + o_c}$ where o_c is the number of odd chords linked with c.

Now note that since the chord c is *odd* its two endpoints have the same signs. For that reason we may define the *sign of an odd chord* c as the sign of an endpoint of this chord. The sign of a chord will be denoted by $s(c)$.

Definition 5.10. Let D be a marked diagram of a free knot K. Then

$$l(D) = \sum_{c \in \{\text{odd chords}\}} s(c).$$

Theorem 5.4. $l(D)$ *is invariant under the Reidemeister moves on marked chord diagrams.*

Proof. To prove this theorem we consider the Reidemeister moves one by one.

1. The first Reidemeister move either adds or deletes a chord unlinked with all other chords of the diagram. This chord is even, and does not contribute to the signs of any odd chords of the diagram. Therefore, the move Ω_1 does not change $l(D)$.

2. The second Reidemeister move either adds or deletes two "close" chords. Both of them are either odd or even. If those chords are even, then they do not contribute to $l(D)$ in any way. If they are odd, then they have opposite signs, because their positions differ by 1 and the number of odd chords linked with them is the same for both chords. Hence they cancel each other in the formula for $l(D)$ and therefore the move Ω_2 does not change $l(D)$.

3. The third Reidemeister move deals with three chords, either all of which are even, or there are two odd ones and an even one among them. Again, if all the chords are even, then they do not contribute to $l(D)$. If two chords are odd, then it is easy to see that the move does not change the signs of those chords. Therefore, the move Ω_3 does not change $l(D)$. $\square$

This theorem essentially means that $l(D)$ is an invariant of oriented long free knots. To obtain an invariant of free knots, we need to examine, what changes when we invert the orientation or move the fixed point.

First, it is obvious that under the orientation change the value of n_c changes for every odd chord c. Indeed, consider an odd chord c with endpoints e_1, e_2 and the arc γ of the base circle of the diagram beginning at e_1 and ending at e_2 containing the fixed point. Since the chord c is odd, there are oddly many endpoints inside the arc γ. Therefore, when walking from the fixed point to the point e_1 and when walking from the fixed point to the point e_2 we obtain values of different parity.

Moreover, the second summand of the power of $s(c)$ does not depend on the orientation of the diagram. Therefore, the orientation change sends $l(D)$ to $-l(D)$.

Now let us move the fixed point. Consider an "elementary" move of the fixed point: let it "jump" over the nearest endpoint of a chord. We can see that this operation changes the parity of n_c for every odd chord and leaves the second summand of $s(c)$ unchanged. Therefore, every move of the fixed point either does not change $l(D)$ or changes its sign.

From those considerations we deduce that even though the value of $l(D)$ may depend on the orientation of the diagram and the fixed point, its absolute value $L(D) = |l(D)|$ is invariant under those transformations. Due to Theorem 5.4 $L(D)$ is also invariant under the Reidemeister moves. For that reason it depends only on the free knot K and we will denote it by $L(K)$. Hence we obtain

Theorem 5.5. $L(K)$ *is an invariant of free knots.*

5.5 Parities on 2-knots and links

Parity theory for 1-knots (classical, virtual, etc.) boils down to the decoration of crossings, that is, codimension 1 singularities of the projection. A natural idea is to perform the same for higher dimensions. In the present section we define parity for 2-knots and surface knots.

5.5.1 *The Gaußian parity*

First of all let us define the Gaußian parity following [Manturov, 2012a]. We begin with the definition of the Gaußian parity for a real 2-knot and then generalise it to abstract 2-knots.

Consider an abstract 2-knot K and its spherical diagram S. Consider a double point $x \in D_0$. It has two preimages and, therefore, two corresponding points on the diagram S; denote them by x_1, x_2. Consider an arbitrary loop η connecting x to itself whose preimage $\tilde{\eta}$ has ends x_1 and x_2 and is in general position with respect to D^*: we suppose all intersections between $\tilde{\eta}$ and D^* to be transverse. All such curves $\tilde{\eta}$ lie on the sphere S and have fixed endpoints, hence homotopic. We only need to fix the behaviour of the curve near x_1 and x_2.

Consider a double line ζ which contains the point x. Orient it arbitrarily and orient its preimages $\zeta_1 \cap U(x_1)$ and $\zeta_2 \cap U(x_2)$ in the coherent way. We impose the following condition: bases $(\dot{\zeta}_1, v_1)$ and $(\dot{\zeta}_2, v_2)$ give different orientation to the sphere S, where $\dot{\zeta}_i$ denotes the unit tangent vector to the curve ζ_i and v_i are the tangent vectors to $\tilde{\eta}$ at the points x_1 and x_2 correspondingly.

Now we call the parity of the point x the parity of the number of elements of the set $Card(\tilde{\eta} \cap D)$. Informally one can say that we count the number of double lines intersecting the curve connecting two preimages of a double point — provided the ends of the curve are properly oriented.

Note that simultaneous change of orientation of the preimages of the double line induced by the orientation change of the double line ζ itself does not change the parity defined this way.

Thus we defined parity of an arbitrary double point x. We call it *the Gaußian parity of the double point.*

Elementary check verifies the following theorem (see [Manturov, 2012a]):

Theorem 5.6. *The Gaußian parity is constant along a double line.*

Now we can define the Gaußian parity of a double line:

Definition 5.11. Consider an arbitrary double point $x \in D_0$ on a double line γ and consider its two preimages x_1, x_2. Connect the point x_1 with the point x_2 by a path $\tilde{\eta}$ so that the behaviour of γ near the endpoints is compatible (as before). The *Gaußian parity* of the double line γ is the parity of the number of intersections between $\tilde{\eta}$ and the preimage of the set D^*.

Now we can define the Gaußian parity for abstract and free 2-knots. Note that the Gaußian parity of a double point (and thus that of a double line) was defined solely by using a spherical diagram. Since abstract (and free) knots are equivalence classes of spherical diagrams (maybe with additional information encoded) modulo moves, the definition of the Gaußian parity for them can be given verbatim.

Due to the above mentioned considerations there is sense in the study of the parity of the double lines appearing in Roseman moves: it is interesting to understand what can be said of their parity regardless of the structure of the rest of a 2-knot. We do not specify in which space the knot lies, it is considered abstractly (as a local link, to be precise, see Definition 4.4). Despite of the fact that the Gaußian parity is trivial for classical 2-knots, the parity can be non-trivial for abstract knots. And since every transformation of a spherical diagram is (locally) a Roseman move, it is important to study the parity of the double lines involved in moves.

First of all note that the parity of double lines not involved in a move should remain unchanged.

Now consider the Roseman moves $\mathcal{R}_1$–$\mathcal{R}_7$, see Fig. 4.2, and the corresponding local links.

(1) Moves $\mathcal{R}_4$ and $\mathcal{R}_6$ yield that every double line ending in a cusp is even.
(2) Two double lines from the second move (located "closely") have the same Gaußian parity.
(3) The double line in the right-hand side of the fourth move is even.
(4) The third move creates a double line; it is always even.
(5) Among the lines meeting at a triple point there is an even number of odd ones.
(6) There are three double lines in the fifth move. There are either two or zero odd double lines among them.

Note that the fifth, sixth and seventh moves do not change the number of double lines. The double lines before and after the move are in a natural one-to-one correspondence. Parities of the corresponding lines are the same.

Later we will see that those properties are not random and stem from general axioms of 2-parity which we deduce from general approach described in the next subsection and are closely related to parities on 2-links.

5.5.2 *General parity principle*

Consider a more general situation: a disjoint union of n-spheres S^n in $\mathbb{R}^{n+2}$. Consider them from a *diagrammatic* viewpoint: as a projection in general position to a subspace $\mathbb{R}^{n+1} \subset \mathbb{R}^{n+2}$. Roseman proved that in any dimension there is a *finite* set of moves with the following property:

Two diagrams represent isotopic links if and only if they can be connected by a finite sequence of diagrams such that each next diagram is obtained using one of those moves or a trivial isotopy (see, for example, [Roseman, 2004]).

For low dimensions those moves are completely described (that is, Reidemeister moves and Roseman moves).

Now we define *parity axioms* corresponding to the moves.

Consider a move M on a diagram D of a local link sending it to a diagram D'. There is a natural bijection between the sheets of the diagram

D and those of the diagram D'. Colour the sheets of the diagram D arbitrarily with two colours. Colour the corresponding sheets of the diagram D' with the same colours (the compatibility condition can limit the choice of colourings). Now we call a codimension 1 set *even* if it is the intersection set of two sheets of the same colour and *odd* otherwise. Finally to define a *parity axiom corresponding to the move M* we take a condition on the parity of the intersection sets which holds for all possible colourings of this local link. Note that different moves can possibly yield the same axiom.

A clarifying example: the third Reidemeister move. Local link which is changed by this move consists of three arcs (we do not care about over- and undercrossings for our purposes). Each can be coloured in one of two colours. That gives eight possible colourings of the local link (see Fig. 5.7). Every colouring yields a parity of each of the three crossings. But there is a property which holds for all eight colourings: The sum of parities modulo 2 equals zero. That is the parity axiom corresponding to the move Ω_3.

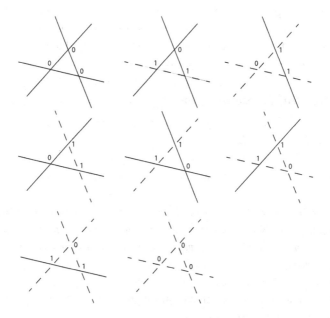

Fig. 5.7   Possible colourings and parity of crossings of the third Reidemeister move

In the case of 2-knots and 2-links the general principle formulated above can be used since all the moves are known: Those are Roseman moves. They yield the following eight parity axioms for double lines:

(0) Parity is constant along double lines;

(1) Double line ending in a cusp is even;

(2) The sum of parities of double lines meeting in a triple point equals 0 mod 2;

(3) All double lines in the move $\mathcal{R}_2$ have the same parity;

(4) Among the three double lines from the move $\mathcal{R}_5$ there are either two or zero odd ones;

(5) The double line appearing in the move $\mathcal{R}_3$ is even;

(6) There is a natural bijection between the lines on the left- and on the right-hand sides of the moves $\mathcal{R}_5, \mathcal{R}_6, \mathcal{R}_7$ induced by the correspondence of the sheets of the diagrams. Parity is preserved by the correspondence;

(7) The double line from the right side of the move $\mathcal{R}_4$ is even.

Besides, it is always supposed that the moves do not change the parity of any double lines not present in the local link.

Note though, that those axioms — obtained from the moves — are redundant. Indeed, the axiom (5) follows from the axiom (1) since the line in question ends in a cusp. The axiom (4) follows from the axioms (2) and (6) since in the left-hand side of the move there are two triple points.

The axioms (3) and (7) seem independent of others. But they can be avoided by using the following trick.

Every double line of a local link either is closed or meets the boundary twice. We will call the connected components of the intersection set between the boundary and the double lines *boundary double points*. There is a natural bijection between the sheets of the local links in the left and the right sides of Roseman moves. It induces the bijection between their boundaries which respects the boundary double points. Therefore, though there is not always a bijection between the double lines, one can always find a bijection between boundary double points.

Thus we can reformulate the axiom (6) in the following way:

(Correspondence axiom) The corresponding boundary double points have the same parity.

This axiom together with axioms (0) and (1) yield axioms (3) and (7). Indeed, on the left-hand side of the fourth move we have two double lines ending with cusps. Due to axiom (1) they are even (and the corresponding boundary double lines are even as well). Therefore due to the correspondence axiom the double line in the right side is even as well. In the same way one can check that the double lines in the second move have the same parity.

In that manner we have a smaller set of parity axioms. Note that they are less dependent of the actual list of Roseman moves: Only the correspondence axiom appeals to them.

(I) *Continuity axiom.* Parity is constant along double lines.

(II) *Selfcrossing axiom.* A double line ending with a cusp is even.

(III) *Triple point axiom.* The sum of parities of three double lines meeting in a triple point equals 0.

(IV) *Correspondence axiom.* There is a natural bijection between boundary double point on the left and right sides of a Roseman move induced by the bijection of the diagram sheets. The parities of the corresponding points are the same.

Definition 5.12. Let $\mathcal{L}$ be a class of 2-links in $\mathbb{R}^4$, and let A be the set of double lines of their diagrams. A mapping $P : A \to \mathbb{Z}_2$ is called *parity* if it satisfies the axioms (I)–(IV).

As one can see, the axioms (0)–(7) above are exactly the same as the properties we got when studying the Gaußian parity. Since the two systems of axioms are equivalent, we got the following

Theorem 5.7. *The Gaußian parity is a parity.*

Using the similar techniques one can define parity axioms for links in arbitrary dimensions.

Remark 5.3. Let p be a parity. It is easy to see that the axioms (I)–(IV) are equivalent to the following set of axioms:

(1) **Continuity axiom:** parity is constant along double lines.
(2) **Correspondence axiom:** there is a natural bijection between boundary double point on the left and right sides of a Roseman move induced by the bijection of the diagram sheets. The parities of the corresponding points are the same;
(3) Ω_1**-axiom:** for a double line γ being the edge of a cylinder over a loop we have $p(v) \equiv 0 \bmod 2$;
(4) Ω_2**-axiom:** for two double lines γ_1, γ_2 being the edges of a cylinder over a bigon we have $p(v_1) + p(v_2) \equiv 0 \bmod 2$;

(5) Ω_3-**axiom:** for three double lines $\gamma_1, \gamma_2, \gamma_3$ of a cylinder over a triangle we have $p(v_1) + p(v_2) + p(v_3) \equiv 0 \mod 2$.

Evidently those axioms exactly coincide with one-dimensional parity axioms applied to the local transversal sections of the knot diagram.

In exactly the same way we can define *parity* for surface knots as a mapping p from the set of the knot (or link) double lines to an abelian group A satisfying the axioms (I)–(IV). The important difference, though, is that the basic parity example — the Gaußian parity — cannot be defined as in case of 2-knots. The reason is that unlike the spherical case, different paths with fixed endpoints on an arbitrary orientable 2-surface are not necessarily homotopic, and thus the notion of Gaußian parity is not necessarily well defined.

5.6 Parity Projection. Weak Parity

5.6.1 *Gaußian parity and parity projection*

For every parity p for virtual knots (or knots in a specific thickened surface), one can consider a *parity projection*, that is a mapping $pr_p : \mathcal{G} \to \mathcal{G}$ from the set of Gauß diagrams $\mathcal{G}$ to itself, defined as follows. For every virtual knot diagram K represented by a Gauß diagram $\mathcal{G}(K)$ we take $pr_p(K)$ to be the virtual knot diagram represented by the Gauß diagram obtained from $\mathcal{G}(K)$ by deleting odd chords with respect to p. At the level of planar diagrams this means that we replace odd crossings by virtual crossings.

Theorem 5.8. *The mapping pr_p is well defined, i.e. if K and K' are equivalent, then so are $pr_p(K)$ and $pr_p(K')$.*

Proof. It is sufficient to show that if two Gauß diagrams K_1 and K_2 are obtained from each other by applying a Reidemeister move, then the Gauß diagrams $pr_p(K_1)$ and $pr_p(K_2)$ differ by a Reidemeister move. In the case of the first Reidemeister move, the chord v of K_1, which is participating in the decreasing first Reidemeister move, has $p_{K_1}(v) = 0$. Then $pr_p(K_1)$ has a chord, which is not linked with other chords and $pr_p(K_2)$ can be obtained by applying the first Reidemeister move to the chord v of $pr_p(K_1)$. In the case of the second Reidemeister move, two crossings v and v', which are participating in the decreasing second Reidemeister move, have either $p_{K_1}(v) = p_{K_1}(v') = 0$ or $p_{K_1}(v) = p_{K_1}(v') = 1$. If $p_{K_1}(v) = p_{K_1}(v') = 0$, then $pr_p(K_1)$ and $pr_p(K_2)$ differ by the second Reidemeister move. If

$p_{K_1}(v) = p_{K_1}(v') = 1$, then chords are deleted by pr_p. Since K_2 already does not have v and v', $pr_p(K_1)$ and $pr_p(K_2)$ coincide. Analogously, in the case of the third Reidemeister move we can obtain either the third Reidemeister move for $pr_p(K_1)$ and $pr_p(K_2)$ if all the three crossings, participating in the third Reidemeister move, are even, or a detour move if the number of even crossings is one. By the definition of the parity all crossings not participating in Reidemeister moves remain fixed, therefore the diagrams $pr_p(K_1)$ and $pr_p(K_2)$ coincide. □

Now consider the Gaußian parity g. For this parity one has a well-defined projection pr_g from the set of virtual knots to itself. Note that if K is a virtual knot diagram, then $pr_g(K)$ might have odd chords: indeed, some crossings which were even in K may become odd in $pr_g(K)$.

However, this map pr_g may take diagrams from one theory to another; for example, if we consider equivalent knots lying in a given thickened surface, then their images should not necessarily be realised in the same surface; they will just be equivalent virtual knots. For virtual knots, this is just a map from virtual knots to virtual knots.

Note that pr_g is not an idempotent map. For example, if we take the Gauß diagram with four chords a, b, c, d where a is linked with b, c, the chord b is linked with a, d, the chord c is linked with a, and the chord d is linked with b, then after applying pr_g we shall get a diagram with two chords a, b, and they will both become odd, see Fig. 5.8.

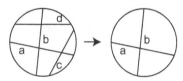

Fig. 5.8   The parity projection is not idempotent

For every Gauß diagram one can decree some chords (crossings) to be *true classical* and remove the other ones, so that the resulting Gauß diagram is classical, and this map will give rise to a well-defined projection from virtual knots to classical knots. For example, in Fig. 5.9 a virtual knot A is drawn in the left part; its band presentation belongs to the thickened torus (see upper part of the right picture); there are four "homologically non-trivial" crossings disappearance of which leads to the diagram D (virtually isotopic to the one depicted in the lower picture of the right half). This is the classical trefoil knot diagram.

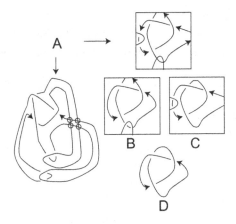

Fig. 5.9   Virtual knot and its classical projection

Theorem 5.9. *For every virtual diagram K there exists a classical diagram $\overline{K}$, such that:*

(1) $\overline{K} < K$;

(2) $\overline{K} = K$ if and only if K is classical;

(3) If K_1 and K_2 are equivalent virtual knots, then so are $\overline{K_1}$ and $\overline{K_2}$;

(4) The map restricted to non-classical knots is a surjection onto the set of all classical knots.

The discrimination between "true classical" crossings and those crossings which will become virtual is of the topological nature, as we shall see in the proof of Theorem 5.9. As usual, we make no distinction between virtually isotopic diagrams: a virtual diagram is said to be *classical* if the corresponding Gauß diagram represents a classical knot. Thus, it makes sense to speak about a map from the set of virtual knots to the set of classical knots. This map will be useful for lifting invariants from virtual knots to classical knots. We shall denote this map by $K \to f(K)$ where K means the knot type represented by K, and $f(K)$ means the resulting knot type of the corresponding classical knots. The only statement of the theorem which deals with diagrams of knots which are not classical, is (4). Otherwise we could just project all diagrams which do not represent classical knots to the unknot diagram (without classical crossings), and the functorial map would be rather trivial. Nevertheless, as we shall see, one can construct various maps of this sort. Different proofs of Theorem 5.9 can be used for constructing various functorial maps and establishing properties of knot invariants.

For example, the following minimality statement holds:

Corollary 5.2. *Let $\mathcal{K}$ be an isotopy class of a classical knot. Then the minimal number of classical crossings for virtual diagrams of $\mathcal{K}$ is realised on classical diagrams (and those obtained from them by the detour move). For every non-classical diagram realizing a knot from $\mathcal{K}$, the number of classical crossings is strictly greater than the minimal number of classical crossings. Moreover, the minimal classical crossing number of a non-classical virtual knot is realised only on minimal genus diagrams.*

Indeed, the projection map from Theorem 5.9 decreases the number of classical crossings, and preserves the knot type.

The proof of Theorem 5.9 is performed in two steps.

Lemma 5.3. *Let K be a virtual diagram, whose underlying diagram genus is not minimal in the class of the knot K. Then there exists a diagram $K' < K$ in the same knot class.*

Lemma 5.4. *There is a map pr from minimal genus virtual knot diagrams to classical knot diagrams such that for every knot K, we have $pr(K) < K$ and if two diagrams K_1 and K_2 are related by a Reidemeister move (performed within the given minimal genus diagram), then their images $pr(K_1)$ and $pr(K_2)$ are related by a Reidemeister move.*

Proof. The proof of Theorem 5.9. We shall construct the projection map in two steps. Let K be a virtual knot diagram. If K is of a minimal genus, then we take $\overline{K}$ to be just $pr(K)$ as in Lemma 5.4. Otherwise take a diagram K' instead of K. It is of the same knot type as K as in Lemma 5.3. If the genus of the resulting diagram is still not minimal, we proceed by iterating the operation K', until we get to a diagram K'' of minimal genus which represents the class K and $K'' < K$. Now, set $\overline{K} = pr(K'')$.

One can easily see that if we insert a small classical knot L inside an edge of a diagram of K, then $f(K \sharp L) = f(K) \sharp f(L)$. So, the last statement of the theorem holds as well and the proof is completed.   □

Proof. The proof of Lemma 5.3. Let K be a virtual knot diagram on a surface S_g of genus g. Assume this genus is not minimal for the knot class of

K. Then by Kuperberg's Theorem it follows that there is a diagram $\tilde{K}$ on S_g representing the same knot as K and a curve γ on S_g such that $\tilde{K}$ does not intersect γ. Indeed, if there were no such diagram $\tilde{K}$, then the knot in $S_g \times I$ corresponding to the diagram K would admit no destabilization, and the genus g would be minimal.

The curve γ gives rise to a (co)homological parity for knots in S_g homotopic to K: a crossing is *even* if the number of intersections of any of the corresponding halves with γ is even, and *odd*, otherwise. This parity, in turn, gives rise to the parity projection pr_γ.

Since K has the underlying diagram genus g, there exists at least one odd crossing of the diagram K. Let $K' = pr_\gamma(K)$ be the result of γ-parity projection applied to K. We have $K' < K$.

By construction, all crossings of $\tilde{K}$ are even; thus, $pr_\gamma(\tilde{K}) = \tilde{K}$.

Let us construct a chain $K = K_0 \to K_1 \to \cdots \to K_n = \tilde{K}$ of Reidemeister moves from K to $\tilde{K}$ and apply the γ−parity projection to it.

We shall get a chain $K' = pr_\gamma(K_0) \to pr_\gamma(K_1) \to \cdots \to pr_\gamma(K_n) = \tilde{K}$. Thus, we get a sequence of Reidemeister moves connecting K' to $\tilde{K}$. So, K' is of the same type of $\tilde{K}$ and K. The claim follows. □

Proof. The proof of Lemma 5.4. Let us construct the projection announced in Lemma 5.4. Fix a surface S_g. Let us consider knots in the thickening of S_g for which genus g is minimal (that is, there is no representative of lower genus for knots in question). Let K be a diagram of such a knot. We shall denote crossings of knot diagrams in S_g and the corresponding points on S_g itself by the same letter (abusing notation).

As above, with every crossing of K we associate the two halves $h_{v,1}$, $h_{v,2}$, now considered as elements of the fundamental group $\pi_1(S_g, v)$, as follows. Let us smooth the diagram K at v according to the orientation of K. Thus, we get a two component oriented link with components $h_{v,1}$, $h_{v,2}$. Consider every component of this link represented as a loop in $\pi_1(S_g, v)$ and denote them again by $h_{v,1}$, $h_{v,2}$.

Let $\gamma_v, \tilde{\gamma}_v$ be the two homotopy classes of the knot K considered as an element of $\pi_1(S_g, v)$: we have two classes because we can start traversing the knot along each of the two edges emanating from v. Note that $h_{v,1} \cdot h_{v,2} = \gamma_v$ and $h_{v,2} \cdot h_{v,1} = \tilde{\gamma}_v$.

Let us now construct a knot diagram $pr(K)$ from K as follows. If for a crossing v we have $h_{v,1} = \gamma_v^k$ for some integer k (or, equivalently, $h_{v,2} = \gamma_v^{1-k}$), then this crossing remains classical for $pr(K)$ (also denoted by K'); otherwise, a crossing becomes virtual. Note that it is immaterial whether we take γ_v or $\tilde{\gamma}_v$ because if $h_{v,1}$ and $h_{v,2}$ are power of the same element of the fundamental groups, then they obviously commute, which means that $\gamma_v = \tilde{\gamma}_v$.

Statement 5.1.

(1) For every K as above, $pr(K)$ is a classical diagram;
(2) $K = pr(K)$ whenever K is classical;
(3) If K_1 and K_2 differ by a Reidemeister move, then $pr(K_1)$ and $pr(K_2)$ differ by either a detour move or by a Reidemeister move.

Proof. Take K as above and consider $pr(K)$. By construction, all "halves" of all crossings for $pr(K)$ are powers of the same homotopy class. We claim that the underlying surface for $pr(K)$ is a 2-sphere. Indeed, when constructing a band presentation for $pr(K)$, we see that the surface with boundary has cyclic homology group. This happens only for a disk or for the cylinder; in both cases, the corresponding compact surface will be S^2.

The situation with the first Reidemeister move is obvious: the new added crossing has one trivial half and the other half equal to the homotopy class of the knot itself.

Now, to prove the last statement, we have to look carefully at the second and the third Reidemeister moves. Namely, if some two crossings A and B participate in a second Reidemeister move, then we have an obvious one-to-one correspondence between their halves such that whenever one half corresponding to A is an power of γ, so is the corresponding half of B.

So, they either both survive in $pr(A)$, $pr(B)$ (do not become virtual) or they both turn into virtual crossings. So, for $pr(A)$, $pr(B)$ we get either the second Reidemeister move, or the detour move. Note that, here, we deal with the second Reidemeister move which does not change the underlying surface.

Now, let us turn to the third Reidemeister move from K to K', and let (A, B, C) and (A', B', C') be the corresponding triples of crossings. We see that the homotopy classes of halves of A are exactly those of A', the same about B, B' and C, C'. So, the only fact we have to check that the number of surviving crossings among A, B, C is not equal to two (the crossings from the list A', B', C' survive accordingly). This follows from Fig. 5.10.

Indeed, without loss of generality assume A and B survive. This means that the class $h_{A,1}$ is a power of the class of the whole knot in the fundamental group with the reference point in A, and $h_{B,1}$ is a power of class of the knot the reference point at B.

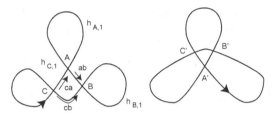

Fig. 5.10   Triviality of two crossings yields the triviality of the third one

Let us now investigate $h_{C,1}$ (for convenience we have chosen $h_{C,1}$ to be the upper right part of the figure).

We see that $h_{C,1}$ consists of the following paths: $(ca)h_{A,1}(ab)h_{B,1}$ $(cb)^{-1}$, where $(ca), (ab), (cb)$ are non-closed paths connecting the points A, B, and C. Now, we can homotop the above loop to $(ca)h_{A,1}(ca)^{-1}$ $(ca)(ab)h_{B,1}(cb)^{-1}$ and then homotop it to the product of $(ca)h_{A,1}(ca)^{-1}$.

We claim that both these loops are homotopic to γ_C^l and γ_C^m for some exponents m, l. Indeed, $h_{A,1}$ is γ_A^k by assumption. Now, it remains to observe that in order to get from γ_A to γ_C, it suffices to "conjugate" by a path along the knot; one can choose (ac) as such a path. The same holds for $h_{B,1}$.

So, if all crossings A, B, C survive in the projection of $pr(K)$ and A', B', C' survive in $pr(K')$, then we see that $pr(K')$ differs from $pr(K)$ by a third Reidemeister move. If no more than one of A, B, C survives, then we have a detour move from $pr(K)$ to $pr(K')$.   □

□

5.6.2   *The notion of weak parity*

Earlier we defined a parity as a way to decorate the crossings of a knot diagram in a way that is compatible with Reidemeister moves. Those compatibility conditions gave rise to the parity axioms.

A natural question is: is it possible to somehow loosen those condition, preserving the good properties of parity? It turns out that the answer is positive. We can partially relax the last parity axiom, allowing either zero, two or three odd crossings among the ones participating in the third Reidemeister move (recall that in Definition 5.6 either zero or two odd crossings were required). The resulting mapping is called a *weak parity*.

In the present section we give rigorous definitions using the same category theory language as before.

Definition 5.13. A *weak parity* p on diagrams of a knot $\mathcal{K}$ is a family of maps $p_K : \mathcal{V}(K) \to \mathbb{Z}_2$, where $K \in ob(\mathfrak{K})$ is an object of the category, such that for any elementary morphism $f : K \to K'$ the following holds:

(1) $p_{K'}(f_*(v)) = p_K(v)$ provided that $v \in \mathcal{V}(K)$ and there exists $f_*(v) \in \mathcal{V}(K')$;

(2) $p_K(v_1) + p_K(v_2) = 0 \pmod 2$ if f is a decreasing second Reidemeister move and v_1, v_2 are the disappearing crossings;

(3) the number of vertices v_i with $p_K(v_i) = 1$ is not equal to one if f is a third Reidemeister move and v_1, v_2, v_3 are the crossings participating in this move.

Remark 5.4. Since the parity axioms or 2-dimensional object turned out to appear from the 1-dimensional parity axioms, one can define weak parity for 2-knots and surface knots in exactly the same way relaxing the Ω_3-axiom.

Definition 5.14. Let p be a parity with coefficients with $\mathbb{Z}_2$ or a weak parity on diagrams of a knot $\mathcal{K}$, let K be any diagram of $\mathcal{K}$. We call a vertex v of K *even* for p if $p_K(v) = 0$, and *odd* otherwise.

Definition 5.15. Let K be a (flat or Gauß) diagram. A *subdiagram of K* is a diagram which is obtained from K by deleting several chords in the case of Gauß diagrams or replacing several classical crossings with virtual crossings in the case of flat diagrams.

Let us define the *closure* of knot theory to be the knot theory obtained from the initial knot theory as follows. Diagrams of the new theory are diagrams of the initial theory and their subdiagrams. Moves in the new theory are defined in the same manner as in the initial theory, i.e. if we have two equivalent diagrams K_1 and K_2, then subdiagrams obtained from K_1 and K_2 by deleting the same chords (replacing the same crossings with virtual ones) not participating in the moves are equivalent in the new theory.

A knot theory is called *closed* if it coincides with its closure.

Theorem 5.10. *Given some knot theory and a weak parity on diagrams of each knot, let us construct the mapping f from the theory to its closure, which sends a diagram K to the diagram obtained from K as follows.*

Any even classical crossing of the diagram K remains classical, and any odd classical crossing is replaced with a virtual one (i.e. when drawing a picture we substitute a virtual crossing for an odd classical crossing).

Then the mapping f is well defined on equivalent classed of knots, i.e. this mapping takes equivalent diagrams to the equivalent diagrams.

Remark 5.5. In the case of free knots, i.e. when framed 4-graphs are considered as diagrams, the mapping f is the operation which deletes all odd crossings and joins two opposite half-edges to form one edge. When knot diagrams are depicted on the plane or a surface, deleted crossings are marked by a circle and considered as virtual crossings.

Proof. Proof of Theorem 5.10. We will show that if K_1 and K_2 are obtained from each other by applying a Reidemeister move, then the diagrams $f(K_1)$ and $f(K_2)$ either coincide up to detour moves or differ by a Reidemeister move. In the case of the first Reidemeister move, the crossing v, which is participating in the first Reidemeister move, has $p_{K_1}(v) = 0$ or $p_{K_1}(v) = 1$. If $p_{K_1}(v) = 0$, then $f(K_1)$ has a loop and $f(K_1)$ and $f(K_2)$ differ by the first Reidemeister move. If $p_{K_1}(v) = 1$, then $f(K_1)$ has a virtual loop and $f(K_1)$ and $f(K_2)$ coincide up to detour moves. In the case of the second Reidemeister move, two crossings v and v', which are participating in the decreasing second Reidemeister move, have either $p_{K_1}(v) = p_{K_1}(v') = 0$ or $p_{K_1}(v) = p_{K_1}(v') = 1$. If $p_{K_1}(v) = p_{K_1}(v') = 0$, then $f(K_1)$ and $f(K_2)$ differ by the second Reidemeister move. If $p_{K_1}(v) = p_{K_1}(v') = 1$, then $f(K_1)$ and $f(K_2)$ coincide up to detour moves. Analogously, in the case of the third Reidemeister move we can obtain either the third Reidemeister move for $f(K_1)$ and $f(K_2)$ if all the three crossings, participating in the third Reidemeister move, are even, or a detour move if the number of even crossings is one or zero. By the definition of the (weak) parity all crossings not participating in Reidemeister moves remain fixed, therefore the diagram K_1 and K_2 are not changed by applying f. $\square$

Now let us describe the mapping f in concrete cases of knot theories.

5.6.3 *Functorial mapping f for Gaußian parity for free, flat and virtual knots*

Let K be a virtual diagram representing a free knot, $D_G(K)$ the corresponding Gauß diagram, and let $At(K)$ be the atom corresponding to the diagram K. Then the following theorem holds.

Theorem 5.11. *The atom $At(K)$ is orientable if and only if all chords of the diagram $D_G(K)$ are even. In particular, the property of an atom to be orientable only depends on the free knot corresponding to K, and does not depend on over/undercrossing information at classical crossings of K.*

Proof. It is known that an atom is orientable if and only if the frame of the atom admits a source-sink structure, i.e. we can orient all edges of its frame in such a way that at each vertex two opposite edges are *emanating*, and the other two opposite edges are *incoming*, see Fig. 5.11 in the center.

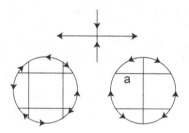

Fig. 5.11   Gauß diagram and the source-sink structures

Recall that edges of the graph, the frame of an atom, correspond to arcs of the Gauß diagram. Besides, adjacent arcs of the Gauß diagram correspond to opposite half-edges. Therefore, a source-sink structure defines an orientation of the Gauß diagram, satisfying the following properties:

- arcs of the Gauß diagram alternate, i.e. each arc oriented in clockwise manner is followed by the arc oriented in counterclockwise manner:
- each chord has two emanating arcs in one of its ends, and two coming arcs in the other end.

From these conditions it easily follows that for each chord a half-circle of the core circle of the Gauß diagram contains an even number of chords' ends. Therefore, the chord is even.   □

Example 5.2. Let us consider the Gauß diagram depicted in Fig. 5.11 (the left part). An orientation of arcs is given in it, this orientation gives rise to the source-sink structure for the frame of the corresponding atom. It is not difficult to check that one of the atoms corresponding to this diagram is spherical (i.e. the surface of the atom is the sphere), therefore, it is orientable, i.e. all atoms with the same frame are orientable.

We cannot define a source-sink structure for the chord diagram as depicted in Fig. 5.11 (the right part) since the chord a is odd (it is linked with one chord). Therefore, an alternating orientation of arrows along the core circle of the chord diagram leads to four incoming edges for the chord a.

Thus, the mapping f preserves those diagrams having orientable atoms (recall that we consider only diagrams of virtual knots, not links). If the atom $At(K)$ corresponding to a diagram K is not orientable, then the diagram $f(K)$ has fewer classical crossings than the diagram K does.

In general, for a given knot K we have $f^2(K) \neq f(K)$ since after the removal of odd chords of the chord diagram the formerly even chords may become odd. Thus, for instance, for the chord diagram with four chords a, b, c, d such that the linked pairs are $(a, b), (a, c), (b, d)$, see Fig. 5.8, the removal of the odd chords c, d yields the diagram with the chords a, b being odd.

It is easy to see that for every diagram K of a virtual knot with n classical crossings there exists a number $m < n$ such that $f^m(K) = f^{m+1}(K)$, i.e. the atom corresponding to the diagram $f^m(K)$ is orientable.

It is evident that the map killing odd crossings is a well-defined map on the set of *atoms*: It suffices to remove all vertices of the atom corresponding to odd crossings and to preserve the white-black structure in the neighbourhoods of the remaining crossings.

This leads us to the following *filtration* on the set of atoms and virtual knots. Let At be an atom whose frame has one unicursal component. Then either At is orientable (in this case we say that At has *grading zero*), or the atom At is not orientable. In the second case there exists a unique natural number $n > 0$ such that $f^n(At)$ is an orientable atom, and $f^{n-1}(At)$ is not. In this case we say that the atom At is *of grading n*.

Analogously, one defines the grading on the set of virtual knots: we say that a virtual knot has grading 0 if it possesses a diagram with orientable atom; a knot K has grading $n > 0$ if the knot $f^n(K)$ has a diagram with orientable atom, whence the knot $f^{n-1}(K)$ has no such diagrams.

Thus we get a natural splitting of the set of virtual knots into subsets: $\mathcal{K}_0 \oplus \mathcal{K}_1 \oplus \mathcal{K}_2 \oplus \cdots \oplus \mathcal{K}_n \oplus \ldots$, where $\mathcal{K}_k$ is the set of virtual knots of grading k.

It is easy to construct examples showing that each of these sets $\mathcal{K}_n$ is non-empty. We construct these examples by induction on n. Let us fix a positive integer k and consider an irreducibly odd diagram on k chords. In the first step we add k pairwise unlinked chords, each linked precisely one chord of the initial diagram. In the lth step we add k pairwise unlinked chords, each of which is linked with exactly one chord added in the previous step. After performing the $n-1$ steps, we obtain a diagram of grading n.

Definition 5.16. We call a map which sets all virtual knots to knots with orientable atoms and takes all virtual knots with orientable atoms to themselves, a *projection*.

The approach above allows one to get a simple proof of the following theorem by Manturov–Viro (first proved in spring 2005 independently by V. O. Manturov and O. Ya. Viro and first published in [Ilyutko and Manturov, 2009]).

Theorem 5.12. *Let K, K' be two equivalent diagrams of virtual knots having orientable atoms. Then there exists a chain of Reidemeister moves*

$$K = K_0 \to K_1 \to \cdots \to K_n = K'$$

such that all atoms corresponding to diagrams K_i are orientable.

Proof. Let us consider a chain of diagrams $K = K_0 \to K_1 \to \cdots \to K_n = K'$ such that the diagram K_i and K_{i+1} are obtained from each other by a classical Reidemeister move (and possibly, a detour move). Denote the maximal grading over all diagrams K_i by l. Let us apply the map f again (in total, we shall apply this map l times starting with the initial chain). Reiterating the process, we get a chain of diagrams

$$f^l(K) = K = K_0 \to f^l(K_1) \to \cdots \to f^l(K_n) = K_n = K',$$

where all atoms corresponding to all diagrams, are orientable, and every two adjacent diagrams either coincide (connected by a detour move) or differ by a Reidemeister move.   □

Remark 5.6. The initial proof of the Manturov–Viro theorem in the general case of *multicomponent links* relies on the geometry of virtual knots and the Kuperberg theorem. Presently, the authors know the proof of this theorem in the general case, relying on the notion on the notion of *relative parity* [Krylov and Manturov, 2011].

5.6.4 *The parity hierarchy on virtual knots*

It follows from the Manturov–Viro theorem that virtual knots with orientable atoms form a natural subclass of all virtual knots with the equivalence being defined only by using those Reidemeister moves which preserve the knot diagram within this subclass.

In particular, this class (let us denote it by $\mathcal{D}^1$) includes all classical knots. We shall show that there exists a natural *filtration* on the set of virtual knots (unrelated to the grading n given above):

$$\mathcal{D}^0 \supset \mathcal{D}^1 \supset \mathcal{D}^2 \supset \ldots \mathcal{D}^n \supset \ldots,$$

which starts with the set $\mathcal{D}^0$ of all virtual knots and has as its limit "index zero knots" $\mathcal{D}^\infty$, the set of knots containing all classical knots.

Thus, let K be a virtual diagram, and let $D_G(K)$ be the corresponding Gauß diagram. We endow the diagram $D_G(K)$ with signs and arrows in the usual way. The positive sign corresponds to a crossing on the left of Fig. 5.12, and the negative sign corresponding to a crossing on the right of Fig. 5.12. The arrow is pointing from the preimage of the overcrossing arc to the preimage of the undercrossing arc.

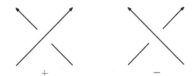

Fig. 5.12 Positive and negative crossings

With each classical crossing we associate its *index*, which will be either a natural number or zero. Let v be a classical crossing of the diagram K, and let $c(v)$ be the corresponding (oriented) chord of the diagram $D_G(K)$. Consider all chords $D_G(K)$ linked with $c(v)$. Let us count for them the sum of signs of those chords linked with the chord $c(v)$ from the left to the right and subtract the sum of signs of those chords linked with the chord $c(v)$ from the right to the left. The absolute value of the obtained number will be called the *index* of the chord $c(v)$ (or the crossing v), and will be denoted by $ind(v) \equiv ind(c(v))$.

It is clear that if the atom corresponding to K is orientable, then the indices of the chords of $D_G(K)$ are all even. Consequently, $\mathcal{D}^1$ consists of those knots having all indices of all chords even.

Let us collect some facts whose proof follows from a simple check.

Proposition 5.1.

(1) If a crossing takes place in the first Reidemeister move, then it has index zero.

(2) In the second Reidemeister move, the indices of the two crossings are equal.

(3) The index of a crossing does not change when the crossing is operated on by the third Reidemeister move

$$ind(v_1) = ind(v_1'), \ ind(v_2) = ind(v_2'), \ ind(v_3) = ind(v_3'),$$

besides, if three crossings v_1, v_2, v_3 participate in a third Reidemeister move, then $ind(v_1) \pm ind(v_2) \pm ind(v_3) = 0$, see Fig. 5.13.

(4) The index of a crossing taking part in a certain Reidemeister move, does not change after this crossing undergoes the Reidemeister move.

(5) All crossings of a classical diagram have index zero.

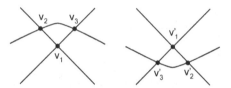

Fig. 5.13   The third Reidemeister move and corresponding crossings

First, note that the property of the index to be equal to zero is a weak parity. The same is true about the congruence to zero modulo some integer number. Thus, the map which eliminates all chords of non-zero indices is a well-defined map in a knot theory with an index.

Besides, the properties of the index described above show that the index can be used in order to define a parity (with coefficients from $\mathbb{Z}_2$).

Indeed, it follows from Proposition 5.1 that one can introduce on diagrams of $\mathcal{D}^1$ the following parity: let K be a knot diagram from $\mathcal{D}^1$; we decree those crossings of K having index divisible by four, to be even, and we decree the remaining ones to be odd.

From Proposition 5.1 it follows that the parity defined in this way satisfies parity axioms.

Let us apply to knots from $\mathcal{D}^1$ the map which kills crossings of index not divisible by four. This operation may take us away from the class $\mathcal{D}^1$.

Nevertheless, if two diagrams K_1, K_2 from $\mathcal{D}^1$ are equivalent, then so are $f(K_1)$ and $f(K_2)$ (even if they do not belong to $\mathcal{D}^1$).

Arguing as above, we can define the sets $\mathcal{D}^k$ of diagrams with all crossings having indices divisible by 2^k, and also the set $\mathcal{D}^\infty$ as the set of diagrams with all crossings having index 0.

There is a set of maps $f_k : \mathcal{D}^k \to \mathcal{D}^0$; each map f_k eliminates all chords of index congruent to 2^k modulo 2^{k+1}. All these maps take equivalent diagrams to equivalent ones. Then the following theorem holds.

Theorem 5.13. *Let K and K' be two diagrams of virtual knots from $\mathcal{D}^k$ (where k is either a natural number or the symbol ∞) corresponding to equivalent virtual knots. Then there exists a chain*

$$K = K_0 \to K_1 \to \cdots \to K_n = K'$$

of diagrams from $\mathcal{D}^k$, where every two adjacent diagrams are obtained from each other by a Reidemeister move or a detour move.

This theorem is proved in the same way as Theorem 5.12 by means of the functorial mapping eliminating all chords of non-zero index (in the case $k = \infty$).

Note that the class $\mathcal{D}^\infty$ is quite interesting: it is an "approximation" of classical knots by virtual knots, and all invariants defined by $\mathcal{D}^\infty$, can be taken to virtual knots by means of the map f.

An example of a non-classical diagram from $\mathcal{D}^\infty$ is shown in Fig. 5.14.

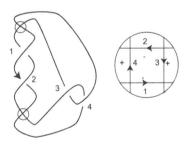

Fig. 5.14   A non-classical diagram from $\mathcal{D}^\infty$

In Fig. 5.14 we depict the Gauß diagram with arrows and signs. It is easy to see that in the Gauß diagram the chords corresponding to the crossings 1 and 2 have opposite directions, that guarantees that the indices of the chords 3 and 4 are both zeros. Analogously, the arrows, corresponding to crossings 3 and 4 are opposite.

One can easily check that the parity hierarchy gives rise to the flat hierarchy: odd crossings will have label 0, those even crossings which become odd after one application of the map f, get label 1, and so on.

Chapter 6

Cobordisms. Applications of Parity Theory to Cobordisms of Knots

In this chapter we review an important and long-studied notion of *cobordism* of knots and extend it to the case of virtual knots and free knots. The latter case can be considered as *graph cobordism*. Those theories also can be studied from picture-valued standpoint. In particular, we give a sliceness (that is, null-cobordance) criterion for odd free knots which presents a finite algorithmic solution to the question whether an odd free knot is slice. This algorithm can be performed given a diagram of the knot. In that sense the diagram itself is the invariant of sliceness.

The key part in the construction of the considered invariant and proofs of theorems in this chapter is played by parity theories on 1-dimensional and 2-dimensional knots and their interconnections.

6.1 Cobordism in knot theories

6.1.1 *Basic definitions*

Consider a pair of classical knots K_1, K_2. Informally one can interpret the *concordance* between them as a possibility to connect those knots with a cylinder in the 4-dimensional space. More precisely, we have the following.

Definition 6.1. Two classical knots $K_1, K_2 \subset \mathbb{R}^3 \times [0, 1]$ such that $K_1 \subset \mathbb{R}^3 \times \{0\}$ and $K_2 \subset \mathbb{R}^3 \times \{1\}$ are called *concordant* if there exists a smooth proper embedding

$$f \colon S^1 \times [0, 1] \to \mathbb{R}^3 \times [0, 1]$$

such that

$$f(S^1 \times \{0\}) = K_1, \ f(S^1 \times \{1\}) = K_2.$$

If we replace the cylinder $S^1 \times [0,1]$ with an orientable 2-surface with two holes $S_g \setminus (D_1 \sqcup D_2)$, then the notion of concordance transforms into the notion of *cobordism*.

Now let us define the notion of cobordism in the case of free knots.

As was defined in Section 3.1, a *free knot* is an equivalence class of chord diagrams (with no orientation and signs on the chords) modulo Reidemeister moves. It turns out fruitful to regard free knots from another perspective: as an equivalence class of framed 4-graphs.

Note that framed graphs are by no means supposed to be planar; we consider them lying in some 3-manifold M^3. When drawing such a graph on a plane, we will depict the union of the opposite edges as a straight line going through the vertex. Informally, the framing can be interpreted as an instruction on how to cross a vertex: when we come to a vertex along an edge, which edge we should choose to leave the vertex.

Now let us choose an arbitrary point on an edge of a framed 4-graph Γ and begin going along the edges respecting the framing. Since the graph is 4-valent, sooner or later we return to the initial point. The subgraph we have covered this way is called a *unicursal component* of the graph Γ.

Consider a framed 4-graph Γ with only one unicursal component. It means that a path starting at an arbitrary point on an edge and respecting the framing is actually an *Euler cycle* on the graph. Consequently, this graph can be interpreted as a circle with some points glued together to form the vertices; the framing is inherited naturally from the circle. Therefore given a framed 4-graph Γ we can construct a chord diagram corresponding to Γ. The process of circumfering the graph gives a natural mapping

$$\varphi \colon S^1 \to \Gamma,$$

which is a bijection everywhere except for the vertices; every vertex v has two preimages $v_1^{-1}, v_2^{-1} \in \varphi^{-1}(v)$. Now, to construct the chord diagram, for every vertex v of the graph Γ we connect the points v_1^{-1}, v_2^{-1} with a chord. Note that this diagram doesn't possess any information on the chords, quite like the chord diagrams of free knots.

Conversely, given a chord diagram one can associate a framed 4-graph with it. To do that we just paste the points connected by a chord together so that the small arcs containing those points intersect transversally. The framing is thus naturally inherited.

In this manner we have established a bijection between the set of framed 4-graphs with one unicursal component and the set of chord diagrams. Given a framed 4-graph Γ, we should denote the corresponding chord diagram by $C(\Gamma)$.

Now we can introduce the following equivalence relation: two framed 4-graphs Γ_1, Γ_2 with one unicursal component are said to be *equivalent* if their chord diagrams $C(\Gamma_1), C(\Gamma_2)$ can be connected by a chain of Reidemeister moves. By definition, an equivalence class of framed 4-graphs is a free knot. That means that every representative of a free knot can be considered as a framed 4-graph.

Remark 6.1. In the construction given above we considered graphs with one unicursal component. If we consider all kinds of framed 4-graphs, then the same construction gives the bijection between framed 4-graphs and generalised chord diagrams with several base circles. Thus, every framed 4-graph can be considered a representative of a free link. The number of the link components is the same as the number of unicursal components of the corresponding graph.

To define the notion of free knot cobordism (or, in the light of the construction given above, framed 4-graph cobordism) we need to introduce *standard complexes*.

Definition 6.2. A finite 2-dimensional CW-complex K is called *standard complex (with boundary)* if every point of K is one of the following:

- an *interior* point of one of the following types: a regular point, a transverse double point, a transverse triple point, or a branch point (cusp) (those types of points are depicted on Fig. 6.1),
- a *boundary point* of one of the types shown on Fig. 6.2.

The union of boundary points is called a *boundary* of the complex K and is denoted by ∂K.

Now we can give the definition of framed graph cobordism.

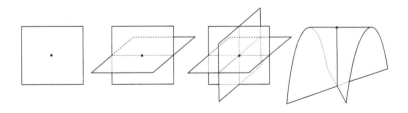

Fig. 6.1 Four allowed types of interior points of a complex K

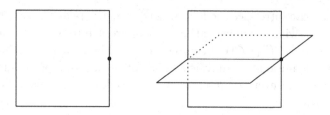

Fig. 6.2   Two allowed types of boundary points of a complex K

Definition 6.3. Two framed 4-graphs Γ_1, Γ_2 with one unicursal component each are called *cobordant* if there exists a triple (M, Σ, f) where M is a 3-manifold, $\Sigma = S_g \setminus (D_1 \sqcup D_2)$ is an oriented 2-manifold (a sphere with g handles) with two discs cut out and f is a mapping $f \colon \Sigma \to M$ of the surface Σ to the manifold M such that

- the image $f(\Sigma)$ is a standard complex with boundary,
- the image of the boundary of the surface is exactly the union of the graphs: $f(\partial \Sigma) = \Gamma_1 \sqcup \Gamma_2$,
- the mapping is compatible with the graphs' framing in the sense that a small neighbourhood of every preimage v_i^{-1} of a vertex v is mapped into the union of the opposite half-edges incident to the vertex v.

The surface Σ is called a *spanning surface* for the pair of graphs Γ_1, Γ_2 and the complex $f(\Sigma)$ is called their *spanning complex*.

Informally, the two graphs being cobordant means the possibility to connect those graphs by a surface such that its projection to some 3-manifold is a standard complex. Naturally, an interesting characteristic of a pair of graphs is the "complexity" of the spanning surface. More precisely:

Definition 6.4. For a pair of framed 4-graphs Γ_1, Γ_2 the minimal genus among the spanning surfaces is called their *cobordism genus*. If the graphs Γ_1, Γ_2 are not cobordant, then the genus is said to be equal to ∞.

One particular case of cobordance is of a special interest:

Definition 6.5. Two framed 4-graphs Γ_1, Γ_2 are called *concordant* if they are cobordant of genus 0.

Among all framed 4-graphs there is one special: the trivial graph given by the trivial chord diagram with no vertices. If a graph Γ is cobordant to the trivial graph, then the trivial graph can be trivially "cupped" by a disc. Thus the graph Γ itself can be spanned by a surface (in the case of concordance — by a disc). This particular case is called *sliceness*:

Definition 6.6. A framed 4-graph Γ is called *slice* if it is cobordant to the trivial graph.

The slice genus is an interesting and important graph characteristic. Moreover, it turns out to be a characteristic of the free knot which representative is the considered graph. To prove this fact we need the following lemma:

Lemma 6.1. *Equivalent framed 4-graphs are genus 0 cobordant.*

Proof. We only need to prove this lemma for graphs which differ by one of the Reidemeister moves. Moreover, evidently, outside a small area where the move is performed, the graphs are connected by the trivial cobordism. Therefore we only need to present the non-trivial cobordisms of the subgraph changed by given Reidemeister moves. Those cobordisms are explicitly shown on Fig. 6.3. □

Fig. 6.3 Cobordisms of graphs connected by a Reidemeister move (left to right: $\Omega_1, \Omega_2, \Omega_3$)

Therefore, since a free knot is essentially a framed 4-graph equivalence class, cobordism and concordance relations are relations of free knots and the cobordism genus can be considered as a characteristic of free knots.

Remark 6.2. Note that the first Reidemeister move produces a cusp and the third Reidemeister move produces a triple point of the spanning complex.

6.1.2 *Cobordism types*

Now let us delve deeper into the possible types of knot and graph cobordisms.

Consider the points of a spanning complex K. Under the map f they have one, two or three preimages on the surface Σ. The points with exactly two preimages are called *double points* of the complex. They are organised into curves, called the *double lines* of the complex. Every intersection of double lines has exactly three preimages on the surface Σ and is called a *triple point*.

Definition 6.7. Let K be a complex with boundary $\partial K = \Gamma$; then we can distinguish between the following types of double lines of K:

(1) cyclic (closed) double line;
(2) double line between two cusps;
(3) double line between two vertices of Γ;
(4) double line between a vertex of Γ and a cusp.

Double lines of the first two types are called *interior*, other double lines are called *boundary*.

We consider different types of spanning surfaces and complexes by imposing different restrictions on the existence of non-regular points of the spanning complex considered. More precisely, we distinguish between the following cases:

(i) **The general case:** We impose no restrictions on the complex K: it can have double lines, cusps, and triple points;
(ii) **The case without cusps:** We forbid cusps, that is, we require that the neigbourhood of each point of K not belonging to Γ looks locally as a union of one, two or three coordinate planes at the origin in $\mathbb{R}^3$;
(iii) **The case without triple points:** We allow cusps, but forbid triple points;
(iv) **The elementary case:** We forbid triple points and cusps;
(v) **The internally trivial case:** We forbid triple points and cusps incident to an interior double line.

Definition 6.8. We shall call these complexes *standard complexes* (with or without boundary). In correspondence with the five cases described above, they should be called

(i) *generic complex;*
(ii) *complex without cusps;*
(iii) *complex without triple points;*
(iv) *elementary complex;*

(v) *internally trivial complex.*

Preimages under the map f (see Definition 6.3) of standard complexes of types (i)–(v) are called

(i) *generic spanning surface;*
(ii) *spanning surface without cusps;*
(iii) *spanning surface without triple points;*
(iv) *elementary spanning surface;*
(v) *internally trivial spanning surface.*

All complexes mentioned in the present work are standard.

Now consider a framed 4-graph Γ and let us fix a type (i)–(v) from Definition 6.8. There exists a set (probably, empty) $\mathfrak{S}$ of spanning surfaces of that type for the graph Γ. Every spanning surface $S \in \mathfrak{S}$ has a genus; if the spanning surface is not connected, its genus is calculated as the sum of the genera of its components. Consider the minimal genus of those spanning surfaces. If there is no spanning surface of the chosen type for the graph Γ, that is, if $\mathfrak{S} = \emptyset$, then the genus is defined as ∞. This genus is a characteristic of the graph Γ.

Remark 6.3. There exist examples of framed 4-graphs with genus of any of the types (i)–(v) not equal to ∞ (that is, with non-empty set $\mathfrak{S}$). The simplest example is the trivial graph given by a circle with no vertices. Evidently, it can be spanned by a disc and thus all five types of genera equal 0. Other examples of slice framed graphs of various types can be found, for example, in [Fedoseev and Manturov, 2019b].

Definition 6.9. For a given framed 4-graph Γ its *slice genus* (*slice genus without cusps, slice genus without triple points, elementary slice genus, internally-trivial genus,* respectively) is the minimal genus of its spanning surfaces of the corresponding type (i)–(v).

The five genera are denoted with $g(\Gamma), g_0(\Gamma), g'(\Gamma), g_{el}(\Gamma), g_{it}(\Gamma)$, respectively.

The following statement follows directly from the definition:

Theorem 6.1. *For each framed 4-graph Γ we have*
$$\infty \geq g_{el}(\Gamma) \geq g_0(\Gamma) \geq g(\Gamma) \geq 0,$$
$$\infty \geq g_{el}(\Gamma) \geq g'(\Gamma) \geq g(\Gamma) \geq 0,$$
and
$$\infty \geq g_{it}(\Gamma) \geq g(\Gamma) \geq 0.$$

Now we can give an equivalent definition of graph sliceness and expand it:

Definition 6.10. We say that a framed 4-graph Γ is *slice* if $g(\Gamma) = 0$. Respectively, we shall use the terms *sliceness without cusps*, *sliceness without triple points*, *elementary sliceness*, and *internally trivial sliceness*.

Evidently, we now can give the definition of graph cobordism and concordance of various types:

Definition 6.11. Two framed 4-graphs are called *cobordant* (*cobordant without cusps*, *cobordant without triple points*, *elementary cobordant*, and *internally trivial cobordant*, respectively) if there exists a spanning surface of the corresponding type (i)–(v) for that pair of graphs.

If two framed 4-graphs are cobordant and the corresponding genus equals zero, then those graphs are called *concordant* (*concordant without cusps*, *concordant without triple points*, *elementary concordant*, and *internally trivial concordant*, respectively).

From the definition we obtain that a framed 4-graph with one unicursal component is slice if and only if it is concordant to a trivial graph consisting of a single circle with no vertices. The same is true for all types of sliceness listed in Definition 6.10.

6.2 Sliceness criteria for certain families of framed graphs

Given a framed graph a natural question arises: which are its slice genera of various types? Additionally, can we determine those genera looking just at the diagram of the graph — that is, its arbitrary representative?

This question is very broad, but in certain particular cases the answer could be given in some nice combinatorial way. In the present book we give two theorems on this matter.

6.2.1 *Odd framed graphs*

First, let us consider *odd* framed graphs. Saying "odd" we mean that every vertex of the corresponding chord diagram is odd in Gaußian sense. The following theorem holds:

Theorem 6.2. *A 1-component odd framed 4-graph Γ is slice if and only if it is elementary slice:*

$$g(\Gamma) = 0 \iff g_{el}(\Gamma) = 0.$$

This theorem is important because the question of elementary sliceness can be solved via a finite combinatorial algorithm given an odd chord diagram of the graph Γ. To prove it we use the Gaußian parity and smoothing techniques, defined in the previous chapters. We will need the following preliminary observation.

Lemma 6.2. *Let Γ be framed 4-graph and K be its spanning complex of genus 0. Then if a double line connects a crossing c of Γ to a cusp, then c is even.*

This lemma essentially means the following. Consider a double line γ of the complex K. Let us denote its endpoints e_1 and e_2. Let the endpoint e_1 coincide with a vertex v of the graph Γ. Since the vertex v is odd, the double line γ is odd as well. Thus, the second endpoint e_2 of the double line γ cannot be a cusp due to Lemma 6.2. Therefore the endpoint e_2 coincides with some other vertex $w \neq v$ of Γ.

Another observation leads to the lemma:

Lemma 6.3 ([Fedoseev and Manturov, 2016]). *Let K be a standard complex with boundary $\partial K = \Gamma$. Let t be a triple point of K such that all double lines are connected to vertices c_1, c_2, c_3, respectively. Then the sum of parities of these vertices is zero modulo 2.*

The reason why Lemma 6.2 and Lemma 6.3 hold originates from the possibility to extend the Gaußian parity from vertices of a graph to double lines of its spanning complex of genus 0.

Lemma 6.3 follows from the fact that at every triple point the sum of the parities of the double lines incident to it equals $0 \bmod 2$. This means that if a double line originates from a triple point and comes to some vertices, then either all of these vertices should be even or there should be one even and two odd vertices.

Now we have all the necessary pieces to prove Theorem 6.2.

Proof. First of all note that one implication of the theorem is trivial: if a graph is elementary slice, then it is slice. Therefore we only need to prove that every slice odd framed 4-graph is elementary slice.

Consider a slice odd graph Γ and let K be the corresponding spanning complex. The complex K can be considered as a free 2-knot with boundary and has two types of double lines: interior ones and exterior ones.

Due to Lemma 4.3 as long as there is at least one interior double line, there exists a smoothing which transforms the complex K into another complex of the same genus 0 and with the same boundary.

Consider such a smoothing of a double line. There are two distinct cases: the smoothed double line can either be incident to triple points or not. If we smooth a double line which is not incident to any triple points, then the number of triple points of the complex doesn't change and the number of double lines decreases by one. If there are triple points incident to the smoothed double line, then the resulting complex has fewer triple points (though the number of double lines can increase).

With every complex K we can associate a pair

$$\mathcal{C}_K = (\#\text{triple points}, \#\text{interior lines})$$

and order those pairs lexicographically. As it was shown in the previous paragraph, smoothing of a double line decreases $\mathcal{C}_K$. That means that the smoothing process is finite. It ends when there are no interior double lines. Let us denote the resulting complex by $\tilde{K}$.

The complex $\tilde{K}$ is also a spanning complex for the graph Γ but it has no interior double lines: every double line starts and ends at a vertex of the graph Γ. Gaußian parity of such double lines coincide with Gaußian parity of their endpoints, therefore every double line of the complex $\tilde{K}$ is odd.

Finally, Lemmas 6.2 and 6.3 yield that the complex $\tilde{K}$ doesn't have either cusps or triple points. Therefore, it provides the elementary spanning complex which completes the proof. □

As was mentioned before, this theorem has a powerful implication: for every diagram of an odd framed 4-graph there is a finite algorithmic procedure of checking whether this graph is slice. To describe this procedure and to prove this corollary, we need to define the following construction. Consider a framed 4-graph Γ and its chord diagram $C(\Gamma)$.

Definition 6.12. *Pairing* $\mathcal{P}$ of the chords of the diagram $C(\Gamma)$ is a partitioning of the chords $\{d_i\}$ into pairwise disjoint sets P_i such that for every i, $|P_i| = 1$ or $|P_i| = 2$ and, for every i such that $P_i = \{c_i, d_i\}$, there is a correspondence between the ends of the chord c_i and those of the chord d_i.

Given a pairing $\mathcal{P}$ we construct a diagram $C(\mathcal{P})$ in the following way: the set of vertices of $C(\mathcal{P})$ coincides with the set of vertices of the diagram $C(\Gamma)$ and every chord of the diagram $C(\mathcal{P})$ is either a chord of the diagram $C(\Gamma)$ belonging to a one-element set P_i or connects the corresponding endpoints of the chords of a two-element set P_i.

Of course, for a given chord diagram there exist different pairings of its chords. Abusing notation a little, we will also say "a pairing of a diagram $C(\Gamma)$" meaning a pairing of the chords of that diagram.

Definition 6.13. We say that a pairing $\mathcal{P}$ of a diagram $C(\Gamma)$ *has no crossings* if the chords of the diagram $C(\mathcal{P})$ are pairwise unlinked and no chord is paired with itself.

Now we can formulate an important implication of Theorem 6.2:

Corollary 6.1. *An odd framed 4-graph Γ is slice if and only if there exists a pairing with no crossings for its chord diagram $C(\Gamma)$.*

Proof. First let us prove that if there exists a pairing with no crossings, then the graph Γ is slice. Let $\mathcal{P}$ be such a pairing of the diagram $C(\Gamma)$. Chord diagram can be regarded as a disc with a set of curves connecting points on its boundary. Note that the chords of the diagram $C(\mathcal{P})$ are divided into pairs in a natural way. Indeed, the pairing $\mathcal{P}$ breaks all the chords of the diagram $C(\Gamma)$ into pairwise disjoint sets P_i. Let us say that two chords c, d of the diagram $C(\mathcal{P})$ form a pair if the set of their endpoints $\{c_1, c_2, d_1, d_2\}$ coincides with the set of endpoints $\{\tilde{c}_1, \tilde{c}_2, \tilde{d}_1, \tilde{d}_2\}$ of the chords of some set $P_i = \{\tilde{c}, \tilde{d}\}$. Without loss of generality we may say that $c_1 = \tilde{c}_1, c_2 = \tilde{c}_2, d_1 = \tilde{d}_1, d_2 = \tilde{d}_2$. In that case we orient the chords in the following way: $c_1 \to c_2, d_1 \to d_2$.

Therefore, the chord diagram $C(\mathcal{P})$ defines a surface diagram of a 2-knot with boundary, whose boundary coincides with the graph Γ. Therefore, the graph Γ is slice.

To prove the inverse implication, consider a slice odd graph Γ and its spanning complex K. Due to Theorem 6.2 we may suppose that the complex K has no cusps, triple points and interior double lines. Consider the surface diagram of the complex K. It consists of a disc and a set of curves, connecting the points on its boundary. Naturally, it defines a chord diagram $C(K)$. The chords of this diagram do not intersect and the set of vertices coincides with the set of vertices of the diagram $C(\Gamma)$. Therefore, the complex K provides a pairing without crossings of the chord diagram $C(\Gamma)$ of the graph Γ. $\square$

Since the number of possible pairings of a given diagram is finite, we obtain an algorithmic procedure of checking the sliceness of an odd framed 4-graph: we just need to go through all the possible pairing one-by-one. If one of them has no crossings, then the graph is slice, otherwise, it is not. In that sense we may say that the diagram itself is an invariant of graph sliceness.

6.2.2 Iteratively odd framed graphs

The next natural class of framed 4-graphs which is bigger than the class of odd graphs is the class of *iteratively odd graphs*. Let us introduce this notion in detail.

First, consider a chord diagram C with the set of vertices $V = \{v_i', v_i''\}$, $i = 1, \ldots, n$ where for every i vertices v_i', v_i'' are connected by a chord e_i.

Definition 6.14. We say that a chord digram C' is obtained from the diagram C by *deletion of a chord* e_j if the set of vertices of the diagram C' is of the form $V' = V \setminus \{v_j', v_j''\}$ and two vertices are connected by a chord if and only if the corresponding vertices are connected by a chord of the diagram C.

Consider a 1-component framed 4-graph Γ and the corresponding chord diagram $C(\Gamma)$. The chord deletion operation on the diagram $C(\Gamma)$ corresponds to "ungluing" operation for the corresponding vertices. Consider a chord diagram C' obtained from $C(\Gamma)$ by parity projection, that is by deleting all odd chords. Let us denote the framed 4-graph corresponding to the chord diagram C' by Γ'.

By definition Γ is odd if and only if Γ' is the circle without vertices. Some even vertices of Γ can become odd for Γ'. Now we can perform the same operation on the graph $\Gamma^1 = \Gamma'$ and its chord diagram. Hence, we can define the graph $\Gamma^2 = (\Gamma^1)'$ and, iteratively, $\Gamma^n = (\Gamma^{n-1})'$.

Definition 6.15. We say that a framed 4-graph Γ is *iteratively odd of order* k if Γ^k has no vertices. Γ is *iteratively odd* if there exists a positive integer k such that Γ is iteratively odd of order k.

Respectively, we say that a vertex c of Γ has order k if it disappears in Γ^{k+1}. For example, odd vertices have order 0. If Γ is not iteratively odd and $\Gamma^m = \Gamma^{m+1}$ has nonempty set of vertices, then all vertices of Γ which persist of $\Gamma^m = \Gamma^{m+1}$ are called *stably even*.

A simple example of an iteratively odd graph is shown on Fig. 5.1 via its chord diagram.

Recall that there are two types of double lines on a complex K with boundary Γ: interior ones and boundary ones (see Definition 6.7). That naturally leads to three types of triple points of such a complex: *interior* triple points (incident to three interior double lines), *exterior* triple points (incident to three boundary double lines) and *mixed* triple points.

The following theorem holds:

Theorem 6.3. *Let Γ be a 1-component iteratively odd framed 4-graph. It is slice if and only if it is slice with only exterior triple points present (that is, with no cusps, interior or mixed triple points).*

To prove it we need some additional facts about spanning complexes and their properties.

Let us define a partial order on spanning complexes in the following way. We define a *complexity* $\mathcal{C}(\mathcal{K})$ of a complex K — the triplet (# triple points, # cusps, # double lines). Those triplets are ordered lexicographically. The concept of complexity is related to the one used in the proof of Theorem 6.2 but includes the number of cusps as well.

Definition 6.16. A complex K_1 is said to be *smaller* than a complex K_2 if

$$\mathcal{C}(K_1) < \mathcal{C}(K_2).$$

A crucial property of parity that we shall need is the following projection lemma:

Lemma 6.4. *For a complex K with boundary $\partial K = \Gamma$, there is a complex K' with the same boundary $\partial K' = \Gamma$ such that K' is obtained from K by deleting all odd interior double lines and which is smaller than the complex K.*

Proof. Consider the surface diagram (S, Γ) of the complex K. Delete all curves corresponding to odd interior double lines. This operation doesn't change the surface S and it's boundary but transforms the set of curves Γ into a set $\tilde{\Gamma}$. Now we need to prove that the pair $(S, \tilde{\Gamma})$ is a surface diagram of a free 2-knot with boundary.

First of all, note that every curve of the set $\tilde{\Gamma}$ is still oriented and paired with exactly one other curve of this set. Therefore, the only condition we need to check is the triple point condition:

If two curves intersect, then the curves paired with them intersect as well (hence a triple point appears on the surface S three times).

Consider a triple point t which was incident to a deleted double line. On the surface diagram it had three corresponding crossings: x_1, x_2, x_3. Let us denote the double lines intersecting at t by α, β, γ and the corresponding curves by $\alpha, \beta, \gamma, \alpha', \beta', \gamma'$. Without loss of generality we may say that $x_1 = \gamma \cap \alpha$, $x_2 = \gamma' \cap \beta$, $x_3 = \alpha' \cap \beta'$.

Let us say that the double line γ was odd and thus deleted by our deletion process. But that means that exactly one of the curves α and β was odd as well, since the sum of parities of α, β, γ must equal $0 \bmod 2$.

That means that exactly two pairs of the curves were deleted and thus no crossings of the set (x_1, x_2, x_3) survive the process.

Therefore, the triple point condition is satisfied and the pair $(S, \tilde{\Gamma})$ is a surface diagram of a knot which is smaller than the initial knot K.   □

In particular, if a cusp point w of K was connected to a vertex c of Γ, then this cusp point will persist in all complexes $K^1 = K', K^2 = (K^1)', \ldots$. By iterating this process, we get

Lemma 6.5. *Let Γ be a slice framed 4-graph, K its spanning complex, that is, $\Gamma = \partial K$. Then each vertex c of Γ which is connected to a cusp is stably even.*

Now we are ready to prove Theorem 6.3.

Proof. Due to Theorem 6.1 we have $g(\Gamma) \leq g_{it}(\Gamma)$. Thus, if a graph Γ is slice with nothing but exterior triple points, that is, $g_{it}(\Gamma) = 0$, then $g(\Gamma) = 0$ and the graph is slice as well. For that reason we only need to prove the converse implication: if an iteratively odd framed graph Γ is slice, then it is slice with only exterior triple points.

Consider a slice iteratively odd graph Γ. Let K be its spanning complex. Apply F-lemma 4.3 to the interior double lines of the spanning complex K. That gives us a spanning complex K' of genus 0 with no interior double lines and hence with no interior or mixed triple points. But K' cannot have any cusps since all the interior cusps were deleted by the F-lemma and no cusps connected to boundary may exist due to Lemma 6.5. It means that only exterior triple point may exist on such the complex K'. Theorem 6.3 is proved.   □

6.2.3 *Multicomponent links*

In the previous two sections we dealt with free knots — that is, framed 4-graphs with one unicursal component. It turns out that Theorem 6.2 holds in the case of multicomponent links if we treat parity in another sense.

Every framed 4-graph Γ having more than one component can be considered as the image of *several circles* $S^1 \sqcup \cdots \sqcup S^1$; we shall distinguish between vertices of Γ which appear by pasting one circle S^1 with itself and those which appear by pasting different circles S^1. We refer to the former as *pure crossings* and to the latter as *mixed crossings*.

Theorem 6.4. *Let Γ be a 2-component framed 4-graph such that all crossings are mixed. Then Γ is slice if and only if it is elementary slice.*

Proof. To prove this theorem we use the following arguments. Let Γ be a slice 2-component framed 4-graph and K its spanning complex. The complex K is an image of the union of two 2-discs D_1, D_2. The interior double lines of K are of the two types: those corresponding to intersections between D_1 with D_2 (mixed double lines) and those corresponding to self-intersections of D_1 or D_2 (pure double lines).

Note that the F-lemma applies to 2-links. Indeed, first consider a pure double line γ. A smoothing of a double line changes only one component of the link, in other words, it can be regarded as a smoothing of a double line of a 2-knot. Thus, due to the F-lemma 4.3 it transforms the component of the link into a new complex of genus 0.

To smooth a mixed double line we use the following trick. Consider the connected sum of the two components. That transforms the link into a 2-knot. Due to the F-lemma 4.3 the smoothing of the double line on this complex produces a complex of genus 0. Now removing the added handle we obtain a 2-component link with the mixed double line smoothed.

As usual, the smoothing process can be applied to interior double lines of a link with boundary verbatim.

To prove Theorem 6.4 we now apply this generalisation of the F-lemma to the complex K. That gives us a new complex K' with the same boundary $\partial K' = \Gamma$ and with no interior double lines. First we note that since all crossings of Γ are mixed, the complex K' has no cusps for every cusp should be a self-crossing of a component. Further, since there are only two components, there are no triple points: one cannot construct a triple point with all double lines incident to it being mixed given only two components of a link.

Therefore the complex K' has neither cusps, nor triple points and Theorem 6.4 is proved. □

6.2.4 *Other results on free knot cobordisms*

In this section we briefly review several further results dealing with some particular cases of knot sliceness and cobordism.

Theorem 6.5. *Let Γ_1 and Γ_2 be two framed 4-graphs. Assume that there is a cobordism of genus g between those graphs such that the corresponding complex K, $\partial K = \Gamma_1 \sqcup \Gamma_2$, has no triple points and no double lines connecting cusps and points on the boundary. Then there exists an elementary cobordism of genus at most g with the same boundary.*

Proof. Since the complex K has no triple points, its double lines do not intersect. Moreover, all the cusps are incident to its interior double lines, and thus can be smoothed due to the generalised F-lemma without genus increase. □

Earlier we proved some properties of odd knots using Gaußian parity on 2-knots and 2-knots with boundary (see Theorem 6.2 and Theorem 6.3). In general, Gaußian parity may not exist on surface knot because different paths on their surface diagrams can be non-homotopic. Nevertheless, in some cases Gaußian parity still exists.

Definition 6.17. A surface diagram on a cylinder C is called *normal* if it satisfies the following conditions:

(1) If a curve γ has both its endpoints on the same boundary component of C, then the curve γ' paired with γ has both its endpoints on the same boundary component of C;
(2) If paired curves γ and γ' have a common endpoint, and their other endpoints lie on the boundary of C, then those endpoints lie on the same boundary component of C;
(3) If a curve γ has endpoints on different boundary components of C, then the curve γ' paired with γ has its endpoints also on different boundary components of C.

In particular, those properties mean that every meridian of the cylinder C not passing through curves' endpoints intersects evenly many curves of the diagram. That means that Gaußian parity is correctly defined on such diagrams.

Now consider two genus zero cobordant free knots. The spanning surface for them is evidently a normal cylinder. Thus the following theorem holds:

Theorem 6.6. *Let K_1 and K_2 be two odd free knots. Then there exists a cobordism of genus 0 between K_1 and K_2 if and only if there exists an elementary genus 0 cobordism between them.*

Proof. The proof of this theorem follows the same pattern as the proof of Theorem 6.2.

First we delete all interior double lines by means of the F-lemma.

Now all the double lines left have their endpoints on the boundary of the spanning surface and thus must be odd. Therefore, the spanning surface has neither cusps nor triple points. □

When dealing with smoothings (either of a crossing of a knot or of a double line of a surface knot) one has two possibilities: either the smoothing yields a connected object (a knot) or not (a link). Let us call the smoothings of the first type "good" and of the second type — "bad".

Lemma 6.6. *Let K be a standard complex of genus 0 whose boundary ∂K is a framed 4-graph $\partial K = \Gamma$. Assume that there is a double line of K connecting a vertex x of ∂K to a cusp point p of K.*

Consider a free knot Γ' obtained from Γ by the one of two smoothings at the vertex x, that gives a free knot (not a link). Then Γ' is slice.

Proof. Consider the complex KK obtained by gluing two copies of K along their boundary $\partial K = \Gamma$. Since K is a knotted disc, the complex KK is a knotted sphere, i.e. a 2-knot, and the graph Γ is a certain transversal section of the complex KK.

The complex KK has a double line γ which connects the two copies of the cusp K. There are two possible smoothing of this double line. One of them gives rise to a connected complex which is a knotted sphere due to the F-lemma.

A smoothing of a double line induces a smoothing on a transversal section of the knot. It is easy to see that the good smoothing of the double line γ induces a one-component smoothing of the knot Γ in the transversal section of KK. Indeed, two possible smoothings of Γ correspond to two possible smoothings of the vertex x. The "bad" smoothing of Γ separates KK into two connected components and thus the transversal section is separated into two connected components as well because the smoothed double line intersects it transversally in one point. Therefore, the good smoothing of γ can only correspond to the good smoothing of x. That completes the proof. □

Finally let us look at the "bad" smoothing of the vertex x from the previous theorem. As it was shown, it corresponds to the "bad" smoothing of the double line γ of the complex KK which breaks it into a 2-component link. But due to the estimate (4.2) the total genus of those two components is at most 1. It means that at least one of them is a knotted sphere. Thus, at least one component of the graph Γ'' obtained from Γ by the "bad" smoothing at the vertex x is capped with a disc, in other words, is slice.

6.3 L-invariant as an obstruction to sliceness

Earlier in Section 5.4 we introduced the notion of L-invariant of free knots. It turns out that this invariant is in fact an *obstruction to sliceness*.

The main result of the present section is the following

Theorem 6.7. *If a free knot K has $L(K) \neq 0$, then K is not slice.*

In particular, this yields the following

Corollary 6.2. *Let γ be a curve immersed in an oriented closed 2-surface S_g. Then if for a free knot Γ corresponding to γ one has $L(\Gamma) \neq 0$, then the underlying γ is not slice as a flat virtual knot.*

Indeed, a general position image of the disc in a 3-manifold is a spanning disc having only three-dimensional singularities. The converse statement is, generally, not true.

The proof of Theorem 6.7 will consist of several steps.

Using the same notation as in Section 5.4 we shall adopt the following notation for the maps: by g we shall mean the map $\mathcal{D} \to D$ corresponding to the cobordism, and by f we shall denote either of the two maps: a Morse function $f : D \to \mathbb{R}$ (see definition ahead) or the composition $g \circ f : \mathcal{D} \to \mathbb{R}$ will be also denoted by f (abusing notation).

Assume the knot K is slice and admits a cobordism $g : \mathcal{D} \to D$ (of genus zero).

Definition 6.18. By a *Morse function* on D we mean a Morse function $g \circ f : \mathcal{D} \to [0, \infty)$ such that if $g(x) = g(y)$, then $f(x) = f(y)$, all Reidemeister move points: triple points, cusp points and tangency points on $\mathcal{D}$, lie on non-critical levels of f, and $f^{-1}(0) = K, f^{-1}(1) = \emptyset$. By abuse of notation we shall denote the function on $\mathcal{D}$ and the function on D by the same letter f. By a *non-singular* value of the function f we mean a noncritical value X such that $f^{-1}(X) \subset \mathcal{D}$ contains no cusps and no triple points. A Morse function on D will be called *simple* if every singular level contains either exactly one critical point or exactly one triple point or exactly one cusp point.

From now on, we require that the Morse function of D is simple and the level 0 is non-singular. It is clear that such Morse functions are *everywhere dense in the class of all functions*. Every Morse function has singular levels of two types: those corresponding to Morse bifurcations (saddles, minima, and maxima) and those corresponding to Reidemeister moves (corresponding to cusps, tangency points and triple points). Denote singular levels of the function f by $c_1 < \cdots < c_k$ and choose non-critical levels a_i: $0 = a_0 < c_1 < a_1 < c_2 < \ldots a_k < c_k < a_{k+1} = 1$.

Let us construct the *Reeb graph* Γ_f (molecule) of the function f as follows. The univalent vertices of the Reeb graph will correspond to minima and maxima of the function f; the vertices of valency three will correspond to saddle points; edges will connect critical points; every edge will correspond to a cylinder $S^1 \times I \subset \mathcal{D}$ which is continuously mapped by f to a closed interval between some two critical point; this cylinder has no Morse critical points inside. One edge will emanate from the point 0 corresponding to the circle $S = \partial \mathcal{D}$, see Fig. 6.4.

Since this graph is a Reeb graph of the Morse function on a disc $\mathcal{D}$, the graph Γ_f is a *tree*.

Our next goal is to endow each edge of the Reeb graph with a non-negative integer *label*. The label of the edge emanating from 0 will coincide with $L(K)$.

For every non-singular level c of f the preimage $K_c = f^{-1}(c) \subset D$ is a free link; when passing through a Reidemeister singular point, the link K_c is operated on by the corresponding Reidemeister move; when passing through a Morse critical point it gets operated on by a Morse-type bifurcation. Every crossing of K_c belongs to some double line of D. Define the *parity* of the crossing to be the Gaußian parity of the double line, it belongs to.

Consider the free link K_c. For every unicursal component $K_{c,j}$ of the free link K_c consider the value $L(K_{c,j})$ and denote them by $l_{c,j}$. Let $L_c = \{l_{c,1}, \ldots, l_{c,m}\}$ be the collection of those integers.

Each $l_{c,i} \in \mathbb{N} \cup \{0\}$ corresponds to a component of the free link K_c and

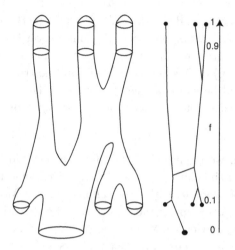

Fig. 6.4   The Reeb graph and circle bifurcations

does not change under Reidemeister move when changing c without passing through Morse critical points. Associate it with the corresponding edge of the graph Γ_f.

Now, let us analyze the behaviour of these labels $l_{c,i}$ at vertices of the graph Γ_f.

Lemma 6.7. *Assume $K_{c-\varepsilon}$ and $K_{c+\varepsilon}$ differ by one Morse bifurcation at the level c.*
   Then:

(1) *If this bifurcation corresponds to a birth of a circle, then $L_{c+\varepsilon}$ is obtained from $L_{c-\varepsilon}$ by an addition of 0;*
(2) *If it corresponds to a death of a circle, then $L_{c+\varepsilon}$ is obtained from $L_{c-\varepsilon}$ by a removal of 0;*
(3) *In the case of fusion all elements of $L_{c-\varepsilon}$ except two ones (m and n) remain the same, and the elements m and n turn into some $k = \pm n \pm m$ to form an element of $L_{c+\varepsilon}$.*
(4) *The fission operation is the inverse to the fusion: instead of one element k one gets a pair of elements m, n such that $\pm m \pm n = k$.*

This leads to the following way of proving Theorem 6.7. The graph Γ_f has all vertices except one (corresponding to the initial knot K) having label 0. At each vertex, the two labels (with signs $\pm$) sum up to give the third label. Thus, taking into account that Γ_f is a tree, we see that the label of the initial vertex is 0, so $L(K) = 0$. A contrapositive completes the proof of Theorem 6.7.

Remark 6.4. Note that the described construction does not work for a cobordism of arbitrary genus, because in that case the Reeb graph is not necessarily a tree.

PART 3

The Groups G_n^k

Chapter 7

General Theory of Invariants of Dynamical Systems and Groups

Usually, invariants of mathematical objects are valued in numerical or polynomial rings, rings of homology groups, etc. In the present chapter, we prove a general theorem about invariants of *dynamical systems* which are valued in groups very close to pictures, the so-called *free k-braids*.

Formally speaking, free k-braids form a group presented by generators and relations; this group has lots of picture-valued invariants. For free 2-braids, the following principle can be realised:

If a braid diagram D is complicated enough, then it realises itself as a subdiagram of any diagram D' contained in D.

In topology, this principle was first demonstrated in terms of *parity* for the case of *virtual knots*, see [Manturov, 2010].

Our invariant of braids is constructed by using horizontal trisecant lines. Herewith, the set of *critical values* (corresponding to these trisecants) leads to a certain picture which appears in all diagrams equivalent to the initial picture.

The main theorem of the present chapter has various applications in knot theory, geometry, and topology. It is based on the following important principle:

If in some science an object has intersections of multiplicity k, then such an object can be studied by k-braids and their generalisations.

7.1 Dynamical systems and their properties

Given a topological space Σ, called the *configuration space*; the elements of Σ will be referred to as *particles*. The topology on Σ^N defines a natural topology on the space of all continuous mappings $[0, 1] \to \Sigma^N$; we shall also study mappings $S^1 \to \Sigma^N$, where $S^1 = [0, 1]/\{0 = 1\}$ is the circle; these mappings will naturally lead to *closures* of k-braids.

Let us fix positive integers n and k.

The *space of admissible dynamical systems* $\mathcal{D}$ is an open subset in the space of all maps $[0, 1] \to \Sigma^n$.

A *state* is an element D of $\mathcal{D}$. By a *dynamical system* of $\mathcal{D}$ we mean the ordered set of particles $D(t) \in \Sigma^n$ for $t \in [0, 1]$. Herewith, $D(0)$ and $D(1)$ are called *the initial state* and *the terminal state*.

We shall also deal with *cyclic dynamical system*, where $D(0) = D(1)$.

As usual, we shall fix the initial state and the terminal state and consider the set of admissible dynamics with such initial and terminal states.

We say that a property $\mathcal{P}$, which is defined for subsets of the set of n particles from Σ, is *k-good*, if the following conditions hold:

(1) if this property holds for some set of particles, then it holds for every subset of this set;
(2) this property holds for every set consisting of $k - 1$ particles among n ones (hence, for every smaller set);
(3) fix $k + 1$ pairwise distinct numbers $i_1, \ldots, i_{k+1}, i_j \in \{1, \ldots, n\}$; if the property $\mathcal{P}$ holds for particles with numbers $i_1, \ldots, i_k$ and for the set of particles with numbers $i_2, \ldots, i_{k+1}$, then it holds for the set of all $k + 1$ particles $i_1, \ldots, i_{k+1}$.

Remark 7.1. Besides *statically good properties* $\mathcal{P}$, which are defined for subsets of the set of particles regardless the state of the dynamics, one can talk about *dynamically good* properties, which can be defined for states considered in time.

Remark 7.2. Our main example deals with the case $k = 3$, where for particles we take different points on the plane, and $\mathcal{P}$ is the property of points to belong to the same line. In general, particles may be more complicated objects than just points.

Let $\mathcal{P}$ be a k-good property defined on a set of n particles $n > k$. For each $t \in [0, 1]$ we shall fix the corresponding state of particles, and pay attention to those t for which there is a set k particles possessing $\mathcal{P}$; we shall refer to these moments as *$\mathcal{P}$-critical* (or just *critical*).

Definition 7.1. We say that a dynamical system D is *pleasant*, if the set of its critical moments is finite whereas each for critical moment there exists exactly one k-index set for which the set $\mathcal{P}$ holds (thus, for larger sets the property $\mathcal{P}$ does not hold). Such an unordered k-tuple of indices will be called a *multiindex* of critical moments.

With a pleasant dynamical system we associate its *type* $\tau(D)$, which will of the set of multiindices $m^1, \ldots, m^N$, written as t increases from $t = 0$ to $t = 1$.

For each dynamics D and each multiindex $m = (m_1, \ldots, m_k)$, let us define the *m-type* of D as the ordered set $t_1 < t_2 < \ldots$ of values of t for which the set of particles $m_1, \ldots, m_k$ possesses the property $\mathcal{P}$.

Notation: $\tau_m(D)$. If the number l of values $t_1 < \cdots < t_l$ is fixed, then the type can be thought of as a point in $\mathbb{R}^l$ with coordinates $t_1, \ldots, t_l$.

Definition 7.2. By the *type* of a dynamical system we mean the set of all its types $\tau_m(D)$. Notation: $\tau(D)$.

If D is pleasant, then these sets are pairwise disjoint.

Definition 7.3. Fix a number k and a k-good property $\mathcal{P}$. We say that D is *$\mathcal{P}$-stable*, if there is a neighbourhood $U(D)$, where for each multiindex m (consisting of k indices), the number l of critical values corresponding to this multiindex is constant, and the type τ_m is a continuous mapping $U(D) \to \mathbb{R}^l$.

We shall often say *pleasant dynamical systems* or *stable dynamical systems* without referring to $\mathcal{P}$ if it is clear from the context which $\mathcal{P}$ we mean.

Definition 7.4. A *deformation* is a continuous path $s : [0, 1] \to \mathcal{D}$ in the space of admissible dynamical systems from a stable pleasant dynamics $s(0)$ to another pleasant stable dynamical system $s(1)$.

Definition 7.5. We say that a deformation s is *admissible* if:

(1) The set of values u, where $s(u)$ is not pleasant or is not stable, is finite, and for those u where $s(u)$ is not pleasant, $s(u)$ is stable.
(2) Inside the stability intervals, the m-types are continuous for each multiindex m.
(3) For each value $u = u_0$, where $s(u_0)$ is not pleasant, exactly one of the two following cases occurs:

(a) There exists exactly one $t = t_0$ and exactly one $(k + 1)$-tuple $m = (m_1, \ldots, m_{k+1})$ satisfying $\mathcal{P}$ for this m (hence, $\mathcal{P}$ does not hold for larger sets).

Let $\tilde{m}_j = m \setminus \{m_j\}, j = 1, \ldots, k + 1$. For types $\tau_{\tilde{m}_j}$, choose those coordinates ζ_j, which correspond to the value $t = t_0$. It is required that for all these values u_0 all functions $\zeta_j(u)$ are smooth, and all derivatives $\frac{\partial \zeta_j}{\partial u}$ are pairwise distinct;

(b) There exists exactly one value $t = t_0$ and exactly two multiindices $m = \{m_1, \ldots, m_k\}$ and $m' = \{m'_1, \ldots, m'_k\}$ for which $\mathcal{P}$ holds; we require that $Card(m \cap m') < k - 1$.

(4) For each value u_0, where the dynamical system $s(u_0)$ is not stable, there exists a value $t = t_0$, which is critical for $s(u_0)$, and a multi-index $\mu = (\mu_1, \ldots, \mu_k)$, for which the following holds. For some small ε all dynamical systems $s(u)$ for $u \in (u_0 - \varepsilon, u_0) \cup (u_0, u_0 + \varepsilon)$ are stable, and for $\delta < \varepsilon$ the type $\tau_\mu(D_{u+\delta})$ differs from the type $\tau_\mu(D_{u-\delta})$ by an addition/removal of two identical multiindices μ in the position corresponding to t_0.

For the space of deformation $\mathfrak{D}$, one defines an induced topology.

Definition 7.6. We say that a k-good property $\mathcal{P}$ is k-*correct* for the space of admissible dynamical systems, if the following conditions hold:

(1) In each neighbourhood of any dynamical system D there exists a pleasant dynamical system D'.
(2) For each deformation s there exists an *admissible* deformation with the same ends $s'(0) = s(0), s'(1) = s(1)$.

Definition 7.7. We say that two dynamical systems D_0, D_1 are *equivalent*, if there exists a deformation $s : [0, 1] \to \mathcal{D}$ such that $s(0) = D_0$ and $s(1) = D_1$.

Thus, if we talk about a correct $\mathcal{P}$-property, then we can talk about an admissible deformation when defining the equivalence.

7.2 Free k-braids

Let us now pass to the definition of the n-*strand free k-braid group* G_n^k.

Consider the following $\binom{n}{k}$ generators a_m, where m runs the set of all unordered k-tuples $m_1, \ldots, m_k$, whereas each m_i are pairwise distinct numbers from $\{1, \ldots, n\}$.

For each unordered $(k+1)$-tuple U of distinct indices $u_1, \ldots, u_{k+1} \in \{1, \ldots, n\}$, consider the $k+1$ sets $m^j = U \setminus \{u_j\}, j = 1, \ldots, k+1$. With U, we associate the relation

$$a_{m^1} \cdot a_{m^2} \ldots a_{m^{k+1}} = a_{m^{k+1}} \ldots a_{m^2} \cdot a_{m^1} \qquad (7.1)$$

for two tuples U and $\bar{U}$, which differ by order reversal, we get the same relation. Thus, we totally have $\frac{(k+1)!\binom{n}{k+1}}{2}$ relations. We shall call them the *tetrahedron relations*.

For k-tuples m, m' with $Card(m \cap m') < k - 1$, consider the *far commutativity relation*:

$$a_m a_{m'} = a_{m'} a_m. \qquad (7.2)$$

Note that the far commutativity relation can occur only if $n > k + 1$.

Besides that, for all multiindices m, we write down the following relation:

$$a_m^2 = 1. \qquad (7.3)$$

Definition 7.8. The *k-free braid group* G_n^k is defined as the quotient group of the free group generated by all a_m for all multiindices m by relations (7.1), (7.2) and (7.3).

Example 7.1. The group G_3^2 is $\langle a, b, c \,|\, a^2 = b^2 = c^2 = (abc)^2 = 1 \rangle$, where $a = a_{12}, b = a_{13}, c = a_{23}$.

Indeed, the relation $(abc)^2 = 1$ is equivalent to the relation $abc = cba$ because of $a^2 = b^2 = c^2 = 1$. This obviously yields all the other tetrahedron relations.

Example 7.2. The group G_4^3 is isomorphic to $\langle a, b, c, d \,|\, a^2 = b^2 = c^2 = d^2 = 1, (abcd)^2 = 1, (acdb)^2 = 1, (adbc)^2 = 1 \rangle$. Here $a = a_{123}, b = a_{124}, c = a_{134}, d = a_{234}$.

It is easy to check that instead of $\frac{4!}{2} = 12$ relations, it suffices to take only $\frac{3!}{2} = 3$ relations.

By the *length* of a word we mean the number of letters in this word, by the *complexity of a free k-braid* we mean the minimal length of all words representing it. Such words will be called *minimal representatives*. The tetrahedron relations (in the case of free 2-braids we call them the *triangle relations*) and the far commutativity relations do not change the complexity, and the relation $a_m^2 = 1$ increases or decreases the complexity by 2.

As usual in the group theory, it is natural to look for minimal length words representing the given free k-braid.

If we deal with conjugacy classes of free k-braids, then one deals with the length of cyclic words.

The number of words of fixed length in a finite alphabet is finite; k-braids and their conjugacy classes are the main invariant of the present work.

Let us define the following two types of homomorphisms for free braids. For each $l = 1, \ldots, n$, there is an *index forgetting homomorphism* $f_l : G_n^k \to G_{n-1}^{k-1}$; this homomorphism takes all generators a_m with multiindex m not containing l to the unit element of the group, and takes the other generators a_m to $a_{m'}$, where $m' = m \setminus \{l\}$; this operation is followed by the index renumbering.

The *strand-deletion homomorphism* d_j is defined as a homomorphism $G_n^k \to G_{n-1}^k$; it takes all generators a_m having multiindex containing j to the unit element; after that we renumber indices.

The free 2-braids (called also *pure free braids*) were studied in [Manturov, 2010; Manturov and Wang, 2012; Manturov, 2015c].

For free 2-braids, the following theorem holds.

Theorem 7.1. *Let b' be a word representing a free 2-braid β. Then every word b which is a minimal representative of β, is equivalent by the triangle relations and the far commutativity relation to some subword of the word b'.*

Every two representatives b_1 and b_2 of the same free 2-braid β are equivalent by the triangle relations and the far commutativity relations.

Thus, for free 2-braids, the recognition problem can be solved by means of considering its minimal representative.

The main idea of the proof of this theorem is similar to the classification of homotopy classes of curves in 2-surfaces due to Hass and Scott [Hass and Scott, 1994], see Chapter 3: in order to find a minimal representative, one looks for "bigon reductions" until possible, and the final result is unique up to third Reidemeister moves for the exception of certain special cases (multiple curves etc.). For free 2-braids "bigon reductions" refer to some cancellations of generators similar generators a_m and a_m in good position with respect to each other, see Fig. 7.2, third Reidemeister moves correspond to the triangle relations, and the far commutativity does not change the picture at all [Manturov, 2015c].

Algebraically, this reduction process stems from the gradient descent algorithm in Coxeter groups, see Theorem 9.3.

For us, it is crucial to know that *when looking at a free 2-braid, one can see which pairs of crossings can be cancelled.* Once we cancel all possible crossings, we get an invariant picture.

Thus, *we get a complete picture-valued invariant of a free 2-braid.*

Theorem 7.1 means that this picture (complete invariant) occurs as a sub-picture in every picture representing the same free 2-braid.

However, various homomorphisms $G_n^k \to G_{n-1}^{k-1}$, whose combination leads to homomorphisms of type $G_n^k \to G_{n-k+2}^2$, allow one to construct lots of invariants of groups G_n^k valued in *pictures*.

In particular, these pictures allow one to get easy estimates for the complexity of braids and corresponding dynamics.

7.3 The main theorem

Let $\mathcal{P}$ be a k-correct property on the space of admissible dynamical systems with fixed initial and final states.

Let D be a pleasant stable dynamical system describing the motion of n particles with respect to $\mathcal{P}$. Let us enumerate all critical values t corresponding to all multiindices for D, as t increases from 0 to 1. With D we associate an element $c(D)$ of G_n^k, which is equal to the product of a_m, where m are multiindices corresponding to the critical values of D as t increases from 0 to 1.

Theorem 7.2. *Let D_0 and D_1 be two equivalent stable pleasant dynamics with respect to $\mathcal{P}$. Then $c(D_0) = c(D_1)$ are equal as elements of G_n^k.*

Proof. Let us consider an admissible deformation D_s between D_0 and D_1. For those intervals of values s, where D_s is pleasant and stable, the word representing $c(D_s)$, does not change by construction. When passing through those values of s, where D_s is not pleasant or is not stable, $c(D_s)$ changes as follows:

(1) Let s_0 be the value of the parameter deformation, for which the property $\mathcal{P}$ holds for some $(k+1)$-tuple of indices at some time $t = t_0$. Note that D_{s_0} is not pleasant, but is stable.

Consider the multiindex $m = (m_1, \ldots, m_{k+1})$ for which $\mathcal{P}$ holds at $t = t_0$ for $s = s_0$. Let $\tilde{m}_j = m \setminus m_j, j = 1, \ldots, k + 1$. For types $\tau_{\tilde{m}_j}$, let us choose those coordinates ζ_j, which correspond to the intersection $t = t_0$ at $s = s_0$. As s changes, these types are continuous functions with respect to s.

Then for small ε, for $s = s_0 + \varepsilon$, the word $c(D_s)$ will contain a sequence of letters $a_{\bar{m}_j}$ in a certain order. For values $s = s_0 - \varepsilon$, the word $c(D_s)$ will contain the same set of letters in the reverse order. This is represented by the relation 7.1 (tetrahedron relation). Here we have used the fact that D_s is stable.

(2) If for some $s = s_0$ we have a critical value with two different k-tuples m, m' possessing $\mathcal{P}$ and $Card(m \cap m') < k - 1$, then the word $c(D_s)$ undergoes the relation 7.2 (far commutativity) as s passes through s_0; here we also require the stability of D_{s_0}.

(3) If at some $s = s_0$, the deformation D_s is unstable, then $c(D_s)$ changes according to the relation 7.3 as s passes through s_0. Here we use the fact that the deformation is admissible.

□

Let us now pass to our main example, the classical braid group. Here particles are distinct points on the plane. We can require that their initial and final positions are uniformly distributed along the unit circle centered at 0.

For $\mathcal{P}$, we take the property to belong to the same line. This property is, certainly, 3-good. Every motion of points where the initial state and the final state are fixed, can be approximated by a motion where no more than 3 points belong to the same straight line at once, and the set of moments where three points belong to the same line, is finite, moreover, no more than one set of 3 points belong to the same line simultaneously. This means that this dynamical system is pleasant.

Finally, the correctness of $\mathcal{P}$ means that if we take two isotopic braids in general position (in our terminology: two pleasant dynamical systems connected by an admissible deformation), then by a small perturbation we can get an admissible deformation for which the following holds. There are only with finitely many values of the parameter s with four points on the same line or two triples of points on the same line at the same moment; moreover, for each such s only one such case occurs exactly for one value of t.

In this example, as well as in the sequel, the properties of being *pleasant* and *correct* are based on the fact that every two *general position* states can be connected by a curve passing through states of codimension 1 (simplest degeneracy) finitely many times, and every two paths with fixed endpoints, which can be connected by a deformation, can be connected by a general position deformation where points of codimensions 1 and 2 occur, the latter happen only finitely many times.

In particular, the most complicated condition saying that the set of some $(k + 1)$ particles satisfies the property $\mathcal{P}$, and the corresponding derivatives are all distinct, is also a general position argument. For example, assuming that some 4 points belong to the same horizontal line (event of codimension 2), we may require that there is no coincidence of any further parameters (we avoid events of codimension 3).

From the definition of the invariant c, one easily gets the following

Theorem 7.3. *Let D be a dynamical system corresponding to a classical braid. Then the number of horizontal trisecants of the braid D is not smaller than the complexity of the free 3-braid $\beta = c(D)$.*

Analogously, various geometrical properties of dynamical systems can be analysed by looking at complexities of corresponding groups of free k-braids, if one can define a k-correct property for these dynamics, which lead to invariants valued in free k-braids.

Let us now collect some situations where the above methods can be applied.

(1) An evident invariant of closed pure classical braids is the conjugacy class of the group G_n^3. To pass from arbitrary braids to pure braids, one can take some power of the braid in question.

(2) Note that the most important partial case for $k = 2$ is the "classical Reidemeister braid theory". Indeed, for a set of points on the plane Oxy, we can take for the property $\mathcal{P}$ that the y-coordinates of points coincide. Then, considering a braid as a motion of distinct points in the plane $z = 1 - t$ as t changes from 0 to 1, we get a set of curves in space whose projection to Oxz will have intersections exactly in the case when the property $\mathcal{P}$ holds. The additional information coming from the x coordinate, leads one to the classical braid theory.

(3) For classical braids, one can construct invariants for $k = 4$ in a way similar to $k = 3$. In this case, we again take ordered sets of n points on the plane, and the property $\mathcal{P}$ means that the set of points belongs to the same circle or straight line; for three distinct points this property always holds, and the circle/straight line is unique.

(4) With practically no changes this theory can be used for the study of weavings [Viro, 1985], collections of projective lines in $\mathbb{R}P^3$ considered up to isotopy. Here $\mathcal{P}$ is the property of a set of points to belong to the same projective line. The main difficulty here is that the general position deformation may contain three lines having infinitely many common horizontal trisecants. Another difficulty occurs when one of our lines becomes horizontal; this leads to some additional relations to our groups G_n^3, which are easy to handle.

(5) In the case of points on a 2-sphere we can define $\mathcal{P}$ to be the property of points to belong to the same geodesic. This theory works with an additional restriction which forbids antipodal points. Some constraints should be imposed in the case of 2-dimensional Riemannian manifolds: for the space of all dynamical systems, we should impose the restrictions which allow one to detect the geodesic passing through two points in a way such that if two geodesics chosen for a, b and for b, c coincide, then the same geodesic should be chosen for a, c.

(6) We can use this paradigm to study configurations of balls in a space. A dynamic of balls induces a dynamic of their centers, so any k-good property assigns a word in some group G_n^k to the dynamic. The difference between dynamics of balls and ones of a set of points is that the centers of balls cannot move too close to each other.

(7) In the case of n non-intersecting projective m-dimensional planes in $\mathbb{R}P^{m+2}$ considered up to isotopy, the theory works as well. Here, in order to define the dynamical system, we take a one-parameter family of projective hyperplanes in general position, for particles we take $(m-1)$-dimensional planes which appear as intersections of the initial planes with the hyperplane.

The properties of being good, correct etc. follow from the fact that in general position "particles" have a unique same secant line (in a way similar to projective lines, one should allow the projective planes not to be straight). For $m = 1$ one should take $k = 3$, in the general case one takes $k = 2m + 1$.

(8) This theory can be applied to the study of fundamental groups of various discriminant spaces, if such spaces can be defined by several equalities and subsets of these equalities can be thought of as property $\mathcal{P}$.

(9) The case of classical knots, unlike classical links, is a bit more complicated: it can be considered as a dynamical system, where the number of particles is not constant, but it is rather allowed for two particles to be born from one point or to be mutually contracted.

The difficulty here is that knots do not possess a group structure, thus, we don't have a natural order on the set of particles. Nevertheless, it is possible to construct a map from classical knots to *free 3-knots* (or free 4-knots) and study them in a way similar to free 3-braids (free 4-braids).

Remark 7.3. In the case of sets of points in a space of dimension 3, the property of some points to belong to a 2-plane (or higher-dimensional plane) is not correct. Indeed, if three points belong to the same line, then whatever fourth point we add to them, the four points will belong to the same plane, thus we can get various multiindices of 4 points corresponding to the same moment.

The "triviality" of such theory taken without any additional constraints has the simple description that the configuration space of sets of points in $\mathbb{R}^3$ has trivial fundamental group.

Remark 7.4. In some cases we can reduce some k-good properties to others. For example, consider the 4-good properties $\mathcal{P}_1$: "points in the plane lie on the same circle (or the same line)", and $\mathcal{P}_2$: "points in the 3-dimensional space lie on the same 2-plane".

Let $f\colon \mathbb{R}^2 \to \mathbb{R}^3$ be the inverse stereographic projection:
$$f(u,v) = \left(\frac{2u}{u^2 + v^2 + 1}, \frac{2v}{u^2 + v^2 + 1}, \frac{u^2 + v^2 - 1}{u^2 + v^2 + 1} \right).$$
The image of the map f is the unit sphere without a point, and the image of any circle or line is a circle on the sphere, i.e. an intersection of the sphere and a 2-plane. Thus, a set $X \subset \mathbb{R}^2$ contains four points which lie on the same circle or line, if and only if the set $f(X) = \{f(x)\}_{x \in X} \subset \mathbb{R}^3$ contains four points which lie on the same 2-plane. In other words, the property $\mathcal{P}_1$ can be reduced to the property $\mathcal{P}_2$.

We can summarise the reasonings above in the following statement.

Proposition 7.1. *Let $C_n(\mathbb{R}^2)$ be the configuration space of ordered sets consisting of n distinct points in $\mathbb{R}^2$, and let $C'_n(\mathbb{R}^3)$ be the configuration space of ordered sets consisting of n distinct points in $\mathbb{R}^3$ such that there are no three points which lie on the same line. Then the following diagram is commutative:*

$$\pi_1(C_n(\mathbb{R}^2)) \xrightarrow{\ \ f_*\ \ } \pi_1(C'_n(\mathbb{R}^3))$$
$$c_{\mathcal{P}_1} \searrow \qquad \swarrow c_{\mathcal{P}_2}$$
$$G_n^4$$

where $c_{\mathcal{P}_i}$ is the homomorphism of Theorem 7.2 determined by the property
$\mathcal{P}_i$, $i = 1, 2$.

The construction above can be used for reduction of other k-good properties to the basic k-property "points lie on the same hyperplane in $\mathbb{R}^{k-1}$".

Remark 7.5. Let M be a complete Riemannian manifold of dimension $k + 1$, $k \geq 2$. We say that M satisfies k-*plane condition* if for any $k + 1$ points of M there exists a k-dimensional plane N (i.e. N is a totally geodesic submanifold) which contains all the $k + 1$ points and for an open dense subset of $(k + 1)$-point configurations the plane is unique.

The k-plane condition implies Cartan's axiom of k-planes: for each point $x \in M$ and each k-dimensional subspace $T \subset T_x M$ there exists a k-dimensional totally geodesic submanifold N containing x such that $T_x N = T$. E. Cartan [Cartan, 1928] proved that real space forms (Riemannian manifolds of constant sectional curvature) are the only Riemannian manifolds of dimension ≥ 3 which satisfy the axiom of k-planes, $k \geq 2$. The completeness of the manifold and the uniqueness of the k-plane in general position mean that M is either S^{k+1}, $\mathbb{R}P^{k+1}$, $\mathbb{R}^{k+1}$ or $\mathbb{H}^{k+1}$.

Note that the uniqueness condition can be discarded if one considers the k-property locally, for points which are near to each other. This approach leads to the theory of groups Γ_n^k, see Part 4.

Let M be a complete Riemannian manifold that satisfies k-plane condition. We consider a k-property $\mathcal{P}$: "points lie on the same hyperplane in M". Denote the configuration space of n points in M such that among them there are no $k + 1$ points which lie on the same $(k - 1)$-plane (i.e. intersection of two hyperplanes), by $C_n'(M)$. Then we have a well defined map $c_M \colon \pi_1(C_n'(M)) \to G_n^k$ induced by the property $\mathcal{P}$.

Let $M = \mathbb{H}^{k+1}$ be the hyperbolic space of dimension $k + 1$. We can use Klein's model and identify M with the open unit ball $U \subset \mathbb{R}^{k+1}$. The hyperplanes in M are the intersections of hyperplanes of $\mathbb{R}^{k+1}$ with U. Hence, $C_n'(\mathbb{H}^{k+1}) = C_n'(U) \subset C_n'(\mathbb{R}^{k+1})$ and $c_{\mathbb{H}^{k+1}} = c_{\mathbb{R}^{k+1}} \circ j_*$ where $j \colon C_n'(U) \to C_n'(\mathbb{R}^{k+1})$ is the inclusion map. Note that j is a homotopy equivalence: we can define a homotopical inverse map by the formula

$$g(x_i) = \frac{x_i}{\max_{i'=1,\ldots,n} \|x_{i'}\| + 1}, \quad (x_1 \ldots, x_n) \in C'(\mathbb{R}^{k+1}).$$

Thus, G_n^k-invariants of the hyperbolic and euclidian spaces coincide.

A natural inclusion $i \colon \mathbb{R}^{k+1} \hookrightarrow \mathbb{R}P^{k+1}$ induces an inclusion $i \colon C_n'(\mathbb{R}^{k+1}) \hookrightarrow C_n'(\mathbb{R}P^{k+1})$ and hence a homomorphism

$i_*\colon \pi_1(C'_n(\mathbb{R}^{k+1})) \to \pi_1(C'_n(\mathbb{R}P^{k+1}))$ compatible with the maps $c_{\mathbb{R}^{k+1}}$ and $c_{\mathbb{R}P^{k+1}}$. The natural covering $p\colon S^{k+1} \to \mathbb{R}P^{k+1}$ induces a 2^n-fold covering $p\colon C'_n(S^{k+1}) \to C'_n(\mathbb{R}P^{k+1})$ and a homomorphism $p_*\colon \pi_1(C'_n(\mathbb{R}^{k+1})) \to \pi_1(C'_n(\mathbb{R}P^{k+1}))$. Thus, we have the following diagram.

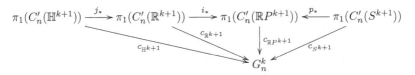

7.4 Pictures

The free k-braids can be depicted by strands connecting points $(1,0), \ldots, (n,0)$ on to points $(1,1), \ldots, (n,1)$; every strand connects $(i,0)$ to $(i,1)$; its projection to the second coordinate is a homeomorphism. We mark crossings corresponding $a_{i,j,k}$ by a solid dot where strands $\#i, \#j, \#k$ intersect transversally.

All other crossings on the plane are artifacts of planar drawing; they do not correspond to any generator of the group, and they are encircled, see Fig. 7.1.

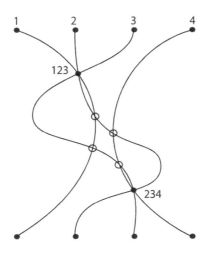

Fig. 7.1 The 3-braid $a_{234}a_{123}$

The clue for the recognition of free 2-braids is the *bigon reduction* shown in Fig. 7.2.

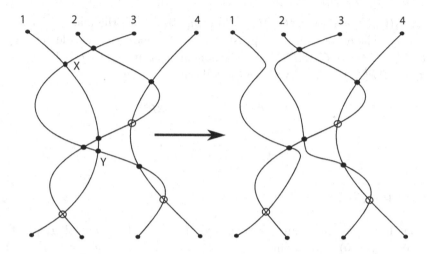

Fig. 7.2 The bigon reduction

Here we reduce the bigon whose vertices are X, Y.

The graph which appears after all possible bigon reductions (with opposite edge structure at vertices) is a complete invariant of the free 2-braid (see also [Manturov, 2015c]).

Chapter 8

Groups G_n^k and Their Homomorphisms. Recognition of Free Braids. Explicit Examples

In the present chapter, we work with groups G_n^k, which generalise classical braid groups and other groups in a very broad sense.

Perhaps, the easiest groups are free products of cyclic groups (finite or infinite). In these groups, the word problem and the conjugacy problem are solved extremely easily: we just contract a generator with its opposite until possible (in a word or in a cyclic word) and stop when it is impossible.

This "gradient descent" algorithm allows one not only to solve the word problem but also to show that "if a word w is irreducible then it is contained in any word w' equivalent to it." For example, $abcab$ is contained in $abaa^{-1}cbb^{-1}ab$ in the free product $\mathbb{Z} * \mathbb{Z} * \mathbb{Z} = \langle a, b, c \rangle$.

Similar phenomena arose in parity theory of knot diagrams in low-dimensional topology discovered by V.O. Manturov:

If a diagram K is odd and irreducible, then it realises itself as a subdiagram of any other diagram K' equivalent to it [Manturov, 2010].

Here "irreducibility" is similar to group-theoretic irreducibility and "oddness" means that all crossings of the diagram are odd or non-trivial in some sense.

In fact, parity theory allows one to endow each crossing of a diagram with a powerful diagram-like information, so that if two crossings (resp., two letters a and a^{-1}) are contracted, then they have the same pictures. Thus, a crossing possessing a non-trivial picture takes responsibility for the non-triviality of the whole diagram (braid, word).

How are these two simple phenomena related to each other?

In the present chapter, we study classical braid groups. However, unlike virtual braids and virtual knots, classical braids and classical knots do not possess any good parity to realise the above principle: for the usual Gaußian parity all crossings turn out to be even and for the component parity there are odd crossings, but they do not lead to non-trivial invariants [Ilyutko, Manturov and Nikonov, 2011, Corollary 4.3].

However, we can change the point of view of what we call "a crossing". There is a natural presentation of the braid group where generators correspond to horizontal trisecants. Besides, there is a similar presentation where generators correspond to horizontal planes having four points lying on the same circle. These two approaches lead to the groups G_n^3 and G_n^4 as shown in Chapter 7.

The groups G_n^3 and G_n^4 share many nice properties with virtual braid groups and free groups. In particular, for these groups, each crossing contains very powerful information.

In Section 8.3, we coarsen this information and restrict ourselves to invariants of classical braids which appear as the image of the map from groups G_n^k to free groups.

One immediate advantage of this approach is that we can give some obvious estimates for various complexities of braids which are easy to calculate.

This chapter is organised as follows. We describe homomorphisms from classical pure braids into free braid groups G_n^3 and G_n^4 in Sections 8.1 and 8.2 respectively. Section 8.3 is devoted to the construction of a homomorphism from even part of free braid groups G_n^k into free products of $\mathbb{Z}_2$. In Section 8.4 we discuss how the defined homomorphisms can be applied to estimates of braids' complexity.

8.1   Homomorphism of pure braids into G_n^3

We recall (see Definition 7.8) that the *k-free braid group with n strands* G_n^k is generated by elements a_m, where m is a k-element subset of $\{1, 2, \ldots, n\}$, and relations $a_m^2 = 1$ for each m, $a_m a_{m'} = a_{m'} a_m$ for any k-element subsets m, m' such that $Card(m \cap m') \leqslant k - 2$, and *tetrahedron relation* $(a_{m_1} a_{m_2} \ldots a_{m_{k+1}})^2 = 1$ for each $(k+1)$-element subset $M = \{i_1, i_2, \ldots, i_{k+1}\} \subset \{1, 2, \ldots, n\}$ where $m_l = M \setminus \{i_l\}, l = 1, \ldots, k+1$.

Groups G_n^k appear naturally as groups describing dynamical systems of n particles in some "general position"; generators of G_n^k correspond to codimension 1 degeneracy, and relations correspond to codimension 2 degeneracy which occurs when performing some generic transformation between two general position dynamical systems. Dynamical systems leading to G_n^3

and G_n^4 are described in Chapter 7; here we describe the corresponding homomorphisms explicitly.

Generators of G_n^3 correspond to configurations in the evolution of dynamical systems which contain a *trisecant*, that is three particles lying in one line. Generators of G_n^4 correspond to configurations which includes four particles that lie in one circle.

Let PB_n be the pure n-strand braid group. It can be presented with the set of generators $b_{ij}, 1 \leqslant i < j \leqslant n$, and the set of relations [Bardakov, 2004]

$$b_{ij}b_{kl} = b_{kl}b_{ij}, \quad i < j < k < l \text{ or } i < k < l < j, \tag{8.1}$$

$$b_{ij}b_{ik}b_{jk} = b_{ik}b_{jk}b_{ij} = b_{jk}b_{ij}b_{ik}, \quad i < j < k, \tag{8.2}$$

$$b_{kl}b_{ik}b_{kl}^{-1}b_{jl} = b_{jl}b_{kl}b_{ik}b_{kl}^{-1}, \quad i < j < k < l. \tag{8.3}$$

It is well known that

Proposition 8.1. *The center $Z(PB_n)$ of the group PB_n is isomorphic to $\mathbb{Z}$.*

For each different indices i, j, $1 \leqslant i, j \leqslant n$, we consider the element $c_{i,j}$ in the group G_n^3 to be the product

$$c_{i,j} = \prod_{k=j+1}^{n} a_{i,j,k} \cdot \prod_{k=1}^{j-1} a_{i,j,k}. \tag{8.4}$$

Proposition 8.2. *The correspondence*

$$b_{ij} \mapsto c_{i,i+1}^{-1} \cdots c_{i,j-1}^{-1} c_{i,j}^2 c_{i,j-1} \cdots c_{i,i+1}, \quad i < j,$$

defines a homomorphism $\phi_n \colon PB_n \to G_n^3$.

Proof. Consider the configuration of n points $z_k = e^{2\pi i k/n}, k = 1, \ldots, n$, in the plane $\mathbb{R}^2 = \mathbb{C}$ where the points lie on the same circle $C = \{z \in \mathbb{C} \,|\, |z| = 1\}$. Pure braids can be considered as dynamical systems whose initial and final states coincide. We can assume that the initial state coincides with the configuration considered above. Then by Theorem 7.2 there is a homomorphism $\phi_n \colon PB_n \to G_n^3$ and we need only to describe explicitly the images of the generators of the group PB_n.

For any $i < j$ the pure braid b_{ij} can be presented as the following dynamical system:

(1) the point i moves along the inner side of the circle C, passes point
    $i+1, i+2, \ldots, j-1$ and lands on the circle before the point j (Fig. 8.1
    upper left);
(2) the point j moves over the point i (Fig. 8.1 upper right);
(3) the point i returns to its initial position over the points $j, j-1, \ldots, i+1$
    (Fig. 8.1 lower left);
(4) the point j returns to its position (Fig. 8.1 lower right).

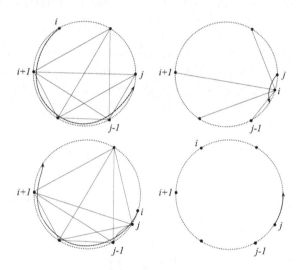

Fig. 8.1   Dynamical system corresponding to b_{ij}

As we check all the situations in the dynamical systems where three
points lie on the same line, and write down these situations as letters in a
word of the group G_n^3 we get exactly the element

$$c_{i,i+1}^{-1} \cdots c_{i,j-1}^{-1} c_{i,j}^2 c_{i,j-1} \cdots c_{i,i+1}.$$

□

Let $\widetilde{G} = \mathbb{Z}_2 * \mathbb{Z}_2 * \mathbb{Z}_2$ and let a_1, a_2, a_3 be the generators of $\widetilde{G}$. The group
$\widetilde{G}$ can be considered as an enhancement of the group $G_3^3 = \mathbb{Z}_2$ if one will
distinguish the middle index in the generators a_{ijk}. Thus, $a_1 = a_{213} = a_{312}$,
$a_2 = a_{123} = a_{321}$ and $a_3 = a_{132} = a_{231}$ will be different generators. Let
$\widetilde{G}_{even}$ be the subgroup in $\widetilde{G}$ which consists of words of even length.

By obvious reason, the 2π rotation of the whole set of points around the
origin has meets no trisecants, thus, the map $PB_3 \to G_n^3$ has an obvious
kernel.

Let us prove that in the case of 3 strings it is the only kernel of our map.

Theorem 8.1. *There is an isomorphism $PB_3/Z(PB_3) \to \widetilde{G}_{even}$.*

Proof. The quotient group $PB_3/Z(PB_3)$ can be identified with the subgroup H in PB_3 that consists of the braids for which the strands 1 and 2 are not linked. In other words, the subgroup H is the kernel of the homomorphism $PB_3 \to PB_2$ that removes the last strand. For any braid in H we can straighten its strands 1 and 2. Looking at such a braid as a dynamical system we shall see a family of states where the particles 1 and 2 are fixed and the particle 3 moves.

The points 1 and 2 split which the line they lie on into three intervals. Let us denote the unbounded interval with the end 1 as a_1, the unbounded interval with the end 2 as a_2 and the interval between 1 and 3 as a_3 (see Fig. 8.2).

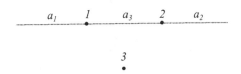

Fig. 8.2   Initial state for a pure braid with 3 strands

We assign a word in letters a_1, a_2, a_3 to any motion of the point 3 as follows. We start with the empty word. Every time the particle 3 crosses the line 12 we append the letter which corresponds to the interval the point 3 crosses. If the point 3 returns to its initial position, then the word will have even number of letters. This construction defines a homomorphism $\phi \colon H \to \widetilde{G}_{even}$.

On the other hand, if we have an even word in letters a_1, a_2, a_3, then we can define a motion of the point 3 up to isotopy in $\mathbb{R}^2 \setminus \{1, 2\}$. Thus, we have a well-defined map from even words to H that induces a homomorphism $\psi \colon \widetilde{G}_{even} \to H \simeq PB_3/Z(PB_3)$. It is easy to see that the homomorphisms ϕ and ψ are mutually inverse. $\qquad \square$

The crucial observation which allows us to prove that this map is an isomorphism, is that we can restore the dynamics from the word in the case of 3 strands. Since relations in G_n^3 correspond to relations in the braid group, this suffices to prove the isomorphism.

When $n > 4$, not all words in G_n^3 correspond to the dynamical systems. Thus, the question whether the mapping $(PB_n)/Z(PB_n) \to G_n^3$ is injective requires additional techniques.

8.2 Homomorphism of pure braids into G_n^4

In the present section, we describe an analogous mapping $PB_n \to G_n^4$; here points $z_1, z_2, \ldots, z_n$ on the plane are in general position of no four points of them belong to the same circle (or line); codimension 1 degeneracies will correspond to generators of G_n^4, where at some moment exactly one quadruple of points belongs to the same circle (line), and relations correspond to the case of more complicated singularities.

Let $a_{\{i,j,k,l\}}, 1 \leqslant i, j, k, l \leqslant n$, be the generators of the group $G_n^4, n > 4$. Let $1 \leqslant i < j \leqslant n$. Consider the elements

$$c_{ij}^I = \prod_{p=2}^{j-1} \prod_{q=1}^{p-1} a_{\{i,j,p,q\}}, \tag{8.5}$$

$$c_{ij}^{II} = \prod_{p=1}^{j-1} \prod_{q=1}^{n-j} a_{\{i,j-p,j,j+q\}}, \tag{8.6}$$

$$c_{ij}^{III} = \prod_{p=1}^{n-j+1} \prod_{q=0}^{n-p+1} a_{\{i,j,n-p,n-q\}}, \tag{8.7}$$

$$c_{ij} = c_{ij}^{II} c_{ij}^{I} c_{ij}^{III}. \tag{8.8}$$

Proposition 8.3. *The correspondence*

$$b_{ij} \mapsto c_{i,i+1} \ldots c_{i,j-1} c_{i,j}^2 c_{i,j-1}^{-1} \ldots c_{i,i+1}^{-1}, \quad i < j, \tag{8.9}$$

defines a homomorphism $\phi_n \colon PB_n \to G_n^4$.

In order to construct this map explicitly, we have to indicate the initial state in the configuration space of n-tuple of points.

By obvious reason, the initial state with all points lying on the circle does not work; so, we shall use the parabola instead.

Let $\Gamma = \{(t, t^2) \,|\, t \in \mathbb{R}\} \subset \mathbb{R}^2$ be the graph of the function $y = x^2$. Consider a rapidly increasing sequence of positive numbers $t_1, t_2, \ldots, t_n$ (precise conditions on the sequence growth will be formulated below) and denote the points $(t_i, t_i^2) \in \Gamma$ by P_i.

Pure braids can be considered as dynamical systems whose initial and final states coincide and we assume that the initial state is the configuration

$\mathcal{P} = \{P_1, P_2, \ldots, P_n\}$. Then by Theorem 7.2 there is a homomorphism $\psi_n \colon PB_n \to G_n^4$. We need to describe explicitly the images of the generators of the group PB_n.

For any $i < j$ the pure braid b_{ij} can be presented as the following dynamical system: the point P_i moves along the graph Γ

(1) the point P_i moves along the graphics Γ and passes points

$$P_{i+1}, P_{i+2}, \ldots, P_j$$

from above (Fig. 8.3 upper left);

(2) the point P_j moves from above the point P_i (Fig. 8.3 upper right);
(3) the point P_i moves to its initial position from above the points

$$P_{j-1}, \ldots, P_{i+1}$$

(Fig. 8.3 lower left);

(4) the point P_j returns to its position (Fig. 8.3 lower right).

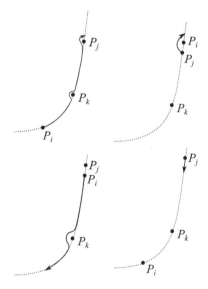

Fig. 8.3 Dynamical system which corresponds to b_{ij}

The proof that this dynamical system leads to the presentation (8.9) is given in Section 8.5.

8.3   Homomorphism into a free group

Let $\mathcal{N} = \{1, 2, \ldots, n\}$ be the set of indices.

Let $H_n^k \subset G_n^k$ be the subgroup whose elements are given by the *even* words, that is words which include any generator $a_m, m \subset \mathcal{N}, |m| = k$, evenly many times. We construct a homomorphism ϕ from the subgroup H_n^k to the free product of $2^{(k-1)(n-k)}$ copies of the group $\mathbb{Z}_2$.

Roughly speaking, any letter a_m in G_n^k will get a collection of "colours" coming from the interaction of indices from m with other indices. These "colours" will remain unchanged when performing the relations $a_m a_{m'} = a_{m'} a_m$ and the tetrahedron relation $\ldots a_m \ldots = \ldots a_m \ldots$; moreover, the two adjacent letters a_m which contracts to 1 inside the word will have the same colour.

Let $a_m, m \subset \mathcal{N}, |m| = k$, be a generator of the group G_n^k. Without loss of generality we can suppose that $m = \{1, 2, \ldots, k\}$. Assume that $p \in \mathcal{N} \setminus m$. For any $1 \leqslant i \leqslant k$ consider the set $m[i] = m \setminus \{i\} \cup \{p\}$. Let us define a homomorphism $\psi_p \colon G_n^k \to \mathbb{Z}_2^{\oplus k-1}$ by the formulas $\psi_p(a_{m[i]}) = e_i, 1 \leqslant i \leqslant k-1, \psi_p(a_{m[k]}) = \sum_{i=1}^{k-1} e_i$ and $\psi_p(a_{m'}) = 0$ for the other $m' \subset \mathcal{N}, |m'| = k$. Here $e_1, \ldots, e_{k-1}$ denote the basis elements of the group $\mathbb{Z}_2^{\oplus k-1}$. The homomorphism ψ_p is well defined since the relations in G_n^k are generated by even words.

Consider the homomorphism $\psi = \bigoplus_{p \notin m} \psi_p$, $\psi \colon G_n^k \to Z$, where $Z = \mathbb{Z}_2^{\oplus (k-1)(n-k)}$. Note that $H_n^k \subset \ker \psi$.

Let $H = \mathbb{Z}_2^{*Z}$ be the free product of $2^{(k-1)(n-k)}$ copies of the group $\mathbb{Z}_2$ and the exponents of its elements are indiced with the set Z. Let $f_x, x \in Z$, be the generators of the group H. Define the action of the group G_n^k on the set $Z \times H$ by the formula

$$a_{m'} \cdot (x, y) = \begin{cases} (x, f_x y) & m' = m, \\ (x + \psi(a_{m'}), y) & m' \neq m. \end{cases}$$

This action is well defined. Indeed, $a_{m'}^2 \cdot (x, y) = (x, y)$ for any $m' \subset \mathcal{N}, |m'| = k$. On the other hand, for any tuple $\tilde{m} = (i_1, i_2, \ldots, i_{k+1})$ such that $m \not\subset \tilde{m}$ one has $\left(\prod_{l=1}^{k+1} a_{\tilde{m} \setminus \{i_l\}} \right)^2 \cdot (x, y) = (x, y)$ since the generators act trivially here. If $m \subset \tilde{m}$, then $\left(\prod_{l=1}^{k+1} a_{\tilde{m} \setminus \{i_l\}} \right)^2 = w_1 a_m w_2 a_m w_3$, where the word w_2 and the word $w_1 w_3$ are products of generators $a_{m[i]}, i = 1, \ldots, k$

and

$$\left(\prod_{l=1}^{k+1} a_{\tilde{m}\setminus\{i_l\}}\right)^2 \cdot (x,y) = w_1 a_m w_2 a_m w_3 \cdot (x,y) =$$

$$w_1 a_m w_2 a_m \cdot (x + \psi(w_3), y) = w_1 a_m w_2 \cdot (x + \psi(w_3), f_{x+\psi(w_3)} y) =$$

$$w_1 a_m \cdot (x + \psi(w_3), f^2_{x+\psi(w_3)} y) = w_1 \cdot (x + \psi(w_3), f^4_{x+\psi(w_3)} y) =$$

$$w_1 \cdot (x + \psi(w_3), y) = (x + \psi(w_1) + \psi(w_3), y) = (x,y).$$

The fourth and the last equalities follow from the equality $\psi(w_1) + \psi(w_3) = \psi(w_1 w_3) = \psi(w_2) = \sum_{i=1}^k \psi(a_{m[i]}) = 0$.

For any element $g \in G_n^k$ define $\phi_x(g)$ from the relation

$$g \cdot (x, 1) = (x + \psi(g), \phi_x(g)1)$$

and let $\phi(g) = \phi_0(g)$. Then $g \cdot (x,y) = (x + \psi(g), \phi_x(g)y)$ for any $(x,y) \in Z \times H$. If $g \in H_n^k$ then $\psi(g) = 0$ and $g \cdot (0,y) = (0, \phi(g)y)$. Hence, for any $g_1, g_2 \in H_n^k$ one has $(g_1 g_2) \cdot (0,1) = g_1 \cdot (g_2 \cdot (0,1)) = g_1 \cdot (0, \phi(g_2)) = (0, \phi(g_1)\phi(g_2))$. On the other hand, $(g_1 g_2) \cdot (0,1) = (0, \phi(g_1 g_2))$. Thus, $\phi \colon H_n^k \to H$ is a homomorphism.

For any element x in H, let $c(x)$ denote the complexity of x, that is the length of the irreducible representative word of x.

Let $b \in PB_n$ be a classical braid and let $\beta = \phi_n(b)$ be the corresponding free braid in G_n^3. Note that the word β belongs to H_n^3. Then for any $m \subset \mathcal{N}, |m| = 3$, the map ϕ defined above is applicable to β. The geometrical complexity of the braid b can be estimated by complexity of the element $\phi(\beta)$.

Proposition 8.4. *The number of horizontal trisecants of the braid b is not less than $c(\phi(\beta))$.*

The proof of the statement follows from the definition of the maps ϕ_n and ϕ.

Using the homomorphism $PB_n \to G_n^4$, we get an analogous estimation for the number of "circled quadrisecants" of the braid.

8.4 Free groups and crossing numbers

The homomorphism described above allows one to estimate the new complexity for the braid group BP_n by using G_n^3 and G_n^4. These complexities have an obvious geometrical meaning as the estimates of the number of horizontal trisecants (G_n^3) and "circled quadrisecants" (G_n^4).

Let $G = \mathbb{Z}_2^{*3}$ be the free product of three copies of $\mathbb{Z}_2$ with generators a, b, c respectively. A typical example of the word in this group is

$$w = abcbabca... \qquad (8.10)$$

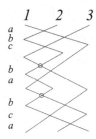

Fig. 8.4   Braid which corresponds to the word w

Note that the word (8.10) is irreducible, so, every word equivalent to it contains it as a subword. In particular, this means that *every word w' equivalent to w contains at least 8 letters.*

The above mentioned complexity is similar to the "crossing number", though, crossings are treated in a non-canonical way.

In [Manturov, 2010], it is proved that for free knots (which are knot theoretic analogues of the group G_n^2) if an irreducible knot diagram K is complicated enough, then it is contained (as a smoothing) in any knot diagram equivalent to it.

In [Manturov, 2015c], similar statements are proved for G_n^2.

Let us now treat one more complicated issue, the unknotting number.

To make the issue simpler, let us start with the toy model. Let $G' = \mathbb{Z}^{*3}$ be the free product of the three copies of $\mathbb{Z}$ with generators a, b, c.

Assume we are allowed to perform the operation of switching the sign $a \longleftrightarrow a^{-1}, b \longleftrightarrow b^{-1}, c \longleftrightarrow c^{-1}$ along with the usual reduction of opposite generators $aa^{-1}, bb^{-1}, cc^{-1}, a^{-1}a, \ldots$.

Given a word v; how many switches do we need to get the word representing the trivial element from v? How to estimate this number from below?

For the word $w = abcbabca$ the answer is "infinity". It is impossible to get 1 from w in G' because w represents a non-trivial element of G, and the operations $a \longleftrightarrow a^{-1}$ do not change the element of G written in generators a, b, c.

Now, let the element of G' be given by the word $w' = a^4 b^2 c^4 b^{-4}$.

This word w' is trivial in G, however, if we look at exponents of a, b, c, we see that they are $+4, -2, +4$. Thus, the number of switchings is bounded below by $\frac{1}{2}(|4| + |2| + |4|) = 5$, and one can easily find how to make the word w' trivial in G' with five switchings.

For classical braids, a crossing switching corresponds to one turn of one string of a braid around another. If i and j are the numbers of two strings of some braid B, then the word in G_n^k, $k = 3, 4$, which corresponds to the dynamical system describing the full turn of the string i around the string j, looks like c_{ij}^2, where $c_{ij} = \prod_{m \supset \{i,j\}} a_m$ and the product is taken over the subsets $m \subset \{1, 2, \ldots, n\}$, which have k element and contain i and j, given in some order. Note that the word c_{ij}^2 is even.

Fix a k-element subset $m \subset \{1, 2, \ldots, n\}$ containing $\{i, j\}$. Consider the homomorphism $\phi \colon H_n^k \to H$ determined by m. The word c_{ij}^2 yields the factor $f_x f_{x+z_{ij}} \in H$, where the index $x \in Z$ depends on the position of c_{ij}^2 inside the word β of G_n^k that corresponds to the braid B, and the element $z_{ij} = \sum_{m' : m' \supset \{i,j\}, Card(m \cap m') = k-1} \psi(m') \in Z$ depends only on i, j, and m and does not depend on the order in the product c_{ij}. Hence, an addition of a full turn of the string i around the string j corresponds to substitution of the subword $1 = f_x^2$ with the subword $f_x f_{x+z_{ij}}$, that is a crossing switch corresponds to the switch of elements f_x and $f_{x+z_{ij}}$ in the image of the braid under the homomorphism ϕ. Thus, the following statement holds.

Proposition 8.5. *The unknotting number of a braid B is estimated below by the number of switches $f_x \mapsto f_{x+z_{ij}}$ which are necessary to make the word $\phi(B) \in H$ trivial.*

The number of switches in the Proposition 8.5 is not an explicit characteristic of a word in the group H, but one can give several rough estimates for it which can be computed straightforwardly. For example, consider the following construction. Let $\pi \colon H \to \mathbb{Z}_2[Z]$ be the natural projection; let Z_0 be the subgroup in Z generated by the elements $z_{ij}, \{i, j\} \subset m$. For any element $\xi = \sum_{z \in Z} \xi_z z \in \mathbb{Z}_2[Z]$ and any $z \in Z$ consider the number $c_z(\xi) = Card(\{z_0 \in Z_0 \,|\, \xi_{z+z_0} \neq 0\})$, and let $c(\xi) = \max_{z \in Z} c_z(\xi)$.

Let $\omega \in H$ be an arbitrary word and $\xi = \pi(\omega)$. Any switch $f_x \mapsto f_{x+z_{ij}}$ in ω corresponds to a switch of ξ. Note that these switches map any element $z \in Z$ into the class $z + Z_0 \subset Z$. After a switch, two summands of ξ, that correspond to generators in $z + Z_0$, can annihilate. Thus, removing all the summands of ξ by the generators in $z + Z_0$ takes at least $\frac{1}{2} c_z(\xi)$ switches.

Thus, we get the following estimate for the unknotting number.

Proposition 8.6. *The unknotting number of a braid B is estimated below by the number* $\frac{1}{2}c\left(\pi(\phi(B))\right)$.

Consider the braid B in Fig. 8.5. Its unknotting number is 2. The braid corresponds to the element

$$\beta = a_{123}a_{234}a_{123}a_{134}a_{123}a_{134}a_{123}a_{234} \in G_4^3.$$

Let $m = \{1,2,3\}$. Then $z_{12} = \psi(a_{124}) = e_1 + e_2 \in Z = \mathbb{Z}_2^{\oplus 2}$, $z_{13} = \psi(a_{134}) = e_2$, $z_{23} = \psi(a_{234}) = e_1$.

The image of the element β is $\phi(\beta) = f_0 f_{e_1} f_{e_1+e_2} f_{e_1}$. It is easily to see that the word $\phi(\beta)$ can not be trivialized with one switch, but it can be made trivial with two switches:

$$f_0 f_{e_1} f_{e_1+e_2} f_{e_1} \xrightarrow{z_{13}} f_0 f_{e_1} f_{e_1} f_{e_1} = f_0 f_{e_1} \xrightarrow{z_{23}} f_0 f_{e_0} = 1.$$

Note that the Proposition 8.6 gives the estimate

$$\frac{1}{2}c(\pi(\phi(\beta))) = \frac{1}{2}c(f_0 + f_{e_1+e_2}) = 1.$$

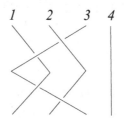

Fig. 8.5   Braid with the unknotting number equal to 2

8.5   Proof of Proposition 8.3

We conclude the present chapter with an explicit proof of Proposition 8.3. This proof also serves as an example of how one can solve the G_n^k groups problems by hand.

Proof. (Proof of Proposition 8.3)

Lemma 8.1. *Let $A_i = (x_i, x_i^2), i = 0, 1, 2, 3$, be different points in the graph Γ. Then A_0, A_1, A_2, A_3 belong to one circle if and only if*

$$x_0 + x_1 + x_2 + x_3 = 0.$$

Proof. Assume that A_0, A_1, A_2, A_3 belong to the circle with the center (a, b) and the radius r for some a, b, r. Then the following equations hold: $(x_i - a)^2 + (x_i^2 - b)^2 = r^2$, $i = 0, \ldots, 3$. Subtracting the first equation from the last three equations, we obtain a linear system on variables a and b with three equations $2(x_0 - x_i)a + 2(x_0^2 - x_i^2)b + (x_i^2 - x_0^2 + x_i^4 - x_0^4) = 0$, $i = 1, 2, 3$. The compatibility equation for this system is $\Delta = 0$ where

$$\Delta = \begin{vmatrix} x_0 - x_1 & x_0^2 - x_1^2 & x_1^2 - x_0^2 + x_1^4 - x_0^4 \\ x_0 - x_2 & x_0^2 - x_2^2 & x_2^2 - x_0^2 + x_2^4 - x_0^4 \\ x_0 - x_3 & x_0^2 - x_3^2 & x_3^2 - x_0^2 + x_3^4 - x_0^4 \end{vmatrix}$$

is the determinant of the system. But

$$\Delta = \begin{vmatrix} x_0 - x_1 & x_0^2 - x_1^2 & x_1^4 - x_0^4 \\ x_0 - x_2 & x_0^2 - x_2^2 & x_2^4 - x_0^4 \\ x_0 - x_3 & x_0^2 - x_3^2 & x_3^4 - x_0^4 \end{vmatrix} =$$

$$(x_0 - x_1)(x_0 - x_2)(x_0 - x_3)(x_1 - x_2)(x_1 - x_3)(x_2 - x_3)(x_0 + x_1 + x_2 + x_3).$$

Hence, the condition $\Delta = 0$ is equivalent to the equality $x_0 + x_1 + x_2 + x_3 = 0$. $\square$

Corollary 8.1. *Let* $\Gamma_+ = \{(t, t^2) \,|\, t \geq 0\}$ *be the positive half of the graphics* Γ *and* $X = \{P_1, P_2, \ldots, P_n\} \subset \Gamma_+, 4 \leqslant n$, *be a finite subset. Denote the circle which contains 3 different points* P_i, P_j, P_k *of* X *by* C_{ijk}. *Then*

$$\Gamma_+ \cap \bigcup_{1 \leqslant i < j < k \leqslant n} C_{ijk} = X.$$

Proof. For any $i < j < k$ one has $C_{ijk} \cap \Gamma = \{P, P_i, P_j, P_k\}$ where $P = (t, t^2)$, $P_i = (t_i, t_i^2)$, $P_j = (t_j, t_j^2)$, $P_k = (t_k, t_k^2)$. Since $t_i, t_j, t_k \geq 0$ and $t + t_i + t_j + t_k = 0$, then $t < 0$ and $P \notin \Gamma_+$. Hence, $C_{ijk} \cap \Gamma_+ = \{P_i, P_j, P_k\}$. $\square$

The corollary means that there is no additional intersections of the circles generated by the points of X with Γ_+. Then in the dynamical systems which corresponds to the pure braid b_{ij}, a configuration of four points lying on one circle can occur only when point i (or j) leaves the curve Γ_+. If that configuration appears when the point i moves above the point $P_k, i < k \leqslant j$, it means that point i, P_k and some other two points P_l and P_m lie on the same circle, that is the point i belongs to the circle C_{klm}. We assume below that $t_l < t_m$.

We need to find the sequence in which the point i intersects circles C_{klm} when moving near P_k. We can substitute circles with their tangent lines at P_k under assumption the point i moves close to P_k.

Given points $P_k, P_l, P_m \in \Gamma_+$ there are three possible cases:

(1) $t_l < t_m < t_k$;
(2) $t_l < t_k < t_m$;
(3) $t_k < t_l < t_m$.

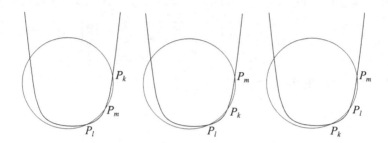

Fig. 8.6   Dynamic system which corresponds to b_{ij}

Case $t_l < t_m < t_k$. The following statement can be obtained by straightforward computation.

Lemma 8.2. *The slope of the tangent line of the circle C_{klm}, which contains the points $P_k = (t_k, t_k^2)$, $P_l = (t_l, t_l^2)$ and $P_m = (t_m, t_m^2)$, is equal to*

$$\kappa_{k,lm} = -\frac{t_k^2(t_l + t_m) + t_k((t_l + t_m)^2 + 2) + t_l t_m(t_l + t_m)}{t_k^2 - t_k(t_l + t_m) - (t_l^2 + t_l t_m + t_m^2 + 1)} \tag{8.11}$$

at the point P_k.

If the sequence $t_i, 1 \leqslant i \leqslant n$, grows rapidly then the coefficient $\kappa_{k,lm}$ is close to the sum $-(t_l + t_m)$. More precisely, we have the following estimate.

Lemma 8.3. *Let $t_1 \geq 1$ and*

$$t_i \geq 100 t_{i-1}^2 \tag{8.12}$$

for any $i, 1 < i \leqslant n$. Then for any $1 \leqslant l < m < k \leqslant n$ we have

$$-(t_l + t_m + 1) < \kappa_{k,lm} < -(t_l + t_m).$$

Proof. The numerator in the formula (8.11) is greater than $t_k^2(t_l + t_m)$ and the denominator is less than t_k^2. Hence, $\kappa_{k,lm} < -(t_l + t_m)$. On the other hand,

$$t_k^2 - t_k(t_l + t_m) - (t_l^2 + t_l t_m + t_m^2 + 1) > t_k^2 - 6t_k t_m \geq t_k^2 \left(1 - \frac{6}{100 t_m}\right),$$

and since $\left(1 - \frac{6}{100t_m}\right)\left(1 + \frac{7}{100t_m}\right) = 1 + \frac{1}{100t_m} - \frac{42}{10000t_m^2} > 1$, we have

$$\left(1 - \frac{6}{100t_m}\right)^{-1} < 1 + \frac{7}{100t_m}.$$

Moreover,

$$t_k^2(t_l + t_m) + t_k((t_l + t_m)^2 + 2) + t_l t_m(t_l + t_m) < t_k^2\left(t_l + t_m + \frac{7t_m^2}{t_k}\right)$$

$$< t_k^2\left(t_l + t_m + \frac{7}{100}\right)$$

since $(t_l + t_m)^2 + 2 < (2t_m)^2 + t_m^2 = 5t_m^2$ and $t_l t_m(t_l + t_m) < 2t_k t_m^2$. Then

$$-\kappa_{k,lm} < \frac{t_k^2\left(t_l + t_m + \frac{7}{100}\right)}{t_k^2\left(1 - \frac{6}{100t_m}\right)} < \left(t_l + t_m + \frac{7}{100}\right)\left(1 + \frac{7}{100t_m}\right) =$$

$$t_l + t_m + \frac{7t_l + 14t_m}{100t_m} + \frac{49}{10000t_m} < t_l + t_m + \frac{21}{100} + \frac{49}{10000} < t_l + t_m + 1.$$

$\square$

Corollary 8.2. *Let $t_1 \geq 1$ and $t_i \geq 100t_{i-1}^2$ for any $i, 1 < i \leq n$. Then for any indices k, l_1, l_2, m_1, m_2 such that $1 \leq l_1 < m_1 < k \leq n$ and $1 \leq l_2 < m_2 < k$ the inequality $\kappa_{k,l_1 m_1} > \kappa_{k,l_2 m_2}$ holds if and only if $m_1 < m_2$ or $m_1 = m_2, l_1 < l_2$.*

Proof. If $m_1 < m_2$ then

$$\kappa_{k,l_2 m_2} < -(t_{l_2} + t_{m_2}) < -100t_{m_1}^2 < -(2t_{m_1} + 1) < \kappa_{k,l_1 m_1}.$$

If $m_1 = m_2$ and $l_1 < l_2$ then

$$\kappa_{k,l_2 m_2} < -(t_{l_2} + t_{m_2}) < -(100t_{l_1}^2 + t_{m_1}) < -(t_{l_1} + t_{m_1} + 1) < \kappa_{k,l_1 m_1}.$$

$\square$

Corollary 8.2 determines the order the point i intersects the circles C_{klm}. Since $\kappa_{k,lm} < 0$ the tangent line to C_{klm} that lies above the graphics Γ_+, belongs to the second quadrant of the plane. The points i intersects the circles according the decreasing order of the coefficients $\kappa_{k,lm}$. Thus, the order on the circles will be

$$C_{k,1,2}, C_{k,1,3}, C_{k,2,3}, C_{k,1,4}, \ldots, C_{k,k-2,k-1}.$$

Case $t_l < t_k < t_m$.

Then the circle C_{klm} goes between the chord $P_l P_k$ and the graphics Γ_+ (see Fig. 8.6 middle). It means the tangent line which lies above Γ_+, belongs to the third quadrant of the plane.

Let r_{klm} be the radius of the circle C_{klm} and let $R_p, 3 \leqslant p \leqslant n$, be equal to $\max_{k,l,m \leqslant p} r_{klm}$. Let $\alpha_p, 3 \leqslant p \leqslant n$, be equal to the minimal angle

$$\min_{u,v,w \leqslant p, u \neq w} \angle P_u P_v P_w.$$

Lemma 8.4. *Let $t_1 \geq 1$ and $t_i \geq t_{i-1}$ for any $i, 1 < i \leqslant n$ and*

$$t_i \geq \max\left(\frac{3}{\sin \alpha_{i-1}} t_{i-1}^2, t_{i-1}^2 + 2R_{i-1} \right) \tag{8.13}$$

for $i > 3$. Then for any indices k, l_1, l_2, m_1, m_2 such that $1 \leqslant l_1 < k < m_1 \leqslant n$ and $1 \leqslant l_2 < k < m_2 \leqslant n$ we have

(1) if $l_1 < k - 1$ then the arc $P_{l_1} P_k$ of the circle $C_{kl_1 m_1}$ lies in the angle $\angle P_{l_1} P_k P_{l_1+1}$;

(2) if $m_1 < m_2$ then $r_{kl_1 m_1} < r_{kl_2 m_2}$;

(3) $\kappa_{k,l_1 m_1} > \kappa_{k,l_2 m_2}$ if and only if $l_1 > l_2$ or $l_1 = l_2, m_1 < m_2$, where $\kappa_{k,l_s m_s}, s = 1, 2$, is the slope of the circle $C_{kl_s m_s}$ at the point P_k.

Proof. 1. Let $r = r_{kl_1 m_1}$ and α be the angle between the circle $C_{kl_1 m_1}$ and the chord $P_{l_1} P_k$ at the point P_k. Then $\sin \alpha = \frac{P_{l_1} P_k}{2r}$. The growth condition implies

$$2r \geq P_k P_{m_1} > t_{m_1}^2 - t_k^2 > \left(\frac{3}{\sin \alpha_k} - 1 \right) t_k^2 > \frac{2t_k^2}{\sin \alpha_k}.$$

On the other hand,

$$P_{l_1} P_k = \sqrt{(t_k - t_{l_1})^2 + (t_k^2 - t_{l_1}^2)^2} < \sqrt{t_k^2 + t_k^4} < 2t_k^2.$$

Then $\sin \alpha < \sin \alpha_k$ and $\alpha < \alpha_k \leqslant \angle P_{l_1} P_k P_{l_1+1}$. Therefore, the arc $P_{l_1} P_k$ lies in the angle $\angle P_{l_1} P_k P_{l_1+1}$.

2. If $m_1 < m_2$ then

$$r_{kl_2 m_2} \geq \frac{P_k P_{m_2}}{2} > \frac{t_{m_2}^2 - t_k^2}{2} \geq \frac{t_{m_2}^2 - t_{m_2-1}^2}{2} \geq R_{m_2-1} \geq r_{kl_1 m_1}.$$

3. If $l_1 > l_2$ then the arc $P_{l_2} P_k$ of the circle $C_{kl_2 m_2}$ lies in the angle $\angle P_{l_2} P_k P_{l_2+1}$ (see Fig. 8.7 left). Then it lies above the chords $P_{l_2+1} P_k$ and $P_{l_1} P_k$. On the other hand, the arc $P_{l_1} P_k$ of the circle $C_{kl_1 m_1}$ lies below the chord $P_{l_1} P_k$. Then the arc $P_{l_2} P_k$ lies above the arc $P_{l_1} P_k$ and $\kappa_{k,l_1 m_1} > \kappa_{k,l_2 m_2}$.

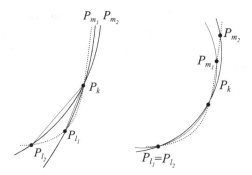

Fig. 8.7   Order of circles in the second case

If $l_1 = l_2$ and $m_1 < m_2$ then the arc $P_{l_1}P_k$ of the circle $C_{kl_1m_1}$ and the arc $P_{l_2}P_k$ of the circle $C_{kl_2m_2}$ have the same chord but the radius of the circle $C_{kl_1m_1}$ is less than the radius of the circle $C_{kl_2m_2}$ (see Fig. 8.7 right). Hence, the arc $P_{l_2}P_k$ lies above the arc $P_{l_1}P_k$ and $\kappa_{k,l_1m_1} > \kappa_{k,l_2m_2}$.

$\square$

The lemma above establishes the order in which the point i crosses the circles C_{klm}: the circle $C_{k,k-1,k+1}$ goes first, then the circles

$$C_{k,k-1,k+2}, C_{k,k-1,k+3}, \ldots, C_{k,k-1,n},$$

then

$$C_{k,k-2,k+1}, C_{k,k-2,k+2}, \ldots, C_{k,k-2,n}$$

and so on till $C_{k,1,n}$.

Case $t_k < t_l < t_m$.

The circle C_{klm} goes between the chord P_lP_k and the graphics Γ_+ (see Fig. 8.6 right). It means the tangent line which lies above Γ_+, belongs to the first quadrant of the plane. The case can be solved in the same manner as the previous case. The order of circles we have here is the following:

$$C_{k,n-1,n}, C_{k,n-2,n}, C_{k,n-2,n-1}, C_{k,n-3,n}, \ldots, C_{k,k+1,k+2}.$$

Assume the sequence $t_i, 1 \leqslant i \leqslant n$, satisfies the growth conditions (8.12) and (8.13). Then we know the order of circles inside each of three cases. Since the first case includes the circles whose tangent lines lie in the second quadrant, the second case includes circles with tangent lines in the third quadrant and the third case gives circles in the first quadrant, the whole order is the following: the circles of the second case go first, then the circles of the first case, then the circles of the third case. In other words, when the point i rounds the point P_k from above we get the word c_{ik} from (8.5). Then the dynamical system which corresponds to the pure braid b_{ij}, leads to the formula (8.9). $\square$

Chapter 9

Generalisations of the Groups G_n^k

9.1 Indices from G_n^3 and Brunnian braids

In classical braid groups, a special role is played by *Brunnian braids on n strands*. A Brunnian braid is a braid such that each braid obtained from it by omitting one strand is trivial for every strand. We can define Brunnian elements in groups G_n^k and we will see that the image of a Brunnian braid is Brunnian in G_n^3 by the homomorphism $\phi_n \colon PB_n \to G_n^3$ described in Section 8.1. In the present section we denote the formula (8.4) by $c_{i,j}^n \in G_n^3$ to clarify the codomain of the homomorphism ϕ_n.

Let us define a mapping $p_m \colon PB_{n+1} \to PB_n$ by

$$
p_m(b_{ij}) = \begin{cases}
1 & \text{if } j = m, \\
b_{ij} & \text{if } i, j < m, \\
b_{i(j-1)} & \text{if } i < m, j > m, \\
b_{(i-1)(j-1)} & \text{if } i, j > m.
\end{cases}
$$

Roughly speaking, this mapping deletes one strand from braids on n strands.

Definition 9.1. *An n-strand Brunnian braid* is a pure braid on n strands such that $p_m(\beta) = 1$ for each index m.

Note that if an element in PB_n is of the form $[\dots [[b_{i_1 j_1}, b_{i_2 j_2}], b_{i_3 j_3}], \dots], b_{i_m j_m}]$ and for every $i \in \{1, \dots, n\}$ there is k such that $i \in \{i_k, j_k\}$, then it is Brunnian, for example, Fig. 9.1.

Let us define a mapping $q_m \colon G_{n+1}^3 \to G_n^3$ by

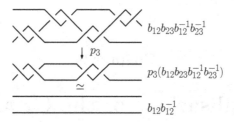

Fig. 9.1   Diagrams of $b_{12}b_{13}b_{12}^{-1}b_{13}^{-1}$ and $p(b_{12}b_{13}b_{12}^{-1}b_{13}^{-1}) = b_{12}b_{12}^{-1}$ in PB_3 and PB_2

$$q_m(a_{ijk}) = \begin{cases} 1 & \text{if } i,j \text{ or } k = m, \\ a_{ijk} & \text{if } i,j,k < m, \\ a_{ij(k-1)} & \text{if } i,j < m, k > m, \\ a_{i(j-1)(k-1)} & \text{if } i < m, j,k > m, \\ a_{(i-1)(j-1)(k-1)} & \text{if } i,j,k > m. \end{cases}$$

Definition 9.2. An element β from G_{n+1}^3 is called *Brunnian* if $q_m(\beta) = 1$ for each $m \in \{1, \ldots, n+1\}$.

Lemma 9.1. *For a Brunnian braid* $\beta \in PB_n$, $\phi_n(\beta)$ *is a Brunnian in* G_n^3.

To prove the above lemma, firstly we prove the following lemma.

Lemma 9.2. *For* $i,j \in \{1, \ldots, n\}$ *and* $i < j$,

$$q_n(c_{i,j}^n) = \begin{cases} 1 & \text{if } j = n, \\ c_{i,j}^{n-1} & \text{if } j \neq n. \end{cases}$$

Proof. If $j = n$, then

$$q_n(c_{i,n}^n) = q_n\left(\prod_{k=1}^{n-1} a_{ink}\right) = \prod_{k=1}^{n-1} q_n(a_{ink}) = 1.$$

If $i,j \neq n$, then

$$q_n(c_{i,j}^n) = q_n\left(\prod_{k=j+1}^{n} a_{ijk} \prod_{k=1}^{j-1} a_{ijk}\right)$$

$$= \prod_{k=j+1}^{n} q_n(a_{ijk}) \prod_{k=1}^{j-1} q_n(a_{ijk}) = \prod_{k=j+1}^{n-1} a_{ijk} \prod_{k=1}^{j-1} a_{ijk} = c_{i,j}^{n-1}.$$

$\square$

Analogously we can show that

$$q_m(c_{i,j}^n) = \begin{cases} 1 & \text{if } i \text{ or } j = m, \\ c_{i,j}^{n-1} & \text{if } i,j < m, \\ c_{i,(j-1)}^{n-1} & \text{if } i < m, j > m, \\ c_{(i-1),(j-1)}^{n-1} & \text{if } i,j > m. \end{cases}$$

Proof of Lemma 9.1. It is sufficient to show that $q_m \circ \phi_n = \phi_{n-1} \circ p_m$, because if $p_m(\beta) = 1$, then $q_m \circ \phi_n(\beta) = \phi_{n-1} \circ p_m(\beta) = \phi_{n-1}(1) = 1$. For $m = n$ and $b_{ij} \in PB_n$, if $j = n$, then $\phi_{n-1} \circ p_n(b_{in}) = 1$ and

$$q_n \circ \phi_n(b_{in}) = q_n((c_{i,i+1}^n)^{-1}(c_{i,i+2}^n)^{-1} \ldots (c_{i,n-1}^n)^{-1}(c_{i,n}^n)^2 c_{i,n-1}^n \ldots c_{i,i+2}^n c_{i,i+1}^n)$$
$$= (c_{i,i+1}^{n-1})^{-1}(c_{i,i+2}^{n-1})^{-1} \ldots (c_{i,n-1}^{n-1})^{-1} q_n((c_{i,n}^n)^2) c_{i,n-1}^{n-1} \ldots c_{i,i+2}^{n-1} c_{i,i+1}^{n-1}$$
$$= (c_{i,i+1}^{n-1})^{-1}(c_{i,i+2}^{n-1})^{-1} \ldots (c_{i,n-1}^{n-1})^{-1} c_{i,n-1}^{n-1} \ldots c_{i,i+2}^{n-1} c_{i,i+1}^{n-1} = 1.$$

If $i,j \neq n$, then

$$\phi_{n-1} \circ p_n(b_{ij}) = \phi_{n-1}(b_{ij})$$
$$= (c_{i,i+1}^{n-1})^{-1}(c_{i,i+2}^{n-1})^{-1} \ldots (c_{i,j-1}^{n-1})^{-1}(c_{i,j}^{n-1})^2 c_{i,j-1}^{n-1} \ldots c_{i,i+2}^{n-1} c_{i,i+1}^{n-1},$$

and

$$q_n \circ \phi_n(b_{ij}) = q_n((c_{i,i+1}^n)^{-1}(c_{i,i+2}^n)^{-1} \ldots (c_{i,j-1}^n)^{-1}(c_{i,j}^n)^2 c_{i,j-1}^n \ldots c_{i,i+2}^n c_{i,i+1}^n)$$
$$= (c_{i,i+1}^{n-1})^{-1}(c_{i,i+2}^{n-1})^{-1} \ldots (c_{i,j-1}^{n-1})^{-1}(c_{i,j}^{n-1})^2 c_{i,j-1}^{n-1} \ldots c_{i,i+2}^{n-1} c_{i,i+1}^{n-1}.$$

Analogously, it is easy to show that $q_m \circ \phi_n = \phi_{n-1} \circ p_m$. $\square$

The aim of this section is to show that the invariants constructed in Section 8.3, which we call the MN invariant, fail to recognize the non-triviality of Brunnian braids (Lemma 9.3). We can enhance these invariants by using the structure of G_n^k; in fact, we shall make only one step allowing us to recognize the commutator, see Example 9.4. But in principle, it is possible to go on enhancing the invariants coming from G_n^k (even from G_n^3) to get invariants which recognize the non-triviality of commutators of arbitrary length: $[[[b_{12},b_{13}],b_{14}],b_{15},\ldots]$. More precisely, the group G_n^3 itself recognizes the non-triviality of such braids, and the corresponding invariants can be derived as maps from G_n^3 to free products of $\mathbb{Z}_2$ (Example 9.4).

Firstly, we reformulate the definition of homomorphisms $\psi_{(i,j,k)} \colon G_n^3 \to \mathbb{Z}_2^{2^{(n-3)}}$ and construct new ones.

Definition 9.3. Let $\beta \in G_n^k$. If the number of the generator a_m in β is even for each multiindex m, then it is called *a word in a good condition*. Analogously, if the number of b_{ij} in $\beta \in PB_n$ is even for every pair i,j, then β is called *a braid in a good condition*.

Remark 9.1. Let β and β' in G_n^k such that $\beta = \beta'$ in G_n^k. Since the relation a_m^2 from G_n^k changes the number of a_m in β by 2 and other relations do not change the number of a_m in β, it is easy to show that if β is in a good condition, then β' is also in a good condition. Let $\beta, \beta' \in PB_n$ such that β and β' are equivalent in PB_n. Analogously, we can show that if β is in a good condition, then β' is also in a good condition.

Remark 9.2. Let H_n be the subset of all elements in a good condition in PB_n. Then H_n is a normal subgroup of PB_n of finite index. Analogously we can show that a subset H_n^k of all elements in a good condition of G_n^k is a normal subgroup of G_n^k of finite index.

Let us start to reformulate the definition of homomorphisms $\psi_{(i,j,k)}\colon G_n^3 \to \mathbb{Z}_2^{2(n-3)}$. Let $\beta \in G_n^3$ be an element in a good condition. For each $c = a_{ijk}$ of β and for $l \in \{1, 2, \ldots, n\} \setminus \{i, j, k\}$, define $i_c(l)$ by

$$i_c(l) = (N_{jkl} + N_{ijl}, N_{ikl} + N_{ijl}) \in \mathbb{Z}_2 \times \mathbb{Z}_2,$$

where N_{ikl} is the number of a_{ikl} from the start of β to the crossing c. Note that i_c can be considered as a map from $\{1, 2, \ldots, n\} \setminus \{i, j, k\}$ to $\mathbb{Z}_2 \times \mathbb{Z}_2$. Fix $i, j, k \in \{1, \ldots, n\}$. Let $\{c_1, \ldots, c_m\}$ be the set of a_{ijk} such that for each $s, t \in \{1, 2, \ldots, m\}$, $s < t$ if and only if we meet c_s earlier than c_t in β. Define a group F_n^3 by the group presentation generated by $\{\sigma \mid \sigma\colon \{1, 2, \ldots, n\} \setminus \{i, j, k\} \to \mathbb{Z}_2 \times \mathbb{Z}_2\}$ with relations $\{\sigma^2 = 1\}$. Note that i_c is a map from $\{1, 2, \ldots, n\} \setminus \{i, j, k\}$ to $\mathbb{Z}_2 \times \mathbb{Z}_2$ and i_c is contained in F_n^3. It is easy to show that F_n^3 is isomorphic to the free product of $2^{2(n-3)}$ copies of $\mathbb{Z}_2$. Define a word $\psi_{(i,j,k)}(\beta)$ in F_n^3 for β by $\psi_{(i,j,k)}(\beta) = i_{c_1} i_{c_2} \ldots i_{c_m}$.

Lemma 9.3. *For a Brunnian $\beta \in PB_n$ and $n \geq 3$, $\psi_{(i,j,k)}(\phi_n(\beta)) = 1$.*

Proof. It is sufficient to show that $\psi_{(ijk)}(\phi_n(b_{lm})) = 1$ for $|\{l, m\} \cap \{i, j, k\}| < 2$. For a Brunnian braid $\beta \in PB_n$ and for $l \notin \{i, j, k\}$, let β_l be the pure braid obtained by omitting b_{lm} from β. Since $\psi_{(ijk)}(\phi_n(b_{lm})) = 1$ for $|\{l, m\} \cap \{i, j, k\}| < 2$,

$$\psi_{(ijk)}(\phi_n(\beta)) = \psi_{(ijk)}(\phi_n(\beta_l)).$$

Since β is Brunnian, β_l is trivial and $\psi_{(ijk)}(\phi_n(\beta)) = \psi_{(ijk)}(\phi_n(\beta_l)) = 1$. Now we shall show that the statement is true. By the definition of ϕ_n,

$$\phi_n(b_{ij}) = (c_{i,i+1}^n)^{-1}(c_{i,i+2}^n)^{-1} \ldots (c_{i,j-1}^n)^{-1}(c_{i,j}^n)^2 c_{i,j-1}^n \ldots c_{i,i+2}^n c_{i,i+1}^n.$$

Note that for $|\{l, m\} \cap \{i, j, k\}| \neq 2$, $c_{l,m}^n$ contains no a_{ijk}. Without loss of generality we may assume that $i < j < k$ and $l < m$. There are 8 subcases:

(1) $\{l, m\} \cap \{i, j, k\} = \emptyset$,
(2) $l \notin \{i, j, k\}, m \in \{i, j, k\}$,
(3) $l = i, i < m < j$,
(4) $l = i, j < m < k$,
(5) $l = i, k < m$,
(6) $l = j, j < m < k$,
(7) $l = j, k < m$,
(8) $l = k, k < m$.

If $\{l, m\} \cap \{i, j, k\} = \emptyset$, then $\phi_n(b_{ij})$ has no $c_{i,j}^n$, $c_{i,k}^n$ and $c_{j,k}^n$. Hence $\psi_{(ijk)}(\phi_n(b_{lm})) = 1$. Analogously $\psi_{(ijk)}(\phi_n(b_{lm})) = 1$ in the cases of 2, 3, 6 and 8.

If $l = i, j < m < k$, then we obtain that

$$\phi_n(b_{im}) = (c_{i,i+1}^n)^{-1} \ldots (c_{i,j}^n)^{-1} \ldots (c_{i,m-1}^n)^{-1}(c_{i,m}^n)^2 c_{i,m-1}^n \ldots c_{i,j}^n \ldots c_{i,i+1}^n$$

has just two a_{ijk}, say $c_1 = a_{ijk}$ in $(c_{i,j}^n)^{-1}$ and $c_2 = a_{ijk}$ in $c_{i,j}^n$, respectively. Since the number of each a_{stu} between c_1 and c_2 is even for every $s, t, u \in \{1, \ldots, n\}$, $i_{c_1} = i_{c_2}$. Therefore $\psi_{(ijk)}(\phi_n(b_{lm})) = 1$. Analogously $\psi_{(ijk)}(\phi_n(b_{lm})) = 1$ in the case of 7.

If $l = i, k < m < n$, then we obtain that

$$\phi_n(b_{im}) = (c_{i,i+1}^n)^{-1} \ldots (c_{i,j}^n)^{-1} \ldots (c_{i,k}^n)^{-1} \ldots (c_{i,m}^n)^2 \ldots c_{i,k}^n \ldots c_{i,j}^n \ldots c_{i,i+1}^n,$$

and it has four a_{ijk}, say $c_1 = a_{ijk}$ in $(c_{i,j}^n)^{-1}$, $c_2 = a_{ijk}$ in $(c_{i,k}^n)^{-1}$, $c_3 = a_{ijk}$ in $c_{i,k}^n$ and $c_4 = a_{ijk}$ in $c_{i,j}^n$, respectively. Since the number of each a_{stu} between c_2 and c_3 is even for every $s, t, u \in \{1, \ldots n\}$, $i_{c_2} = i_{c_3}$. Similarly, $i_{c_1} = i_{c_4}$. Therefore $\psi_{ijk}(\phi_n(b_{im})) = i_{c_1} i_{c_2} i_{c_3} i_{c_4} = 1$. The proof is complete. □

By the above Lemma the MN-invariant for G_n^3 does not recognize the non-triviality of Brunnian braids in PB_n. In the next paragraph we are going to make the "indices" stronger and we will use the parity for elements of the group G_n^2.

9.2 Groups G_n^2 with parity and points

In the present section we consider the groups G_n^2 with additional enhancements: parity and points. Essentially, we assign additional information to the strands of the braids following the ideas originally appearing in [Fedoseev and Manturov, 2015]. Note that in the work [Fedoseev and Manturov, 2015] elements of the group, corresponding to G_n^2 with points, were called *dotted braids*.

When mapping G_n^3 to G_{n-1}^2, we lose a lot of information. It turns out that we can save some of that information in the form of parity.

Definition 9.4. For a positive integer n, let us define $G_{n,p}^2$ as the group presentation generated by $\{a_{ij}^\epsilon \mid \{i,j\} \subset \{1,\ldots,n\}, i < j, \ \epsilon \in \{0,1\}\}$ subject to the following relations:

(1) $(a_{ij}^\epsilon)^2 = 1$, $\epsilon \in \{0,1\}$ and $i,j \in \{1,\ldots,n\}$,
(2) $a_{ij}^{\epsilon_{ij}} a_{kl}^{\epsilon_{kl}} = a_{kl}^{\epsilon_{kl}} a_{ij}^{\epsilon_{ij}}$ for $\{i,j\} \cap \{k,l\} = \emptyset$,
(3) $a_{ij}^{\epsilon_{ij}} a_{ik}^{\epsilon_{ik}} a_{jk}^{\epsilon_{jk}} = a_{jk}^{\epsilon_{jk}} a_{ik}^{\epsilon_{ik}} a_{ij}^{\epsilon_{ij}}$, for distinct i,j,k, where $\epsilon_{ij} + \epsilon_{ik} + \epsilon_{jk} \equiv 0$ mod 2.

The relations of $G_{n,p}^2$ are closely related to the axioms of parity. We call $a_{ij}^0 (a_{ij}^1)$ an *even generator* (*an odd generator*). If a word β in $G_{n,p}^2$ has no odd generators, then we call β an *even word* (or *an even element*). G_n^2 can be considered as a subgroup of $G_{n,p}^2$ by a homomorphism i from G_n^2 to $G_{n,p}^2$ defined by $i(a_{ij}) = a_{ij}^0$. It is easy to show that i is well-defined. Moreover, the following statement can be proved.

Lemma 9.4. *The homomorphism* $i: G_n^2 \to G_{n,p}^2$ *is a monomorphism.*

Proof. To prove this statement, consider the projection map $p : G_{n,p}^2 \to G_n^2$ defined by

$$p(a_{ij}^\epsilon) = \begin{cases} a_{ij} & \text{if } \epsilon = 0, \\ 1 & \text{if } \epsilon = 1. \end{cases}$$

It is easy to show that p is well-defined homomorphism. Let β and β' be two words in G_n^2 such that $i(\beta) = i(\beta')$. By the definition of i, $p(i(\beta)) = \beta$ and $p(i(\beta')) = \beta'$. That is, $\beta = \beta'$ in G_n^2, therefore, the proof is completed. $\square$

Corollary 9.1. *If two words β and β' in G_n^2 are equivalent in $G_{n,p}^2$, then they are equivalent in G_n^2.*

Now, we define a group $G_{n,d}^2$ and we call it G_n^2 *with points.*

Definition 9.5. For a positive integer n, define $G_{n,d}^2$ by the group presentation generated by $\{a_{ij} \mid \{i,j\} \subset \{1,\ldots,n\}, i < j\}$ and $\{\tau_i \mid i \in \{1,\ldots,n\}\}$ with the following relations;

(1) $a_{ij}^2 = 1$ for $\{i,j\} \subset \{1,\ldots,n\}, i < j$,

(2) $a_{ij}a_{kl} = a_{kl}a_{ij}$ for distinct $i, j, k, l \in \{1, \ldots, n\}$,

(3) $a_{ij}a_{ik}a_{jk} = a_{jk}a_{ik}a_{ij}$ for distinct $i, j, k \in \{1, \ldots, n\}$,

(4) $\tau_i^2 = 1$ for $i \in \{1, \ldots, n\}$,

(5) $\tau_i\tau_j = \tau_j\tau_i$ for $i, j \in \{1, \ldots, n\}$,

(6) $\tau_i\tau_j a_{ij}\tau_j\tau_i = a_{ij}$ for $i, j \in \{1, \ldots, n\}$,

(7) $a_{ij}\tau_k = \tau_k a_{ij}$ for distinct $i, j, k \in \{1, \ldots, n\}$.

We call τ_i a *generator for a point on i-th component* or simply, *a point on i-th component*. Geometrically, a_{ij} corresponds to a 4-valent vertex and τ_i corresponds to a point on the i-th strand of the free braid, see Fig. 9.2.

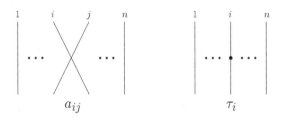

Fig. 9.2 Geometrical meanings of a_{ij} and τ_i

9.2.1 *Connection between* $G_{n,p}^2$ *and* $G_{n,d}^2$

In fact, two groups $G_{n,p}^2$ and $G_{n,d}^2$ are closely related to each other. Define a homomorphism ϕ from $G_{n,p}^2$ to $G_{n,d}^2$ by

$$\phi(a_{ij}^\epsilon) = \begin{cases} a_{ij} & \text{if } \epsilon = 0, \\ \tau_i a_{ij}\tau_i & \text{if } \epsilon = 1. \end{cases}$$

Geometrically, the image of a_{ij}^1 is a crossing between i-th and j-th strands of the braid with two points just before and after the crossing on i-th strand, see Fig. 9.3. Notice that, since we can get $\tau_i a_{ij}\tau_i = \tau_j a_{ij}\tau_j$ from the relation $\tau_i\tau_j a_{ij}\tau_j\tau_i = a_{ij}$, two diagrams with points in Fig. 9.3 are equivalent. On the other hand, the number of τ_i and τ_j before a given crossing $\phi(a_{ij}^\epsilon)$ is equal to ϵ modulo 2, because every image of a_{kl}^ϵ before $\phi(a_{ij}^\epsilon)$ has two points or no points.

Lemma 9.5. *The map ϕ from $G_{n,p}^2$ to $G_{n,d}^2$ is well defined.*

$$a_{ij}^1 \xrightarrow{\ \phi\ }$$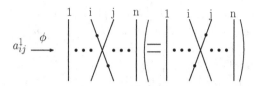

Fig. 9.3   Image of a_{ij}^1

Proof. To show that ϕ is well defined, it is enough to show that every relation is preserved by ϕ. For relations $(a_{ij}^\epsilon)^2 = 1$, if $\epsilon = 0$, then $\phi((a_{ij}^0)^2) = a_{ij}^2 = 1$. If $\epsilon = 1$, then

$$\phi((a_{ij}^1)^2) = \tau_i a_{ij} \tau_i \tau_i a_{ij} \tau_i = \tau_i a_{ij} a_{ij} \tau_i = \tau_i \tau_i = 1,$$

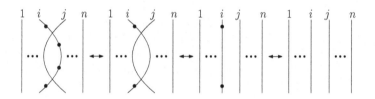

Fig. 9.4 $\phi(a_{ij}^2) = 1$

which can be presented geometrically as Fig. 9.4. For relations $a_{ij}^{\epsilon_{ij}} a_{kl}^{\epsilon_{kl}} = a_{kl}^{\epsilon_{kl}} a_{ij}^{\epsilon_{ij}}$ with distinct i, j, k, l, since $i, j, k,$ and l are different indices, clearly the commutativity holds. For relations $a_{ij}^{\epsilon_{ij}} a_{ik}^{\epsilon_{ik}} a_{jk}^{\epsilon_{jk}} = a_{jk}^{\epsilon_{jk}} a_{ik}^{\epsilon_{ik}} a_{ij}^{\epsilon_{ij}}$, where $\epsilon_{ij} + \epsilon_{ik} + \epsilon_{jk} = 0 \bmod 2$, there are only two cases: all ϵ's are equal to 0 or only two of them are equal to 1. If every ϵ is 0, then

$$\phi(a_{ij}^0 a_{ik}^0 a_{jk}^0) = a_{ij} a_{ik} a_{jk} = a_{jk} a_{ik} a_{ij} = \phi(a_{jk}^0 a_{ik}^0 a_{ij}^0).$$

Suppose that only two of them are equal to 1, say $\epsilon_{ij} = 1$, $\epsilon_{ik} = 1$ and $\epsilon_{jk} = 0$. Then

$$\phi(a_{ij}^1 a_{ik}^1 a_{jk}^0) = \tau_i a_{ij} \tau_i \tau_i a_{ik} \tau_i a_{jk} = \tau_i a_{ij} a_{ik} \tau_i a_{jk} = \tau_i a_{ij} a_{ik} a_{jk} \tau_i =$$
$$\tau_i a_{jk} a_{ik} a_{ij} \tau_i = a_{jk} \tau_i a_{ik} \tau_i \tau_i a_{ij} \tau_i = \phi(a_{jk}^0 a_{ik}^1 a_{ij}^1),$$

which can be presented geometrically as Fig. 9.5.

Therefore, ϕ is well-defined. $\square$

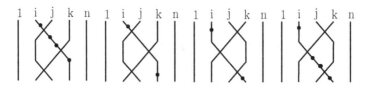

Fig. 9.5 $\phi(a_{ij}^1 a_{ik}^1 a_{jk}^0) = \phi(a_{jk}^0 a_{ik}^1 a_{ij}^1)$

Let $H_{n,d}^2 = \{\beta \in G_{n,d}^2 \mid N_i(\beta) = 0 \bmod 2, \; i \in \{1,\ldots,n\}\}$, where $N_i(\beta)$ is the number of τ_i in β. Note that, by the definition of ϕ, it is clear that the image of $G_{n,p}^2$ by ϕ goes into the set $H_{n,d}^2$. Since every relation of $G_{n,d}^2$ preserves the number of τ's modulo 2, $H_{n,d}^2$ is a subgroup of $G_{n,d}^2$.

Lemma 9.6. *Let* $\beta = T_0 a_{i_1 j_1} T_1 \ldots T_{k-1} a_{i_k j_k} T_k \ldots T_{m-1} a_{i_m j_m} T_m \in H$, *where* T_k *is a product of* τ's. *Define a function* $\chi : H_{n,d}^2 \to G_{n,p}^2$ *by*

$$\chi(\beta) = a_{i_1 j_1}^{\epsilon_1} \ldots a_{i_k j_k}^{\epsilon_k} \ldots a_{i_m j_m}^{\epsilon_m},$$

where N_{i_k} *is the number of* τ_{i_k} *in* $T_0,\ldots,T_{k-1}$ *and* $\epsilon_k = N_{i_k} + N_{j_k}$. *Then* χ *is well-defined.*

Proof. Consider $\beta = T_0 a_{i_1 j_1} T_1 \ldots T_{k-1} a_{i_k j_k} T_k \ldots T_{m-1} a_{i_m j_m} T_m \in H$. Assume that $a_{i_k j_k}$ is contained in one of relations, which can be applied to β. If $a_{i_k j_k}$ is contained in the relation (6) $\tau_i \tau_j a_{ij} \tau_j \tau_i = a_{ij}$, then the number of τ_{i_k} and τ_{j_k} in T_{k-1} is preserved modulo 2 and then the sum of the numbers of τ_{i_k} and τ_{j_k} in $T_0,\ldots,T_{k-1}$ is not changed modulo 2. Therefore $\epsilon_k = N_{i_k} + N_{j_k}$ is compatible with 2. If $a_{i_k j_k}$ is contained in the relation (7) $\tau_k a_{ij} = a_{ij} \tau_k$, then the number of τ_{i_k} and τ_{j_k} in T_{k-1} is preserved, since in the relation (7) i,j,k are different. Therefore the number of τ_{i_k} and τ_{j_k} in $T_0,\ldots,T_{k-1}$ is not changed, and $\epsilon_k = N_{i_k} + N_{j_k}$ is preserved modulo 2. The relation (2) $a_{ij} a_{kl} = a_{kl} a_{ij}$ is preserved by χ by relation $a_{ij}^{\epsilon_1} a_{kl}^{\epsilon_2} = a_{kl}^{\epsilon_2} a_{ij}^{\epsilon_1}$ for different i,j,k,l. Suppose that $a_{i_k j_k}$ is appeared in the relation (1) $a_{ij}^2 = 1$. Then $i_k = i_{k+1}$, $j_k = j_{k+1}$ and there are no τ's in T_k. Therefore $\epsilon_k = \epsilon_{k+1}$ and the relation is preserved by $(a_{ij}^\epsilon)^2 = 1$. Suppose that $a_{i_k j_k}$ is appeared in the relation (3) $a_{ij} a_{ik} a_{jk} = a_{jk} a_{ik} a_{ij}$. Suppose that $a_{i_k j_k} a_{i_{k+1} j_{k+1}} a_{i_{k+2} j_{k+2}} = a_{i_{k+2} j_{k+2}} a_{i_{k+1} j_{k+1}} a_{i_k j_k}$ and $i_k = i_{k+1}$, $j_k = i_{k+2}$ and $j_{k+1} = j_{k+2}$. It is sufficient to show that $\epsilon_k + \epsilon_{k+1} + \epsilon_{k+2} \equiv 0$ modulo 2. Then there are no τ's in T_k, T_{k+1}, T_{k+2}, and hence $N_{i_k} = N_{i_{k+1}}$, $N_{j_k} = N_{i_{k+2}}$, $N_{j_{k+1}} = N_{j_{k+2}}$. Therefore

$$\epsilon_k + \epsilon_{k+1} + \epsilon_{k+2} = N_{i_k} + N_{j_k} + N_{i_{k+1}} + N_{j_{k+1}} + N_{i_{k+2}} + N_{j_{k+2}} \equiv 0 \bmod 2$$

and the proof is completed. $\square$

Theorem 9.1. $G_{n,p}^2$ *is isomorphic to* $H_{n,d}^2$.

Proof. We will show that $\chi \circ \phi = 1_{G_{n,p}^2}$ and $\phi \circ \chi = 1_{H_{n,d}^2}$. To show that $\chi \circ \phi = 1_{G_{n,p}^2}$ let $\beta = a_{i_1 j_1}^{\epsilon_1} \ldots a_{i_m j_m}^{\epsilon_m} \in G_{n,p}^2$. Then

$$\chi(\phi(\beta)) = \chi(\tau_{i_1}^{\epsilon_1} a_{i_1 j_1} \tau_{i_1}^{\epsilon_1} \ldots \tau_{i_m}^{\epsilon_m} a_{i_m j_m} \tau_{i_m}^{\epsilon_m}) = a_{i_1 j_1}^{\theta_1} \ldots a_{i_m j_m}^{\theta_m}.$$

For each k, the number of all τ's which are appeared from $a_{i_1 j_1}^{\epsilon_1} \ldots a_{i_{k-1} j_{k-1}}^{\epsilon_m}$ is even. Since θ_k is equal to the number of all τ_{i_k} and τ_{j_k} modulo 2, $\theta_k = \epsilon_k$. Therefore $\chi \circ \phi = 1_{G_{n,p}^2}$.

Now we will show that $\phi \circ \chi = 1_{H_{n,d}^2}$. Let

$$\beta = T_0 a_{i_1 j_1} T_1 \ldots T_{k-1} a_{i_k j_k} T_k \ldots T_{m-1} a_{i_m j_m} T_m \in H_{n,d}^2.$$

Firstly we will show that β is equivalent to an element in the form

$$\tau_{i_1}^{\epsilon_1} a_{i_1 j_1} \tau_{i_1}^{\epsilon_1} \ldots \tau_{i_m}^{\epsilon_m} a_{i_m j_m} \tau_{i_m}^{\epsilon_m}.$$

By the relations $\tau_i \tau_j = \tau_j \tau_i$ and $\tau_k a_{ij} = a_{ij} \tau_k$, we can assume that β has the form

$$T_0 a_{i_1 j_1} T_1 \ldots T_{k-1} a_{i_k j_k} T_k \ldots T_{m-1} a_{i_m j_m} T_m$$

such that T_k only has τ_{i_k}, τ_{j_k}, $\tau_{i_{k+1}}$ and $\tau_{j_{k+1}}$. In point of $a_{i_k j_k}$, it is a product of the following;

$$A_{i_k j_k} = \tau_{i_k}^{\theta_{i_k}^f} \tau_{j_k}^{\theta_{j_k}^f} a_{i_k j_k} \tau_{i_k}^{\theta_{i_k}^b} \tau_{j_k}^{\theta_{j_k}^b}.$$

Now we claim that there is an element β' equivalent to β which is a product of

$$A'_{i_k j_k} = \tau_{i_k}^{\theta_{i_k}^f} \tau_{j_k}^{\theta_{j_k}^f} a_{i_k j_k} \tau_{i_k}^{\theta_{i_k}^b} \tau_{j_k}^{\theta_{j_k}^b},$$

where $\tau_{i_k}^{\theta_{i_k}^f} = \tau_{i_k}^{\theta_{i_k}^b}$ and $\tau_{j_k}^{\theta_{j_k}^f} = \tau_{j_k}^{\theta_{j_k}^b}$. Suppose that i is an index such that $A_{ij} = \tau_i^{\theta_i^f} \tau_j^{\theta_j^f} a_{ij} \tau_i^{\theta_i^b} \tau_j^{\theta_j^b}$ is the first part in which $\theta_i^f \neq \theta_i^b$, say $\theta_i^f = 1$ and $\theta_i^b = 0$. Note that if $\theta_i^b = 1$ and $\theta_i^f = 1$, then $\tau_i^{\theta_i^b}$ can be move to the very next A_{il} and it is same to the case of $\theta_i^f = 1$ and $\theta_i^b = 0$. Since β is in $H_{n,d}^2$, the number of τ_i should be even and there is $A_{ik} = \tau_i^{\theta_i'^f} \tau_k^{\theta_k^f} a_{ik} \tau_i^{\theta_i'^b} \tau_k^{\theta_k^b}$ such that $\theta_i'^f = 1$ or $\theta_i'^b = 1$ as (9.1) and (9.2).

$$\beta = \ldots A_{ij} \ldots A_{ik} \ldots \tag{9.1}$$

$$= \ldots \tau_j^{\theta_j^f} a_{ij} \tau_i^{\theta_i^b} \tau_j^{\theta_j^b} \ldots \tau_i^{\theta_i^f} \tau_k^{\theta_k^f} a_{ik} \tau_k^{\theta_k^b} \ldots \tag{9.2}$$

If $\theta_i'^f = 1$ and there are no A_{il}, then by the relations $a_{ij}\tau_k = \tau_k a_{ij}$ and $\tau_i \tau_j = \tau_j \tau_i$, τ_i can be move to the next of a_{ij} as (9.3) and (9.4).

$$\beta = \ldots \tau_i \tau_j^{\theta_j^f} a_{ij} \tau_j^{\theta_j^b} \ldots A_{**} \ldots \tau_i \tau_k^{\theta_k^f} a_{ik} \tau_k^{\theta_k^b} \ldots \tag{9.3}$$

$$= \ldots \tau_i \tau_j^{\theta_j^f} a_{ij} \tau_i \tau_j^{\theta_j^b} \ldots A_{**} \ldots \tau_k^{\theta_k^f} a_{ik} \tau_k^{\theta_k^b} \ldots \tag{9.4}$$

If $\theta_i'^f = 1$ and there is A_{il}'s, then by the relation $\tau_i^2 = 1$ new τ's can appear and by the relations $a_{ij}\tau_k = \tau_k a_{ij}$ and $\tau_i \tau_j = \tau_j \tau_i$, τ_i's can be moved to the next of a_{ij} and to the front of a_{il} as (9.5), (9.6) and (9.7).

$$\beta = \ldots \tau_i \tau_j^{\theta_j^f} a_{ij} \tau_j^{\theta_j^b} \ldots A_{il} \ldots \tau_i \tau_k^{\theta_k^f} a_{ik} \tau_k^{\theta_k^b} \ldots \tag{9.5}$$

$$= \ldots \tau_i \tau_j^{\theta_j^f} a_{ij} \tau_j^{\theta_j^b} \ldots \tau_i \tau_i A_{il} \ldots \tau_i \tau_k^{\theta_k^f} a_{ik} \tau_k^{\theta_k^b} \ldots \tag{9.6}$$

$$= \ldots \tau_i \tau_j^{\theta_j^f} a_{ij} \tau_i \tau_j^{\theta_j^b} \ldots \tau_i A_{il} \tau_i \ldots \tau_k^{\theta_k^f} a_{ik} \tau_k^{\theta_k^b} \ldots \tag{9.7}$$

Analogously the case $\theta_i'^b = 1$ also can be solved.

Now assume that $\beta = \prod_{k=1}^m = \tau_{i_k}^{\theta_{i_k}} \tau_{j_k}^{\theta_{j_k}} a_{i_k j_k} \tau_{i_k}^{\theta_{i_k}} \tau_{j_k}^{\theta_{j_k}}$.

Then

$$\phi \circ \chi(\beta) = \phi \circ \chi \left(\prod_{k=1}^m \tau_{i_k}^{\theta_{i_k}} \tau_{j_k}^{\theta_{j_k}} a_{i_k j_k} \tau_{i_k}^{\theta_{i_k}} \tau_{j_k}^{\theta_{j_k}} \right)$$

$$= \phi \left(\prod_{k=1}^m a_{i_k j_k}^{\theta_{i_k j_k}} \right) = \prod_{k=1}^m \left(\tau_{i_k}^{\theta_{i_k j_k}} a_{i_k j_k} \tau_{i_k}^{\theta_{i_k j_k}} \right),$$

where $\theta_{i_k j_k} \equiv \theta_{i_k} + \theta_{j_k} \bmod 2$. If $\theta_{i_k} = \theta_{j_k} = 1$, by the relation $\tau_i \tau_j a_{ij} \tau_j \tau_i = a_{ij}$, β and $\phi \circ \chi(\beta)$ are same elements in $G_{n,d}^2$. If $\theta_{i_k} = 1$ and $\theta_{j_k} = 0$ (or $\theta_{i_k} = 0$ and $\theta_{j_k} = 1$) by the relation $\tau_i \tau_j a_{ij} \tau_j \tau_i = a_{ij}$ (in other words, $\tau_i a_{ij} \tau_i = \tau_j a_{ij} \tau_j$), β and $\phi \circ \chi(\beta)$ are same elements in $G_{n,d}^2$ and hence $\phi \circ \chi = 1_{G_{n,d}^2}$. $\square$

9.2.2 Connection between $G_{n,d}^2$ and G_{n+1}^2

We recall that maps from G_n^k to G_{n-1}^k and from G_n^k to G_{n-1}^{k-1} are constructed by deleting one subindex. In these maps, each generator is mapped to a generator. Here we justify these maps and extend to the case of mappings from G_n^2 to $G_{n-1,d}^2$. We obtained the group monomorphisms $i\colon G_n^2 \to G_{n,p}^2$ and $\phi\colon G_{n,p}^2 \to G_{n,d}^2$. To show that they are group monomorphisms, we found the right inverses for them. In this section we consider relations between $G_{n,d}^2$ and G_{n+1}^2.

Let us define a homomorphism ψ from G_{n+1}^2 to $G_{n,d}^2$ by

$$\psi(a_{ij}) = \begin{cases} a_{ij} & \text{if } n+1 \notin \{i,j\}, \\ \tau_i & \text{if } j = n+1, \\ \tau_j & \text{if } i = n+1. \end{cases}$$

To this end, we prove the following:

Lemma 9.7. *The mapping ψ from G_{n+1}^2 to $G_{n,d}^2$ is well-defined.*

Proof. It is enough to show that every relation for G_{n+1}^2 is preserved by ψ. If every index is different with $n+1$, it is clear. Consider the relation $a_{ij}a_{kl} = a_{kl}a_{ij}$ for distinct $i,j,k,l \in \{1,\ldots,n+1\}$. If one of i,j,k and l is $n+1$, say $i = n+1$, then

$$\psi(a_{(n+1)j}a_{kl}) = \tau_j a_{kl} = a_{kl}\tau_j = \psi(a_{kl}a_{(n+1)j}).$$

For the relations $a_{ij}a_{ik}a_{jk} = a_{jk}a_{ik}a_{ij}$, if one of i,j and k is $n+1$, say $i = n+1$, then

$$\psi(a_{ij}a_{ik}a_{jk}) = \tau_j \tau_k a_{jk} = a_{jk}\tau_k \tau_j = \psi(a_{jk}a_{ik}a_{ij}).$$

Clearly, the relation $a_{ij}^2 = 1$ is preserved by ψ and the statement is proved. $\qquad\square$

Now, let us define the inverse mapping. We define ω from $G_{n,d}^2$ to G_{n+1}^2 by $\omega(a_{ij}) = a_{ij}$ and $\omega(\tau_i) = a_{i(n+1)}$, for example, see Fig. 9.6.

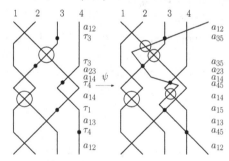

Fig. 9.6   Example of an image of the mapping ω

But this mapping is not well-defined because the relation $\tau_i \tau_j = \tau_j \tau_i$ in $G_{n,d}^2$ goes to $a_{i(n+1)}a_{j(n+1)} = a_{j(n+1)}a_{i(n+1)}$, but $a_{i(n+1)}$ and $a_{j(n+1)}$ do not commute. However, we can modify the mapping and get the following lemma.

Lemma 9.8. *The mapping* $\omega\colon G_{n,d}^2 \to G_{n+1}^2/\langle a_{i(n+1)}a_{j(n+1)} = a_{j(n+1)} a_{i(n+1)} \rangle$, *defined by* $\omega(a_{ij}) = a_{ij}$ *and* $\tau_i = a_{i(n+1)}$, *is well-defined.*

Proof. It is sufficient to show that every relation for $G_{n,d}^2$ goes to an identity in $G_{n+1}^2/\langle a_{i(n+1)}a_{j(n+1)} = a_{j(n+1)}a_{i(n+1)} \rangle$. For relations (1), (2), (3) and (4), it is clear. Consider the relation $\tau_i \tau_j = \tau_j \tau_i$. Then

$$\omega(\tau_i \tau_j) = a_{i(n+1)}a_{j(n+1)} = a_{j(n+1)}a_{i(n+1)} = \omega(\tau_j \tau_i).$$

The relation (6) $\tau_i \tau_j a_{ij} \tau_j \tau_i = a_{ij}$ can be rewritten as $\tau_i \tau_j a_{ij} = a_{ij} \tau_j \tau_i$. Then

$$\omega(\tau_i \tau_j a_{ij}) = a_{i(n+1)}a_{j(n+1)}a_{ij} = a_{ij}a_{j(n+1)}a_{i(n+1)} = \omega(a_{ij}\tau_j \tau_i).$$

Finally, we can show that the relations (5) and (7) are preserved by using the relation $a_{i(n+1)}a_{j(n+1)} = a_{j(n+1)}a_{i(n+1)}$ and the proof is completed. $\quad\square$

Geometrically the relations $a_{i(n+1)}a_{j(n+1)} = a_{j(n+1)}a_{i(n+1)}$ mean that two consecutive classical crossings pass a virtual crossing, see Fig. 9.7. Generally, this move in this case, is called *a forbidden move by* $(n+1)$-*th strand*.

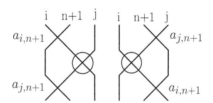

Fig. 9.7   Forbidden move

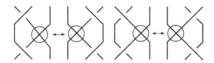

Fig. 9.8   Forbidden moves for virtual links

In the virtual knot theory *forbidden moves* consists of two moves in Fig. 9.8. The following proposition is well known:

Proposition 9.1. *Every virtual knot is equivalent to the trivial knot under the virtual Reidemeister moves and the forbidden moves.*

It follows that the forbidden moves for free knots unknot free knots. But in the case of (free or virtual) links, it is not true.

Corollary 9.2. *The two groups*

$$G^2_{n,d} \text{ and } G^2_{n+1}/\langle a_{i(n+1)}a_{j(n+1)} = a_{j(n+1)}a_{i(n+1)}\rangle$$

are isomorphic.

Proof. We will show that

$$\omega \circ \psi = 1_{G^2_{n,d}} \text{ and } \psi \circ \omega = 1_{G^2_{n+1}/\langle a_{i(n+1)}a_{j(n+1)}=a_{j(n+1)}a_{i(n+1)}\rangle}.$$

For generators a_{ij} in $G^2_{n+1}/\langle a_{i(n+1)}a_{j(n+1)} = a_{j(n+1)}a_{i(n+1)}\rangle$,

$$\psi(a_{ij}) = \begin{cases} a_{ij} & \text{if } n+1 \in \{i,j\}, \\ \tau_i & \text{if } j = n+1, \\ \tau_j & \text{if } i = n+1. \end{cases}$$

By definition of ω, $\omega(\psi(a_{ij})) = a_{ij}$. Hence $\omega \circ \psi = 1_{G^2_{n,d}}$. Analogously, it can be shown that $\psi \circ \omega = 1_{G^2_{n+1}/\langle a_{i(n+1)}a_{j(n+1)}=a_{j(n+1)}a_{i(n+1)}\rangle}$. □

Now define a mapping $\psi_m \colon G^2_{n+1} \to G^2_{n,d}$ by

$$\psi_m(a_{ij}) = \begin{cases} \tau_i & \text{if } j = m, \ i < m, \\ \tau_{i-1} & \text{if } j = m, \ i > m, \\ a_{ij} & \text{if } i,j < m, \\ a_{i(j-1)} & \text{if } i < m, j > m, \\ a_{(i-1)(j-1)} & \text{if } i,j > m. \end{cases}$$

Analogously ψ_m is well-defined (see Lemma 9.7). Then, for each braid β on $(n+1)$ strands such that the number of crossings between i-th and m-th strands is even for each index $i \neq m$, that is, $\beta \in \psi_m^{-1}(H^2_{n,d})$, and a braid with parity can be obtained.

Example 9.1. Let $\beta = a_{12}a_{23}a_{13}a_{23}a_{13}a_{23}a_{12}a_{23}$ in G^2_3. Note that $\psi_m(\beta) \in H^2_{2,d}$ for each index m. Then

- $\chi(\psi_1(\beta)) = \chi(\tau_1 a_{12}\tau_2 a_{12}\tau_2 a_{12}\tau_1 a_{12}) = a^1_{12}a^0_{12}a^1_{12}a^0_{12} \neq 1$,
- $\chi(\psi_2(\beta)) = \chi(\tau_1\tau_2 a_{12}\tau_2 a_{12}\tau_2\tau_1\tau_2) = \chi(a_{12}\tau_1 a_{12}\tau_1) = a^0_{12}a^1_{12} \neq 1$,
- $\chi(\psi_3(\beta)) = \chi(a_{12}\tau_2\tau_1\tau_2\tau_1\tau_2 a_{12}\tau_2) = \chi(a_{12}\tau_2 a_{12}\tau_2) = a^0_{12}a^1_{12} \neq 1$.

Since ψ_m is well defined, β is not trivial in G^2_3.

9.3 Parity for G_n^2 and invariants of pure braids

Let $\beta \in G_{n,p}^2$. For a fixed pair $i, j \in \{1, \ldots, n\}$ and for $k \in \{1, \ldots, n\} \backslash \{i, j\}$, define $i_{a_{ij}^\epsilon}^p(k)$ for each a_{ij}^ϵ in β by

$$i_{a_{ij}^\epsilon}^p(k) = \begin{cases} N_{ik}^0 + N_{jk}^0 \mod 2 \text{ if } \epsilon = 0, \\ N_{ik}^0 + N_{jk}^1 \mod 2 \text{ if } \epsilon = 1, \end{cases}$$

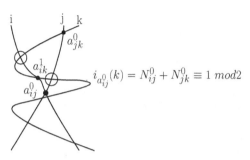

$$i_{a_{ij}^0}(k) = N_{ij}^0 + N_{jk}^0 \equiv 1 \bmod 2$$

Fig. 9.9   The value of $i_{a_{ij}}^p(k)$

where N_{ik}^ϵ is the number of a_{ik}^ϵ, which appears before a_{ij}^ϵ, for example, see Fig. 9.9. Let $\{c_1, \ldots, c_m\}$ be the ordered set of $a_{ij}^{\epsilon_m}$'s in β such that the order agrees with the order of position of a_{ij}^ϵ's. Define $w_{ij}^p \colon G_{n,p}^2 \to F_n^2$ by $w_{ij}^p(\beta) = \prod_{s=1}^m i_{c_s}^p$. Here, the superscript p means "parity".

Lemma 9.9. w_{ij}^p *is well defined.*

Proof. It suffices to show that the image of w_{ij}^p does not change when relations of $G_{n,p}^2$ are applied to β. In the cases of $(a_{ik}^\epsilon)^2 = 1$ and $a_{ij}^{\epsilon_1} a_{kl}^{\epsilon_2} = a_{kl}^{\epsilon_2} a_{ij}^{\epsilon_1}$, it is easy. For relations $a_{ij}^{\epsilon_{ij}} a_{ik}^{\epsilon_{ik}} a_{jk}^{\epsilon_{jk}} = a_{jk}^{\epsilon_{jk}} a_{ik}^{\epsilon_{ik}} a_{ij}^{\epsilon_{ij}}$, where $\epsilon_{ij} + \epsilon_{ik} + \epsilon_{jk} \equiv 0 \bmod 2$, suppose that the relation is not applied on c_s. Then the number of a_{ik}^ϵ and a_{jk}^ϵ before c_s remains the same and then $w_{ij}^p(\beta)$ does not change. Suppose that c_s is in the applied relation, say $c_s^{\epsilon_1} a_{il}^{\epsilon_2} a_{jl}^{\epsilon_3} = a_{jl}^{\epsilon_3} a_{il}^{\epsilon_2} c_s^{\epsilon_1}$, where $\epsilon_1 + \epsilon_2 + \epsilon_3 \equiv 0 \bmod 2$. If $l \neq k$, then the number of a_{ik}^ϵ and a_{jk}^ϵ before c_s is not changed. Suppose that $l = k$. If $\epsilon_1 = 0$, then $\epsilon_2 = \epsilon_3 = 0$ or $\epsilon_2 = \epsilon_3 = 1$. Then the sum of the number of a_{ik}^0 and a_{jk}^0 remains the same modulo 2. If $\epsilon_1 = 1$, then $\epsilon_2 = 1, \epsilon_3 = 0$ or $\epsilon_2 = 0$, $\epsilon_3 = 1$. Then $i_k^p(c_s) = N_{ik}^0 + N_{jk}^1$ is not changed modulo 2 and $w_{ij}^p(\beta)$ is not changed. Therefore w_{ij}^p is an invariant. $\qquad\square$

Definition 9.6. Let $\psi_{ij}^l \colon G_{n+1}^2 \to F_n^2$ be defined by $\psi_{ij}^l = w_{ij}^p \circ \psi_l$.

For example, for $\beta = a_{12}a_{34}a_{13}a_{34}a_{13}a_{12}$, we obtain $\psi_4(\beta) = a_{12}^0 a_{13}^1 a_{13}^0 a_{12}^0$. Then $\psi_{12}^4(\beta) = w_{12}^p \circ \psi_4(\beta) = 01 \in F_3^2$, see Fig. 9.10.

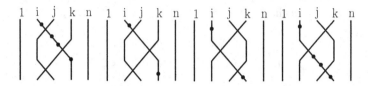

Fig. 9.10 $\beta = a_{12}a_{34}a_{13}a_{34}a_{13}a_{12}$ and $\psi_{12}^4 = 01 \in F_3^2$

Corollary 9.3. *ψ_{ij}^l is an invariant for $\beta \in G_{n+1}^2$.*

Proof. Since ψ_l is a homomorphism, by Lemma 9.9, ψ_{ij}^l is an invariant for $\beta \in G_{n+1}^2$.   □

Example 9.2. Let $X = a_{12}a_{13}a_{12}a_{13}$ and $Y = a_{23}a_{35}a_{23}a_{35}$ and $\beta = [X, Y]$ in G_5^2. Let us show now that β is not trivial. To this end, we consider the element w_{12}^5 where the parity is obtained from 5-th strand. For the pair $(1, 2)$, the value of the MN invariant for G_n^2 of β is trivial, because Y is in a good condition and Y contains no a_{12}. But $\psi_{12}^5(\beta)$ is not trivial. Now we calculate it. Firstly,

$$\psi_5(\beta) = a_{12}^0 a_{13}^0 a_{12}^0 a_{13}^0 a_{23}^0 a_{23}^1 a_{13}^0 a_{12}^0 a_{13}^0 a_{12}^0 a_{23}^1 a_{23}^0.$$

Let $c_1 = a_{12}^0$, $c_2 = a_{12}^0$, $c_3 = a_{12}^0$, $c_4 = a_{12}^0$ such that the order of c_i agrees with the order of a_{12}^ϵ in β. Then $i_{c_1}^p(3) = 0$, $i_{c_2}^p(3) = 1$, $i_{c_3}^p(3) = 0$, $i_{c_4}^p(3) = 1$ and $i_{c_s}(4) = 0$. Then $\psi_{12}^5(\beta) = \zeta_{(0,0)}\zeta_{(1,0)}\zeta_{(0,0)}\zeta_{(1,0)}$ and it cannot be canceled in F_4^2, where $\zeta_{a,b}$ is defined by $\zeta_{a,b}(3) = a$ and $\zeta_{a,b}(4) = b$.

This invariant can be used for $\beta \in G_{n+1}^3$ by the homomorphism $r_m \colon G_{n+1}^3 \to G_n^2$ defined by

$$r_m(a_{ijk}) = \begin{cases} a_{ij} & \text{if } k = m, i, j < m, \\ a_{i(j-1)} & \text{if } k = m, i < m, j > m, \\ a_{(i-1)(j-1)} & \text{if } k = m, i > m, j > m, \\ 1 & \text{if } i, j, k \neq m. \end{cases}$$

Example 9.3. Let

$$\beta = a_{124}a_{123}a_{135}a_{134}a_{124}a_{134}a_{135}a_{123}a_{134}a_{135}a_{134}a_{123}a_{135}a_{134}a_{124}a_{134}$$
$$a_{135}a_{123}a_{124}a_{134}a_{135}a_{134},$$

in G_5^3. Then

$$\beta_1 = r_1(\beta)$$
$$= a_{24}a_{23}a_{35}a_{34}a_{24}a_{34}a_{35}a_{23}a_{34}a_{35}a_{34}a_{23}a_{35}a_{34}a_{24}a_{34}a_{35}a_{23}a_{24}a_{34}a_{35}a_{34}.$$

For index 5,

$$\psi_5(\beta_1) = a_{24}^0 a_{23}^0 a_{34}^1 a_{24}^0 a_{34}^1 a_{23}^0 a_{34}^0 a_{34}^1 a_{23}^1 a_{34}^0 a_{24}^0 a_{34}^0 a_{23}^1 a_{24}^0 a_{34}^1 a_{34}^0.$$

Let $c_1 = a_{24}^0$, $c_2 = a_{24}^0$, $c_3 = a_{24}^0$ and $c_4 = a_{24}^0$ such that the order of c_i agrees with the order of a_{24}^ϵ in $\psi_5(\beta_1)$. Then

- $i_{c_1}(3) = N_{23}^0 + N_{34}^0 = 0 + 0 \equiv 0 \mod 2,$
- $i_{c_2}(3) = N_{23}^0 + N_{34}^0 = 1 + 0 \equiv 1 \mod 2,$
- $i_{c_3}(3) = N_{23}^0 + N_{34}^0 = 2 + 2 \equiv 0 \mod 2,$
- $i_{c_4}(3) = N_{23}^0 + N_{34}^0 = 2 + 3 \equiv 1 \mod 2.$

Therefore $\psi_{24}^5(\beta_1) = 0101 \neq 1$ and hence β is not trivial in G_5^3.

Example 9.4. Let

$$\beta = [[[b_{12}, b_{14}], b_{16}], [b_{13}, b_{15}]]$$
$$= b_{12}b_{14}b_{12}^{-1}b_{14}^{-1}b_{16}b_{14}b_{12}b_{14}^{-1}b_{12}^{-1}b_{16}^{-1}b_{13}b_{15}b_{13}^{-1}b_{15}^{-1}b_{16}b_{12}b_{14}b_{12}^{-1}b_{14}^{-1}b_{16}^{-1}$$
$$b_{14}b_{12}b_{14}^{-1}b_{12}^{-1}b_{15}b_{13}b_{15}^{-1}b_{13}^{-1} \in PB_6.$$

Note that for each $k \in \{1, 2, 3, 4, 5, 6\}$, $p_k(\beta) = 1$, that is, β is Brunnian in PB_n. Then
$$\psi_6(r_1(\phi_6(\beta))) = a_{23}^0 a_{24}^0 a_{25}^0 a_{23}^1 a_{35}^1 a_{34}^1 a_{45}^0 a_{24}^0 a_{34}^0 a_{45}^1 a_{24}^1 a_{35}^1 a_{23}^0 a_{25}^0 a_{24}^0 a_{23}^0$$
$$a_{25}^1 a_{24}^1 a_{24}^0 a_{25}^0 a_{23}^0 a_{24}^0 a_{25}^0 a_{23}^1 a_{35}^1 a_{24}^1 a_{45}^1 a_{34}^0 a_{24}^0 a_{45}^0 a_{34}^1 a_{35}^1 a_{23}^1 a_{25}^0 a_{24}^0 a_{23}^0 a_{25}^1$$
$$a_{24}^1 a_{35}^1 a_{24}^1 a_{35}^1 a_{25}^0 a_{25}^1 a_{35}^1 a_{24}^1 a_{35}^1 a_{23}^0 a_{34}^0 a_{35}^0 a_{23}^0 a_{34}^1 a_{24}^1 a_{45}^1 a_{25}^0 a_{35}^0 a_{45}^0 a_{25}^1 a_{35}^1$$
$$a_{24}^1 a_{34}^1 a_{23}^0 a_{35}^0 a_{24}^0 a_{35}^1 a_{25}^0 a_{45}^1 a_{35}^0 a_{25}^1 a_{45}^0 a_{24}^0 a_{35}^1 a_{24}^1 a_{25}^1 a_{23}^0 a_{24}^0 a_{25}^0 a_{23}^0 a_{35}^0 a_{34}^0$$
$$a_{45}^0 a_{24}^0 a_{34}^1 a_{45}^1 a_{24}^1 a_{35}^0 a_{23}^0 a_{25}^0 a_{24}^0 a_{23}^1 a_{25}^1 a_{24}^0 a_{35}^1 a_{24}^1 a_{45}^1 a_{34}^0 a_{24}^0 a_{45}^0 a_{34}^1 a_{24}^1 a_{35}^0$$
$$a_{25}^1 a_{25}^0 a_{35}^0 a_{24}^1 a_{34}^0 a_{45}^0 a_{24}^0 a_{34}^1 a_{45}^1 a_{24}^1 a_{35}^0 a_{24}^1 a_{23}^1 a_{23}^0 a_{24}^0 a_{25}^0 a_{23}^0 a_{35}^1 a_{24}^1 a_{34}^1$$
$$a_{24}^0 a_{35}^0 a_{34}^0 a_{35}^0 a_{23}^0 a_{25}^0 a_{24}^0 a_{23}^1 a_{25}^1 a_{24}^1 a_{35}^0 a_{24}^1 a_{45}^1 a_{25}^0 a_{35}^1 a_{45}^0 a_{25}^1 a_{35}^0 a_{24}^1 a_{35}^0 a_{23}^0$$
$$a_{34}^1 a_{35}^1 a_{23}^0 a_{34}^1 a_{24}^1 a_{35}^0 a_{25}^0 a_{45}^1 a_{35}^0 a_{25}^1 a_{45}^1 a_{24}^1 a_{34}^0 a_{23}^1 a_{24}^1 a_{25}^1$$
and the number of a_{24}^ϵ is 40. We obtain that

$$\psi_{24}^6 = \zeta_{(0,0)}\zeta_{(0,1)}\zeta_{(1,1)}\zeta_{(0,0)}\zeta_{(0,1)}\zeta_{(1,1)}\zeta_{(0,0)}\zeta_{(0,1)}\zeta_{(0,0)}\zeta_{(0,1)}\zeta_{(1,1)}\zeta_{(0,1)},$$

where $\zeta_{a,b}$ is defined by $\zeta_{a,b}(3) = a$ and $\zeta_{a,b}(5) = b$ and ψ_{24}^p is not trivial in F_5^2.

9.4 Group G_n^3 with imaginary generators

9.4.1 *Homomorphisms from classical braids to $\widetilde{G}_n^3$*

In this section, by using a simple intuitive construction we show how to associate with any word in Artin's generators a certain word in a larger set of generators containing the initial word inside (Theorem 9.2). We get a phenomenon of *imaginary generators* which allows one to *read between* *"classical crossings"*: for a word of a classical pure braid we can "see" the letters a_{ijk}, placed between classical crossings, so that the equivalence of classical pure braids leads to the equivalence of the resulting words. In the algebraic language this is described by means of an injection of a smaller group to a larger one; the composition of this homomorphism with an obvious projection is the identity homomorphism.

This allows one to use the larger group (a small modification of the group G_n^3) as a modification of the small group (actually, the braid group on n strands).

Definition 9.7. The group $\widetilde{G}_n^3$ is given by the group presentation generated by $\{a_{\{i,j,k\}} \mid \{i,j,k\} \subset \{1,\ldots,n\}, |\{i,j,k\}| = 3\}$ and $\{\sigma_{ij} \mid i,j \in \{1,\ldots,n\}, |\{i,j\}| = 2\}$ subject to the following relations:

(a) $a_{\{i,j,k\}}^2 = 1$ for $\{i,j,k\} \subset \{1,\ldots,n\}, |\{i,j,k\}| = 3$,

(b) $a_{\{i,j,k\}}a_{\{s,t,u\}} = a_{\{s,t,u\}}a_{\{i,j,k\}}$, if $|\{i,j,k\} \cap \{s,t,u\}| < 2$,

(c) $a_{\{i,j,k\}}a_{\{i,j,l\}}a_{\{i,k,l\}}a_{\{j,k,l\}} = a_{\{j,k,l\}}a_{\{i,k,l\}}a_{\{i,j,l\}}a_{\{i,j,k\}}$ for distinct i,j,k,l,

(d) $\sigma_{ij}\sigma_{kl} = \sigma_{kl}\sigma_{ij}$ for distinct i,j,k,l,

(e) $\sigma_{ij}a_{\{s,t,u\}} = a_{\{s,t,u\}}\sigma_{ij}$, if $|\{i,j\} \cap \{s,t,u\}| < 2$,

(f) $a_{\{i,j,k\}}\sigma_{ij}\sigma_{ik}\sigma_{jk} = \sigma_{jk}\sigma_{ik}\sigma_{ij}a_{\{i,j,k\}}$ for distinct i,j,k,

(g) $\sigma_{ij}a_{\{i,j,k\}}\sigma_{ik}\sigma_{jk} = \sigma_{jk}\sigma_{ik}a_{\{i,j,k\}}\sigma_{ij}$ for distinct i,j,k.

We denote $a_{ijk} \sim a_{\{i,j,k\}}$; note that $\sigma_{ij} \neq \sigma_{ji}$.

As we already noticed in Chapter 8, pure braids can be presented by dynamical systems of moving points. Let us now consider pure braids on n strands as n moving points with one additional fixed (infinite) point. Let us define a homomorphism from PB_n to $\widetilde{G}_n^3$. Now we consider pure braids as moving n points on upper semi-disc $\{z \in \mathbb{C} \mid |z| \leq 2, Imz \geq 0\}$. As above, the homomorphism from PB_n to $\widetilde{G}_n^3$ will be defined by "reading" those moments when some three points are collinear, which is analogous to the construction of the homomorphism from PB_n to G_n^3. Let enumerated

n points $Z = \{z_1(0), \ldots, z_n(0)\}$ be placed on the semicircle $\{z \in \mathbb{C} \mid |z| = 1, Im\, z \geq 0\}$ in the order of enumeration of points with respect to the clockwise orientation. Let us place one more (infinite) point z_∞ in the center of the semicircle. When the points $Z(t) = \{z_1(t), \ldots, z_n(t)\}$ move as $t \in [0, 1]$ changes, if three points $\{z_i(t), z_j(t), z_k(t)\} \subset Z(t)$ are collinear when $t = t_0$, then we write down the generator a_{ijk}. If points z_j, z_i, z_∞ are on the same ray emanating from z_∞ in this order and the point z_i passes the directed line from the left to the right, then we write down the generator σ_{ij}, see Fig. 9.11. For $i, j \in \{1, \ldots, n\}, i < j$ let us define

$$c_{i,j} = \left(\prod_{k=j+1}^{n} a_{ijk} \right) \sigma_{ij}^{-1} \left(\prod_{k=1}^{j-1} a_{ijk} \right), \tag{9.8}$$

$$c_{j,i} = \left(\prod_{k=j+1}^{n} a_{ijk} \right) \sigma_{ji}^{-1} \left(\prod_{k=1}^{j-1} a_{ijk} \right). \tag{9.9}$$

In particular, for the generator b_{ij} of PB_n, the image of $\widetilde{\phi}_n$ has the form of

$$\widetilde{\phi}_n(b_{ij}) = c_{i,i+1}^{-1} c_{i,i+2}^{-1} \cdots c_{i,j-1}^{-1} c_{j,i} c_{i,j} c_{i,j-1} \cdots c_{i,i+2} c_{i,i+1}. \tag{9.10}$$

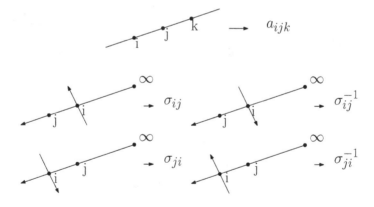

Fig. 9.11 Generators related for each cases

Example 9.5. Let $n = 6$, $i = 2, j = 4$ and points are placed on the semi-circle shown in Fig. 9.12. By reading every set of triple points, we obtain the following word in the group $\widetilde{G}_6^3$:

$$\widetilde{\phi}_6(b_{24}) = a_{123}\sigma_{23}a_{236}a_{235}a_{234}a_{245}a_{246}\sigma_{42}^{-1}a_{124}a_{234}a_{245}a_{246}\sigma_{24}^{-1}a_{124}a_{234}$$
$$a_{234}a_{235}a_{236}\sigma_{23}^{-1}a_{123},$$

see Fig. 9.13. By definition, we obtain

$$\widetilde{\phi}_6(b_{24}) = c_{23}^{-1}c_{42}c_{24}c_{23}. \tag{9.11}$$

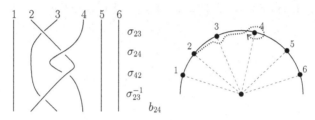

Fig. 9.12   A generator b_{24} in PB_6 and its dynamical system

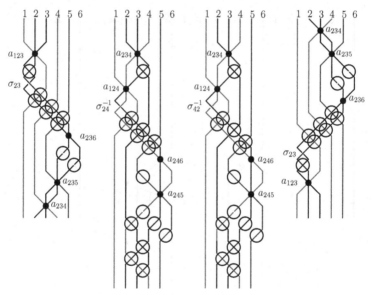

Fig. 9.13   The diagram of the image of b_{24} along $\widetilde{\phi}_6$

As usual the sets of codimension 1 are related to generators, and the sets of codimension 2 are related to relations. They are analogous to those from Chapter 8 in the case of $n + 1$ points. In our setting, if a point passes to a triple point, which contains the infinite point (∞), then it is possible to give an additional information and write one of generators not as $a_{ij\infty}$, but as $\sigma_{ij}, \sigma_{ij}^{-1}, \sigma_{ji}$ or σ_{ji}^{-1}. This is the main idea of how to prove the theorem given below, but this theorem differs from Proposition 8.2, roughly speaking, the image has more generators and relations than the image in Proposition 8.2.

Theorem 9.2. *The homomorphism $\widetilde{\phi}_n$ from PB_n to $\widetilde{G}_n^3$ is well-defined. In other words, if two pure braids β, β' are equivalent as pure braids, then $\widetilde{\phi}_n(\beta)$ and $\widetilde{\phi}_n(\beta')$ present the same element in $\widetilde{G}_n^3$. In particular,*

$$\widetilde{\phi}_n(b_{ij}) = c_{i,i+1}^{-1}c_{i,i+2}^{-1}\cdots c_{i,j-1}^{-1}c_{i,j}c_{j,i}c_{i,j-1}\cdots c_{i,i+2}c_{i,i+1}. \tag{9.12}$$

Proof. Let $z_k = e^{\pi i(n-k)/(n-1)}$, $k = 1, \ldots, n$, be points on the semicircle $C = \{z = a + bi \in \mathbb{C} \mid |z| = 1, b \geq 0\}$. Pure braids can be considered as dynamical systems, in which points $Z(t) = \{z_1(t), \ldots, z_n(t)\}$ move on the plane as $t \in [0, 1]$ changes such that the position of points in the beginning and in the end are same as indicated in the above. Now we clearly formulate the image of each generator in PB_n. For $i < j$ the generator b_{ij} can be considered as the following dynamical system (see Fig. 9.14):

(1) The points $Z(t) \backslash \{z_i(t)\}$ do not move as t changes.
(2) The point $z_i(t)$ moves along the semi-circle C passes beside
$$z_{i+1}(t), z_{i+2}(t), \ldots, z_{j-1}(t)$$
to the point $z_j(t)$.
(3) The point $z_i(t)$ turns around $z_j(t)$ in the clockwise orientation.
(4) The point $z_i(t)$ comes back to the initial position beside $z_{j-1}(t), \ldots, z_{i+1}(t)$.

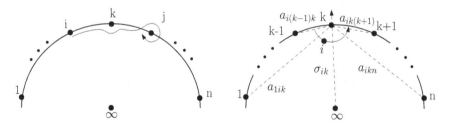

Fig. 9.14   The dynamical systems associating with b_{ij} and c_{ik}

When three points are placed on the same straight line, we write down one of generators of $\widetilde{G}_n^3$ with respect to the rules, which are shown in

Fig. 9.11. In the end of the process, we obtain the word

$$\widetilde{\phi_n}(b_{ij}) = c_{i,i+1}^{-1} c_{i,i+2}^{-1} \cdots c_{i,j-1}^{-1} c_{j,i} c_{i,j} c_{i,j-1} \cdots c_{i,i+2} c_{i,i+1}.$$

Now we will show that if two pure braids β and β' are equivalent in PB_n, then $\widetilde{\phi_n}(\beta)$ and $\widetilde{\phi_n}(\beta')$ are equivalent in $\widetilde{G_n^3}$. Without loss of generality we may assume that the pure braids have the forms of pleasant stable dynamical systems, see Section 7.1.

Let $\{\beta_r\}_{r\in I=[0,1]}$ be an admissible deformation between β and β'. Without loss of generality, we may assume that $\{\beta_r\}_{r\in I}$ satisfies the followings:

(1) For every $r \in (r_0, r_1) \subset I$, if β_r is stable and pleasant, then the set of critical moments of β_r changes continuously.

(2) In I there are finite $s_0 \in I$ such that β_{s_0} is not pleasant or not stable. In other words, β_{s_0} satisfies one of the followings;

A Suppose that there are three points $\{z_i(t_0), z_j(t_0), z_k(t_0)\} \subset Z(t_0)$ such that the points $\{z_i(t_0), z_j(t_0), z_k(t_0)\}$ are collinear in the moment $t = t_0$ and they are still collinear in the first approximation centered at $t = t_0$ (that is, β_{t_0} is not stable). For some $\epsilon > 0$, the word $\widetilde{\phi_n}(\beta_{s_0-\epsilon})$ has the form of $F a_{ijk} a_{ijk} B$, but $\widetilde{\phi_n}(\beta_{s_0+\epsilon})$ has the form of FB (see Fig. 9.15). That is, when β_r passes the moment β_{s_0}, $\widetilde{\phi_n}(\beta_{s_0+\epsilon})$ is obtained from $\widetilde{\phi_n}(\beta_{s_0-\epsilon})$ by the relation (a) in Definition 9.7. Here we use the fact that the deformation is admissible.

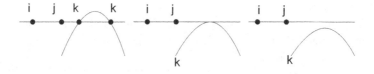

Fig. 9.15   Case A

B Suppose that four points are collinear in the moment $t = t_0$, that is, β_{t_0} is not pleasant, but stable. If the four points are contained in $Z(t_0)$ (see Fig. 9.16), then for some $\epsilon > 0$, $\widetilde{\phi_n}(\beta_{s_0-\epsilon})$ contains a product of $a_{ijk}, a_{ijl}, a_{ikl}, a_{jkl}$ in some order, $\widetilde{\phi_n}(\beta_{s_0+\epsilon})$ have the product of $a_{ijk}, a_{ijl}, a_{ikl}, a_{jkl}$ in the reverse order. As t is changed, $\widetilde{\phi_n}(\beta_{s_0+\epsilon})$ is obtained from $\widetilde{\phi_n}(\beta_{s_0-\epsilon})$ by the relation (c) in Definition 9.7.

If one of the four points is the infinite point, then for some $\epsilon > 0$ (see Fig. 9.17), the word $\widetilde{\phi_n}(\beta_{s_0-\epsilon})$ contains a product of $a_{ijk}, \sigma_{ij}, \sigma_{ik}, \sigma_{jk}$ in some order, and the word $\widetilde{\phi_n}(\beta_{s_0+\epsilon})$ has the

Fig. 9.16 Case B-1: Four points from Z are collinear

product of $a_{ijk}, \sigma_{ij}, \sigma_{ik}, \sigma_{jk}$ in the reverse order. When β_t passes the moment β_{s_0}, the word $\widetilde{\phi_n}(\beta_{s_0+\epsilon})$ is obtained from the word $\widetilde{\phi_n}(\beta_{s_0-\epsilon})$ by the relations (f), (g), (h) of the group $\widetilde{G}_n^3$ in Definition 9.7. Here we have used the fact that β_{t_0} is stable.

Fig. 9.17 Case B-2: Three points from Z are collinear on the same ray emanating from z_∞

C Suppose that two sets m and m' of three points on the lines l and l' respectively in β_{s_0} at $t = t_0 \in [0, 1]$, such that $|m \cap m'| < 2$ (see Fig. 9.18). Then $\widetilde{\phi_n}(\beta_r)$ is changed according to one of the relations (b),(d),(e) of relations of the group $\widetilde{G}_n^3$ in Definition 9.7. In this case we also use the stability of β_{t_0}.

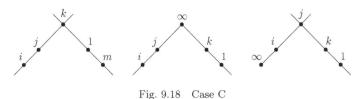

Fig. 9.18 Case C

As before, we can rewrite every type of deformations (codimension 2), which correspond to general position isotopy between two braids. Passing those moments which are either not good or not stable, the word is deformed by one of relations of the group $\widetilde{G}_n^3$ which completes the proof. □

Since every word of PB_n can be presented as a pure braid, we consider the homomorphism $\widetilde{\phi_n}$ as the mapping from pure braids words to words of $\widetilde{G}_n^3$.

9.4.2 Homomorphisms from $\widetilde{G}_n^3$ to G_{n+1}^3

Define the homomorphism $pr : \widetilde{G}_n^3 \to G_n^3$ by

$$pr(a_{ijk}) = \begin{cases} 1, & \text{if } \infty \in \{i,j,k\}, \\ a_{ijk}, & \text{if } \infty \notin \{i,j,k\}. \end{cases} \tag{9.13}$$

The proposition below follows from the definition of pr:

Proposition 9.2. $pr \circ \widetilde{\phi}_n = \phi.$

Now we define the homomorphism i from G_n^3 to $\widetilde{G}_n^3$ by $i(a_{ijk}) = a_{ijk}$. Then $i \circ pr = Id_{G_n^3}$.

Besides, we define the homomorphism π from $\widetilde{G}_n^3$ to G_{n+1}^3 by $\pi(a_{ijk}) = a_{ijk}$ and $\pi(\sigma_{ij}) = a_{ij(n+1)}$. We have $\pi \circ \widetilde{\phi}_n = \phi_{n+1} \circ j_n$ where $j_n \colon PB_n \to PB_{n+1}$ is the natural inclusion. In Chapter 8 we studied homomorphisms from G_n^k to the free product of copies of groups $\mathbb{Z}_2$. This, in turn, leads to the homomorphism $w_{ijk} : G_n^3 \to F_n^3$, where

$$F_n^3 = \langle \{\sigma \mid \sigma : \{1,2,\dots,n\}\backslash\{i,j,k\} \to \mathbb{Z}_2 \times \mathbb{Z}_2\} \mid \{\sigma^2 = 1\}\rangle \cong \mathbb{Z}_2^{*2^{2(n-3)}}.$$

For each generator a_{ijk} in $\beta = Fa_{ijk}B \in G_n^k$, let us define the mapping $i_{a_{ijk}} : \{1,2,\dots,n\}\backslash\{i,j,k\} \to \mathbb{Z}_2 \times \mathbb{Z}_2$ by

$$i_c(l) = (N_{jkl} + N_{ijl}, N_{ikl} + N_{ijl}) \in \mathbb{Z}_2 \times \mathbb{Z}_2,$$

for $l \in \{1,2,\dots,n\}\backslash\{i,j,k\}$, where N_{jkl} is the number of occurrences of a_{ikl} in F. We call $i_{a_{ijk}}$ the index of a_{ijk} in β. Let $\{c_1,\dots,c_m\}$ be the ordered set of all a_{ijk} in β such that if $\beta = T_l c_l B_l$, then $c_s \in T_l$ for $s < l$. Define $w_{ijk} : G_n^3 \to F_n^3$ by

$$w_{(i,j,k)}(\beta) = i_{c_1} i_{c_2} \dots i_{c_m}.$$

Note that if generators on the right hand side have indices, then we can define indices for generators on the left hand side. In other words, if we get a homomorphism $x : G \to H$ for generators of the group G and H, and if indices i for generators in the group H are defined,

$$\dots a \dots \to \dots x(a) \dots$$

then the indices j for generators in G can be defined as

$$j(a) := i(x(a)).$$

That is, for a generator b in $\widetilde{G_n^3}$, the index j_b can be defined by means of G_{n+1}^3 as follow:

$$j_b(l) := i_{\pi(b)}(l),$$

for each $l \in \{1, 2, \ldots, n+1\} \setminus \{i, j, k\}$. Analogously we define the homomorphism from $\widetilde{G_n^3}$ to the free product of copies of groups $\mathbb{Z}_2$ as follow:

$$\widetilde{w}_{ijk} := w_{(i,j,k)} \circ \pi : \widetilde{G_n^3} \to G_{n+1}^3 \to \mathbb{Z}_2^{*2^{2((n+1)-3)}}.$$

Example 9.6. Let $\beta = [b_{12}, b_{13}]$ in PB_3.

$$\widetilde{\phi}_3(b_{12}) = a_{123}\sigma_{21}^{-1}a_{123}\sigma_{12}^{-1},$$

$$\widetilde{\phi}_3(b_{13}) = \sigma_{12}a_{123}\sigma_{31}^{-1}a_{123}\sigma_{13}^{-1}\sigma_{12}^{-1}.$$

Then

$$\widetilde{w}_{123}(\widetilde{\phi}_3(\beta)) = (0,0)(1,1)(1,0)(0,0)(1,1)(1,0) \neq 1$$

in F_4^3, where $i_{a_{123}}(4) = (N_{124} + N_{234}, N_{134} + N_{234}) \bmod 2$, and β is nontrivial in PB_3. Note that the braid β is Brunnian, and $w_{ijk} \circ \phi(\beta) = 1$ for Brunnian braids β, that is, $w_{123}(\phi(\beta)) = 1$ (Theorem 9.3).

9.5 G_n^k-groups for simplicial complexes and the word problem on $G^2(K)$

In the present section we introduce a natural G_n^k-type construction of groups on any given simplicial complex K as a generalisation of the notion of groups G_n^k, which gives a functor from the category of simplicial complexes and injective simplicial maps to the category of groups. The construction depends on the combinatorial data of the simplicial complex K, which might have some connections with knot theory via G_n^k-type groups.

9.5.1 G_n^k-groups for simplicial complexes

Let K be a simplicial complex with its vertex set $V(K)$ to be linearly ordered. For each integer k with $1 \leq k \leq \dim(K) + 1$, define the group $G^k(K)$ to be generated by a_σ for each $(k-1)$-simplex $\sigma \in K$ with defining relations:

(A) $a_\sigma^2 = 1$
(B) **Commuting Rule.** $a_\sigma a_\tau = a_\tau a_\sigma$ if $\dim(\sigma \cap \tau) < k - 2$.

(C) **Tetrahedron Relations on Boundaries.** For each k-simplex σ,

$$a_{d_0\sigma}a_{d_1\sigma}\ldots a_{d_k\sigma} = a_{d_k\sigma}a_{d_{k-1}\sigma}\ldots a_{d_0\sigma},$$

where, for $\sigma = [u_0 u_1 \ldots u_k]$,

$$d_i\sigma = [u_0 u_1 \ldots u_{i-1}u_{i+1}\ldots u_k],$$

and the relation is obtained for every ordering of the set of the vertices of the simplex σ.

Let us consider the special cases of groups $G^k(K)$. Suppose that $K = \Delta^{n-1}$ is an $(n-1)$-simplex with vertices labeled as $u_1, u_2, \ldots, u_n$. Then $G^k(K) = G_n^k$, hence our construction is a generalisation of groups G_n^k. Other special cases are as follows.

If $k = 1$, then relation (B) does not occur. Group $G^1(K)$ is generated by a_v for each vertex v of K with defining relations: $a_v^2 = 1$ from relation (A), and $a_v a_w = a_w a_v$ if v and w are joined by a 1-simplex of K from relation (C). Hence $G^1(K)$ is exactly the right-angled Coxeter group on the graph given by the 1-skeleton of K.

Now let $K = \partial\Delta^k$ be the boundary of a k-simplex. Then tetrahedron relations do not occur in $G^k(K)$. Let σ and τ be two different k-simplices in K given by removing i and j, respectively. Then $\sigma \cap \tau$ is given by removing i and j. Hence the commuting rule does not occur in $G^k(K)$. It follows that $G^k(\partial\Delta^k)$ is a free product of $k+1$ copies of $\mathbb{Z}/2$.

The functorial property of groups $G^k(K)$ is guaranteed by the following proposition.

Proposition 9.3. *Any injective simplicial map* $f \colon K \to L$ *induces a group homomorphism* $f_* \colon G^k(K) \to G^k(L)$ *with* $f_*(a_\sigma(K)) = a_{f(\sigma)}(L)$.

Proof. The proof is immediate. □

Corollary 9.4. *Let K be a finite simplicial complex. Then*

(1) *Each m-simplex σ^m with $m \geq k$ induces a group homomorphism*
    $i_\sigma \colon G_{m+1}^k \to G^k(K)$,
(2) *There is a group homomorphism* $G^k(K) \to G_{\mathrm{Card}(V(K))}^k$.

Proof. The inclusion $\sigma^m \hookrightarrow K$ is an injective simplicial map, and there is a canonical injective simplicial map $K \hookrightarrow \Delta^{\mathrm{Card}(V(K))-1}$. □

As mentioned above, group $G^1(K)$ is a right-angled Coxeter group. The next interesting groups are $G^2(K)$. The main result in this section is to solve the word problem on groups $G^2(K)$ as our preliminary progress on this construction given in Theorem 9.4.

9.5.2 The word problem for $G^2(K)$

In [Manturov, 2015b], the word problem in G_n^2 was solved by means of a bijection between elements of the group G_n^2 and a Coxeter group we call $C(n, 2)$.

Below, we recall the above construction and show how it leads to the solution of the word problem for $G^2(K)$, where K is an arbitrary 2-dimensional complex (and hence, it works for a complex for any dimension since the group G^2 depends only on the 2-frame of the complex).

Let $C(n, 2)$ be the Coxeter group having $\binom{n}{2}$ generators b_{ij}, where (i, j) run all unordered pairs of elements $\{i, j\} \subset \{1, \ldots, n\}$. The relations in the Coxeter group are:

(1) $b_{ij}^2 = 1$ (A)
(2) $(b_{ij}b_{kl})^2 = 1$ (B)
(3) $(b_{ij}b_{ik})^3 = 1$ (C)

Here in (A), (B), (C) i, j, k, l range over all elements from $\{1, \ldots, n\}$ and different letters correspond to different numbers.

In other words, we have the Coxeter group with $\binom{n}{2}$ generators and connections with label 2 in the case (B) and connections of label 3 in the case (C).

The correspondence between elements of G_n^2 and $C(n, 2)$ goes as follows (cf. Section 10.2). Let $w = a_{i_1 j_1} a_{i_2 j_2} \ldots a_{i_k j_k}$ be a word in G_n^2. We start rewriting it to get a word in $C(n, 2)$ step-by-step. First, we take $b_{i_1 j_1}$ with the same pair of indices (i_1, j_1) as the first letter for $a_{i_1 j_1}$. After that, in all subsequent words $a_{i_2 j_2}, \ldots, a_{i_k j_k}$ we act on all indices by the transposition (i_1, j_1). We get a word $b_{i_1 j_1} a_{i_2' j_2'} \ldots a_{i_k' j_k'}$. After that we write $b_{i_2' j_2'}$ and act on all remaining indices $i_2, j_2, i_3, j_3, \ldots, i_k, j_k$ by the transposition (i_2', j_2'). We get a word $b_{i_1 j_1} b_{i_2' j_2'} a_{i_3'' j_3''} \ldots a_{i_k'' j_k''}$.

Then we write $b_{i_3'' j_3''}$ and act on the remaining letters by the transposition (i_3'', j_3''). We continue the process until we rewrite the word completely.

We denote the resulting word by $f(w)$.

We summarise the results from [Manturov, 2015b] as follows

Theorem 9.3. *The map f is a bijection between words in a_{ij} and words in b_{ij}.*

Two words w, w' in letters a_{ij} are equal as elements of G_n^2 if and only if the resulting words $f(w), f(w')$ are equal as elements of $C(n, 2)$. Moreover, if two words w, w' are related by a relation from G_n^k, then the resulting words $f(w), f(w')$ are related by the corresponding relation in the Coxeter group $C(n, 2)$. Namely, (A) comes from the relation $a_{ij}^2 = 1$, (B) comes from $a_{ij}a_{kl} = a_{kl}a_{ij}$, and (C) comes from the relation $a_{ij}a_{ik}a_{jk} = a_{jk}a_{ik}a_{ij}$.

In particular, this means that in G_n^2 the word problem is solved by using the gradient descent algorithm: *a word w is equal to 1 if and only if we can trivialise w by using those relations which do not increase the length of it.*

The inverse rewriting algorithm (from b_{ij} to a_{ij}) is more or less the same.

We are going to generalise this result for the case of $G^2(K)$ for arbitrary complex K. Denote the vertices of K by $1, \ldots, n$. Not every 2-complex is a frame of a full simplex. Namely, it may happen that some edge (ij) is missing or some 2-face (ijk) is missing despite all three edges $(ij), (ik), (jk)$ are present.

Since edges correspond to generators, and faces correspond to relations, this suggests the following definition of the analogue of $C(n, 2)$. For the simplicial complex K on vertices $1, \ldots, n$ we define

$$C(K) = \langle b_{ij}, ij \text{ are connected by an edge} | (D), (E), (F) \rangle,$$

where (D) means that for generators b_{ij} we have $b_{ij}^2 = 1$, (E) means that b_{ij} commutes with b_{kl} for pairwise distinct i, j, k, l, and (F) means that $(b_{ij}b_{jk})^3 = 1$ whenever K contains a 2-simplex spanned by i, j, k. Note that in this case we also have $(b_{ij}b_{jk})^3 = (b_{ik}b_{jk})^3 = 1$.

Note that the above group is indeed a Coxeter group, however, some labels may be infinite. Namely, if two edges (ij) and (ik) are present, but there is no 2-simplex (ijk), then we have no relation of the type $(b_{ij}b_{ik})^N = 1$ for any integer N.

Now, we want to formulate the analogue of Theorem 9.3. We shall do it in two steps. First, assume every two vertices of K are connected by an edge. Then the formulation is literally the same. Namely, starting from a word w in generators a_{ij}, we apply the rewriting process as above, and get a word in generators b_{ij}.

Theorem 9.4. *Theorem 9.3 works verbatim when we replace G_n^2 with $G^2(K)$, $C(n,2)$ with $C(K)$ and apply relations (C) only when the simplex (ijk) is present.*

To handle the general case when some edges are missing, it suffices to make the following observation.

Statement 9.1. *Let K be a 2-complex on vertices $1, 2, \ldots, n$ and let K' be the complex obtained from K by adding the missing edges. Then the natural map $G^2(K) \to G^2(K')$ which takes generators to the generators with the same names, is an inclusion.*

Proof. Indeed, the inverse map is obtained by taking all generators corresponding to the "new edges" to 1. It is crucial that these edges do not appear as boundaries of 2-simplices. □

Thus, we get the following

Corollary 9.5. *Let K be a 2-dimensional cell complex on vertices $1, \ldots, n$, and let K' be the complex obtained from K by adding all missing edges. Then the rewriting map f in Theorem 9.3 restricted to $G^2(K)$ is an injection $G^2(K) \to C(K')$.*

In particular, the word problem in $G^2(K)$ can be solved by a gradient descent algorithm.

Interestingly, the image $f(G^2(K))$ may contain all letters in the alphabet b_{ij}. For example, if K consists of points $1, 2, 3$ and 1-simplices $(12), (13)$, then when rewriting the elements from the group $\mathbb{Z}_2 * \mathbb{Z}_2 = \langle a_{12}, a_{13} | a_{12}^2 = a_{13}^2 = 1 \rangle$, we can obtain all letters b_{12}, b_{13}, b_{23}. Nevertheless, the only decreasing relations in $C(K')$ are $b_{12}^2 \to 1, b_{13}^2 \to 1, b_{23}^2 \to 1$, so starting with a word $f(w)$ originating from $w \in G^2(K)$, we will be always dealing with words of the form $f(w)$.

9.6 Yet another braid invariant: codimension one properties arising from tangent circles

In the present section we give one more geometric construction, which gives rise to invariants of classical braids. As the constructions, presented above, it relies upon *the general position codimension 1 property*.

Namely, we consider a motion of n points $z_j(t), j = 1, \ldots, n$ inside the unit circle $D = \{|z| < 1\} = \{x, y \mid x^2 + y^2 < 1\}$. One can readily check that

through any two points inside the circle one can draw exactly two circles
tangent to the absolute $|z| = 1$. We are interested in those moments, where

some three points $z_i(t), z_j(t), z_k(t)$ lie on a circle tangent to the absolute.

Later on, when considering the circle tangent to the absolute, we shall
enumerate points on this circle starting from the tangency point in the counterclockwise direction. We shall say that the point a *precedes* the point b if
when passing the circle starting from the tangency point counterclockwise,
we first encounter a and then b.

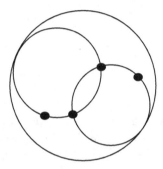

Fig. 9.19   Two circles tangent to the absolute

Unlike the k-properties considered before ("three points are collinear"
for $k = 3$, "four points are on the same circle" for $k = 4$), this property has
the following subtlety. There are two circles passing through the two points
and tangent to the absolute, not one. More precisely, this disadvantage can
be formulated as follows:

*From the fact that three points a, b, c belong to a circle tangent to the
absolute and three points a, b, d belong to a circle tangent to the absolute, it
does not follow that all four points a, b, c, d belong to a circle tangent to the
absolute.*

One can easily see that the following statement holds:

Statement 9.2. *If points a, b, c belong to the same circle tangent to the
absolute and points a, b, d belong to the same circle tangent to the absolute
and in both circles a precedes b, then a, b, c, d belong to the same circle
tangent to the absolute.*

The ability to work with such situations is quite important, since when studying dynamical systems of motions of several points generic codimension 1 properties can be related to more complicated curves than just lines or circles, so that there are a fixed number of curves of the given type passing through fixed $k - 1$ points. This leads to the generalisation of the G_n^k-approach and to the following group (in our case $k = 3$). Now let us fix the number of points n.

Definition 9.8. The group "G_n^3 is defined by the presentation

$$\text{``}G_n^3 = \langle a''_{ijk} | (4), (5), (6) \rangle,$$

having $n(n-1)(n-2)$ generators a''_{ijk}, where i, j, k range all possible *ordered* triples of points from 1 to n, and three types of relations:

$$a''_{ijk}{}^2 = 1 \text{ for all pairwise distinct } i, j, k \tag{4}$$

$$a''_{ijk} a''_{pqr} = a''_{pqr} a''_{ijk}, \tag{5}$$

if none of the ordered pairs $\{i,j\}, \{i,k\}, \{i,l\}$ coincides with any of the ordered pairs $\{p,q\}, \{p,r\}, \{q,r\}$.

$$(a''_{ijk} a''_{ijl} a''_{ikl} a''_{jkl})^2 = 1, i, j, k, l \text{ are pairwise distinct.} \tag{6}$$

Note that the condition (5) significantly differs from (2). For example, a''_{123} commutes with a''_{421}, but does not commute with a''_{134}.

In the case of "G_n^3, there is no evident homomorphism to G_n^3. Indeed, in the group "G_n^3 for $n \geq 4$ one has, for example, the relation $a''_{123} a''_{421} = a''_{421} a''_{123}$; the relation $a_{123} a_{124} = a_{124} a_{123}$ in G_n^3 does not hold. Moreover, if we take the quotient of the group G_n^3 by all commutativity relations of such sort, the resulting quotient group will be isomorphic to a direct product of groups $\mathbb{Z}_2$.

Now let dynamics of motion of n points inside the unit circle be given. Imposing natural "general position" conditions with respect to the circles tangent to the absolute, we can associate a word in letters a''_{ijk} to this dynamics, as follows. At each critical moment t_l we have three points on the circle (i_l, j_l, k_l), which are enumerated in the counterclockwise direction starting from the tangency point. With such a moment, we associate the generator a''_{i_l, j_l, k_l}. The word $g(\beta)$ is the product of all generators corresponding to all critical moments, as t increases.

Analogously to the main theorem of [Manturov, 2015a], one can prove

Theorem 9.5. *The map $\beta \mapsto g(\beta)$ constructed above, is a homomorphism from the pure braid group PB_n to "G_n^3.*

Proof. Let us enumerate all those events of codimension 2 which will lead us to relations. The case of "unstable triple point" is treated as follows: at some moment, we have three points on a circle tangent to the absolute, and this disappears after a small perturbation of dynamics. This corresponds to the relation "$a_{ijk}^2 = 1$.

In the same manner, one can deal with "quadruple points". Let our dynamics be such that at some moments some four points are on the same circle tangent to the absolute. Let us enumerate these points in the counterclockwise direction starting from the tangency point: i, j, k, l. After a small perturbation, the quadruple point splits into four instances with triple points and corresponding generators $a_{ijk}'', a_{ijl}'', a_{ikl}'', a_{jkl}''$. The "opposite" small perturbation gives rise to the product of the same four generators in the inverse order in such a way that a_{ijk}'' can not be next to a_{ikl}''.

The most important feature of "G_n^3 is the situation with two "independent simultaneous" triples of points.

Assume that for some moment t for the n points, there are two triples of indices $m = \{a, b, c\}, m' = \{d, e, f\}$, such that $(z_a(t), z_b(t), z_c(t))$ belong to a circle tangent to the absolute and $(z_d(t), z_e(t), z_f(t))$ belong to another circle tangent to the absolute and in each triple the indices are pairwise distinct. If $Card((a, b, c) \cap (d, e, f)) \leq 1$, then we get a commutativity relation (5) analogous to the relation (2). If the intersection consists of two indices, we get (unordered) triples (a, b, c) and (a, b, f), then, according to Statement 9.2, for the same triple, the point a precedes b when counting counterclockwise, and in the second triple, b precedes a, which is also described by relation (5).   □

The group "G_n^3 admits a natural homomorphism to the group G_N^2, where $N = n(n-1)$, and each of the indices p, q of generators $a_{p,q}$ is an ordered pair of distinct elements from 1 to n.

Let us define the map $h : \text{``}G_n^3 \to G_N^2$ by the formula

$$h : a_{ijk}'' \mapsto a_{ij,ik}. \tag{7}$$

Theorem 9.6. *The map constructed above is a homomorphism* $h : \text{``}G_n^3 \to G_N^2$.

The check is left to the reader.

Remark 9.3. The composite map $PB_n \mapsto G_N^2$ can be constructed without mentioning the group "G_n^3. Namely, when considering a braid $\alpha \in PB_n$ as a dynamics of motion of n points inside the unit circle, to each moment when some three points i, j, k belong to the same circle tangent to the absolute, we associate the product $a_{ij,ik} a_{ij,ik} a_{ik,jk}$, if when walking along the circle starting from the tangent point, we encounter these points in the order i, j, k.

Chapter 10

Representations of the Groups G_n^k and Their Connections with Permutahedra

10.1 Faithful representation of Coxeter groups

The purpose of this section is to prove that there exist faithful linear presentations of Coxeter groups. For more details see [Cohen, 2008].

Definition 10.1. Let $M = (m_{ij})_{1 \leq i,j \leq n}$ be a symmetric $n \times n$ matrix with entries from $\mathbb{N} \cup \{\infty\}$ such that $m_{ii} = 1$ for all $i \in \{1, 2, \ldots, n\}$ and $m_{ij} > 1$ whenever $i \neq j$. *The Coxeter group of type M* is the group

$$W(M) = \langle S = \{s_1, \ldots, s_n\} \mid \{(s_i s_j)^{m_{ij}} = 1 \mid i, j \in \{1, 2, \ldots, n\}, m_{ij} < \infty\}\rangle.$$
$$(10.1)$$

Here we agree that if $m_{ij} = \infty$, then there are no relation $(s_i s_j)^{m_{ij}} = 1$. The pair $(W(M), S)$ is called *the Coxeter system of type M*.

Example 10.1.

(1) If $n = 1$, then $M = (1)$ and $W(M) = \langle s_1 \mid s_1^2 = 1 \rangle$, the cyclic group of order 2.

(2) If $n = 2$, then $M = \begin{bmatrix} 1 & m \\ m & 1 \end{bmatrix}$ and $W(M)$ is isomorphic to the dihedral group D_{2m} for some $m \in \mathbb{N} \cup \{\infty\}$. We denote such Coxeter matrix by $I_2(m)$. If $m = \infty$, the Coxeter group $W(M)$ is called the *infinite dihedral group*.

The Coxeter matrix $M = (m_{ij})_{1 \leq i,j \leq n}$ is often described by a labeled graph $\Gamma(M)$ with vertices $\{1, \ldots, n\}$ such that two vertices i and j are joined by an edge labeled m_{ij} if and only if $m_{ij} > 2$. If $m_{ij} = 3$, then the label 3 of the edge $\{i, j\}$ is often omitted. If $m_{ij} = 4, 5, 6, \ldots$, then instead of the labels at the edge $\{i, j\}$ one often draws a double edge, a triple edge, a quadruple edge and so on. Such labeled graphs are called the *Dynkin diagrams* of M.

Example 10.2. The left diagram in Fig. 10.1 is the Dynkin diagram for the dihedral group D_{2m}. The middle and right diagrams in Fig. 10.1 are the Dynkin diagrams for the group $C(3,2)$ and $C(4,2)$, which is introduced in Section 9.5.2.

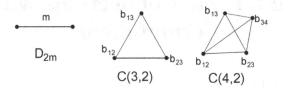

Fig. 10.1   Dynkin diagrams

Example 10.3. It is well known that the symmetric group of a n-simplex is the symmetric group S_{n+1} and it is isomorphic to the Coxeter group, which has the Dynkin diagram of type A_n as described in Fig. 10.2. The symmetric group of the n-cube is known as the Coxeter group with the Dynkin diagram of type B_n in Fig. 10.2. The Coxeter groups of type n are isomorphic to symmetric groups of certain semiregular polytopes. Every finite irreducible Coxeter groups is classified, including Coxeter groups with Dynkin diagrams in Fig. 10.2. For more details see [Cohen, 2008].

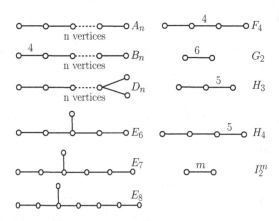

Fig. 10.2   Dynkin diagrams for finite Coxeter groups

Definition 10.2. Let $(W(M), S)$ be a Coxeter system of type M. A word u of the minimal length with $u \equiv w$ in $W(M)$ is called *a minimal expression of w*. The length of u is called *the length of w*. We denote it by $l(w)$.

Let us see some properties of the length of words.

Lemma 10.1. *Let $(W(M), S)$ be a Coxeter system of type M. For $s \in S$ and $w \in W(M)$, we have $l(sw) = l(w) \pm 1$.*

Proof. It is easy to see that $l(sw) \leq 1 + l(w)$ and $l(w) = l(s(sw)) \leq 1 + l(sw)$, that is, $l(w) - 1 \leq l(sw) \leq l(w) + 1$. We will show that $l(w) \not\equiv l(sw)$ mod 2. Let us consider the mapping $wr : W(M) \to \mathbb{Z}_2$ defined by $wr(s) = 1$ for $s \in S$. It is easy to see that wr is a well-defined group homomorphism. Then for $w \in W(M)$ we obtain that $wr(w) \equiv l(w)$ mod 2. It follows that $l(sw) \equiv wr(sw) = wr(s) + wr(w) \equiv l(s) + l(w) = 1 + l(w)$ and hence $l(sw) \equiv 1 + l(w)$ mod 2. $\qquad\qquad\square$

Definition 10.3. Let $(W(M), S)$ be a Coxeter system of type M. Let $T \subset S$. An element $w \in W(M)$ is called *left T-reduced*, if $l(tw) > l(w)$ for all $t \in T \cup T^{-1}$. An element $w \in W(M)$ is called *right T-reduced*, if $l(wt) > l(w)$ for all $t \in T \cup T^{-1}$.

Lemma 10.2. *Let $(W(M), S)$ be a Coxeter system of type M and $T \subset S$.*

(1) *For $w \in W(M)$ there exist $u \in \langle T \rangle$ and a left T-reduced element v such that $w = uv$ and $l(w) = l(u) + l(v)$.*
(2) *If $w \in \langle T \rangle$, then $l(w) = l_T(w)$, where l_T is the minimal length in $\langle T \rangle$.*

Proof. For the statement (1). Let $w \in W(M)$. Let us consider the subset D of $\langle T \rangle \times W(M)$ consisting all pairs (u, v) such that $w = uv$, $l(w) = l(u) + l(v)$ and $l(u) = l_T(u)$. Since $(1, w) \in D$, $D \neq \emptyset$. Let (u_0, v_0) be an element of D such that $l(u_0)$ is maximal in D. We will show that v_0 is left T-reduced. Suppose that v_0 is not left T-reduced, that is, $l(tv_0) < l(v_0)$. Then $v_0 = tv'$ for some $v' \in W(M)$ such that $l(v_0) = l(v') + 1$. Since $w = u_0 v_0 = u_0 tv'$, we obtain the inequality

$$l(w) \leq l(u_0 t) + l(v') \leq (l(u_0) + 1) + (l(v_0) - 1) = l(w)$$

and hence $l(u_0 t) = l(u_0) + 1$. Additionally, we obtain that

$$l(u_0 t) \leq l_T(u_0 t) \leq l_T(u) + 1 = l(u) + 1 = l(u_0 t),$$

that is, $l(u_0 t) = l_T(u_0 t)$. By definition of D, $(u_0 t, v') \in D$ and $l(u_0 t) > l(u_0)$ and it is contradiction to $l(u_0)$ is maximal. That is, v_0 is left T-reduced and the statement (1) is proved. For the statement (2) let $w \in \langle T \rangle$. By the previous proof there exist $u \in \langle T \rangle$ and left T-reduced element v in $W(M)$ such that $w = uv$ and $l(w) = l(u) + l(v) = l_T(u) + l(v)$. Note that if $v \in \langle T \rangle$ is left T-reduced, then $v = 1$. Therefore $l(w) = l_T(u) + l(v) = l_T(u)$. $\qquad \square$

10.1.1   *Coxeter group and its linear representation*

Let $(W(M), S)$ be a Coxeter system of type M. From now on we assume that $|S| = n < \infty$. We will construct a real linear representation of $W(M)$ of degree n such that the images of the elements of S are reflections in $\mathbb{R}^n$.

Definition 10.4. By *a reflection* in a real vector space V we mean a linear transformation on V fixing a subspace of V of codimension 1, called its *mirror*, and having a nontrivial eigenvector with eigenvalue -1, called a *root of the reflection*.

Definition 10.5. Fix a Coxeter matrix $M = (m_{ij})_{1 \le i,j \le n}$. Let V be a real vector space with basis $\{e_i\}_{i \in \{1,2,\dots,n\}}$. Let us define a symmetric bilinear form B_m on V by formulas

$$B_M(e_i, e_j) = -2\cos\left(\frac{\pi}{m_{ij}}\right) \tag{10.2}$$

for $m_{ij} < \infty$, $i, j \in \{1, 2, \dots, n\}$ and

$$B_M(e_i, e_j) = -2, \tag{10.3}$$

if $m_{ij} = \infty$. It is easy to see that the form is symmetric, since $m_{ij} = m_{ji}$ for $i, j \in \{1, 2, \dots, n\}$. We call B_M *the symmetric bilinear form associated with M*.

By simple calculations we obtain the following corollary.

Corollary 10.1. *Let B_M be the symmetric bilinear form associated with a Coxeter matrix M. Then*

(1) $B_M(e_i, e_i) = 2$ for all $i \in \{1, 2, \dots, n\}$ by definition of Coxeter matrix.
(2) $B_M(e_i, e_j) \le 0$ for $i, j \in \{1, 2, \dots, n\}$ with $i \ne j$. The equality holds if and only if $m_{ij} = 2$.

Let Q_M be the quadratic form determined by B_M, i.e.

$$Q_M(x) = B_M(x, x)$$

for all $x \in V$. We call Q_M *the quadratic form associated with M*.

For $x = -2\Sigma_i x_i e_i$ we obtain $Q_M(x) = \Sigma_{i,j \in \{1,2,...,n\}} x_i x_j \cos(\pi/m_{ij})$. The bilinear form B_M is related to Q_M via the formula

$$Q_M(x+y) = Q_M(x) + Q_M(y) + 2B_M(x,y).$$

We use the symmetric bilinear form B_M to define reflection in $GL(V)$ preserving B_M. We shall see some general properties of B_M and these reflections.

Let us define a linear transformation $\rho_i : V \to V$ for $i \in \{1, \ldots, n\}$ by

$$\rho_i(x) = x - B_M(x, e_i)e_i,$$

for $x \in V$.

Lemma 10.3. *Let B_M be the symmetric bilinear form associated with the Coxeter matrix M. The symmetric bilinear form B_M and the linear transformations ρ_i satisfy the following properties:*

(1) For each $i \in \{1, \ldots, n\}$, the transformation ρ_i is a reflection in V. Its mirror is $e_i^{\perp} := \{x \in V \mid B(x, e_i) = 0\}$.
(2) For all $x, y \in V$ we have $B_M(\rho_i x, \rho_i y) = B_M(x, y)$.
(3) The order of $\rho_i \rho_j$ is m_{ij}.

Proof. From simple computations the statement (1) follows. For the statement (2), by computation, we get

$$
\begin{aligned}
B_M(\rho_i x, \rho_i y) &= B_M(x - B_M(x, e_i)e_i, y - B_M(y, e_i)e_i) \\
&= B_M(x,y) - B_M(x, e_i)B_M(e_i, y) - B_M(x, e_i)B_M(y, e_i) \\
&\quad + B_M(x, e_i)B_M(y, e_i)B_M(e_i, e_i) \\
&= B_M(x,y) - B_M(x, e_i)B_M(e_i, y) - B_M(x, e_i)B_M(y, e_i) \\
&\quad + 2B_M(x, e_i)B_M(y, e_i) \\
&= B_M(x,y).
\end{aligned}
$$

For the statement (3), we notice that the linear subspace $\mathbb{R}e_i + \mathbb{R}e_j$ of V is invariant under ρ_i and ρ_j. Put $b = B_M(e_j, e_i)$. Linear transformations ρ_i and ρ_j on the basis e_i, e_j can be expressed by the matrices as

$$
\rho_i : \begin{bmatrix} -1 & -b \\ 0 & 1 \end{bmatrix} \text{ and } \rho_j : \begin{bmatrix} 1 & 0 \\ -b & -1 \end{bmatrix}.
$$

It follows that $\rho_i \rho_j$ has matrix

$$
\rho_i \rho_j : \begin{bmatrix} -1 + b^2 & b \\ -b & -1 \end{bmatrix}.
$$

The characteristic polynomial of this matrix is $\lambda^2 - (b^2 - 2)\lambda + 1$, which factors as $(\lambda - e^{2\pi i/m_{ij}})(\lambda - e^{-2\pi i/m_{ij}})$, where $i = \sqrt{-1}$.

Suppose that $m_{ij} = \infty$, so $b = B_M(e_i, e_j) = -2$. Then the above matrix is $\begin{bmatrix} 3 & -2 \\ 2 & -1 \end{bmatrix}$, but its minimal polynomial is $(\lambda - 1)^2$. From Linear algebra it follows that the matrix has infinite order and hence $\rho_i \rho_j$ has the infinite order.

Suppose $m_{ij} < \infty$. The restriction of the quadratic form Q_M associated with M to $\mathbb{R}e_i + \mathbb{R}e_j$ is

$$
\begin{aligned}
Q_M(x_i e_i + x_j e_j) &= Q_M(x_i e_i) + Q_M(x_j e_j) + 2B_M(x_i e_i, x_j e_j) \\
&= 2x_i^2 + 2x_j^2 - 4x_i x_j \cos(x/m_{ij}) \\
&= 2(x_i - x_j \cos(\pi/m_{ij}))^2 + 2x_j^2 \sin^2(\pi/m_{ij}),
\end{aligned}
$$

that is, Q_M is a positive definite quadratic form on $\mathbb{R}e_i + \mathbb{R}e_j$, and so $B_M(x, y)$ is an inner product on V. That is, we obtain that $V = (\mathbb{R}e_i + \mathbb{R}e_j) \oplus (e_i^\perp \cap e_j^\perp)$. From the statement (1) it follows that the order of $\rho_i \rho_j$ is the order of its restriction to $\mathbb{R}e_i + \mathbb{R}e_j$. The above formula for the characteristic polynomial of this restriction of $\rho_i \rho_j$ shows that its eigenvalues on that subspace are $e^{2\pi i/m_{ij}}$ and $e^{-2\pi i/m_{ij}}$, which are primitive m_{ij}-th roots of unity. Therefore, the order of $\rho_i \rho_j$ is equal to m_{ij}. $\square$

Theorem 10.1. *Let M be a Coxeter matrix of dimension n.*

(1) *The mapping $w \mapsto \rho_w$ given by $\rho_w = \rho_{r_1} \rho_{r_2} \dots \rho_{r_t}$ for $w = r_1 \dots r_t$ with $r_i \in S$ define a linear transformation of $W(M)$ on V preserving B_M.*

(2) *The mapping from $\{1, \dots, n\}$ to $\{\rho_{s_1}, \dots, \rho_{s_n}\}$ given by $i \mapsto \rho_{s_i}$ is a bijection.*

(3) *The restriction of ρ to the subgroup $\langle s_i, s_j \rangle$ of $W(M)$ is faithful for all $i, j \in \{1, \dots, n\}$.*

Proof. For the statement (1), by Lemma 10.3 (1) and (3), the subspace of $GL(V)$ generated by the $\rho_{s_i}, i \in \{1, \dots, n\}$, satisfies relations of $W(M)$. It follows that the mapping $s_i \mapsto \rho_{s_i}$ determines a unique group homomorphism $\rho \colon W(M) \to GL(V)$ satisfying the given equations. The symmetric bilinear form B_M is preserved by ρ due to Lemma 10.3 (2).

Since $m_{ij} > 1$ for $i, j \in \{1, \dots, n\}$ with $i \neq j$, statements (2) and (3) of this theorem follow from Lemma 10.3 (3). $\square$

Definition 10.6. If $(W(M), S)$ is a Coxeter system of type M, the corresponding linear representation $\rho \colon W \to GL(V)_B$ defined in Theorem 10.1 is called *the reflection representation of W*.

10.1.2 *Faithful representation of Coxeter groups*

In the previous section we constructed the reflection representation of Coxeter groups. To prove that the reflection representation is faithful, we consider the affine space associated with a vector space.

Definition 10.7. Let V be a vector space over a field $\mathbb{F}$. *The affine space* $A(V)$ *associated with V* is the set V with the collection of distinguished subsets, which are called *affine subspaces*, and the relation on the set of affine subspaces, which is called *parallelism*. An affine subspace and the parallelism are defined as follows:

- an affine subspace of $A(V)$ is a coset $v + U$ of a linear subspace U of V;
- two affine subspaces are *parallel* if they are cosets of the same linear subspace of V. This relation is called *parallelism*.

We call elements of $A(V)$ *points*.

It is easy to see that parallelism is an equivalence relation on affine subspaces. We denote two parallel affine subspaces X and Y by $X \parallel Y$. *A real affine space* is an affine space of a real vector space. *An affine line* is a parallel of 1-dimensional linear space. *An hyperplane* is a parallel of $(n-1)$-dimensional linear subspace.

Definition 10.8. For a linear subspace U of V, if U has a dimension d, then we say that $Y = y + U$ is *of the dimension d*. We decree that the empty set has dimension -1.

For a subset X of V, let us define *the affine subspace $\langle X \rangle$ generated by X* to be the smallest of all affine subspaces of $A(V)$ containing X. This definition makes sense, because, if $\{X_i\}_{i \in I}$ is a collection of affine subspaces of $A(V)$, the intersection $\cap_{i \in I} X_i$ is again an affine subspace.

Definition 10.9. A permutation f of V is *an affine automorphism of $A(V)$* if it satisfies the followings:

- For affine subspaces X, Y of $A(V)$, if $X \subset Y$, then $f(X) \subset f(Y)$.
- For affine subspaces X, Y of $A(V)$, if $X \parallel Y$, then $f(X) \parallel f(Y)$.

The group of all affine automorphism of $A(V)$ is denoted by $Aut(A(V))$.

It is easy to see that $f \in GL(V)$ is an affine automorphism of the affine space $A(V)$. That is, $GL(V) \subset Aut(A(V))$ is a subgroup. For $v \in V$, let us define the function $t_v : A(V) \to A(V)$ by $t_v(x) = x + v$ for $x \in V$. Then the function t_v is an affine automorphism of the affine space $A(V)$ and we call t_v *translation by v*. We denote the group of all translations on $A(V)$ by $T(V)$.

Proposition 10.1. *Let V be a vector space. The subgroup of $Aut(A(V))$ generated by $T(V)$ and $GL(V)$ is a semi-direct product with normal subgroup $T(V)$.*

Proof. Since non-trivial translations do not fix the origin 0, but linear transformations of V do fix the origin 0, it is easy to see that $GL(V) \cap T(V) = \{id_V\}$. Suppose that $v \in V$. If $g \in Aut(A(V))$, then

$$gt_v g^{-1}(x) = g(g^{-1}(x) + v) = g(g^{-1}(x)) + g(v) = t_{g(v)}(x),$$

that is, $gT(V)g^{-1} \subset T(V)$ and $T(V)$ is a normal subgroup of $Aut(A(V))$. From the above calculation it follows that $T(V)GL(V) = GL(V)T(V)$, and hence the subgroup of $Aut(A(V))$ generated by $T(V)$ and $GL(V)$ is a semi-direct product of $T(V)$ and $GL(V)$. $\square$

Definition 10.10. The subgroup of $Aut(A(V))$ generated by $T(V)$ and $GL(V)$ is called *the affine linear group of V*, denote by $AGL(V)$.

Definition 10.11. An element of $AGL(V)$ of order 2 fixing an affine hyperplane is called *an affine reflection* on a real affine space $A(V)$. The fixed affine hyperplane is called its *mirror*.

Definition 10.12. Let G be a group acting on a set E. A nonempty subset F of E is called *a prefundamental domain for G* if $F \cap gF = \emptyset$ for all $g \in G \backslash \{1\}$.

For a group action of G on a set E, if there is a prefundamental domain for G, then it follows that the group action is faithful.

Example 10.4. In Fig. 10.3 let us define three reflections on $\mathbb{R}^2$ ρ_l, ρ_m and ρ_n by reflections with respect to lines l, m, n, respectively. That is, the mirrors of ρ_l, ρ_m and ρ_n are the lines l, m, n, respectively. Put $G = \langle \rho_l, \rho_m, \rho_n \rangle \subset AGL(\mathbb{R}^2)$. It is easy to see that G acts on the plane and the area inside of the hexagons, which is not contained in mirrors l, m, n is a prefundamental domain for G.

Theorem 10.2. *Let $\{H_i \mid i \in I\}$ be a family of affine hyperplanes of the affine space $A(V)$ of the real vector space V. For each $i \in I$ let ρ_i be an*

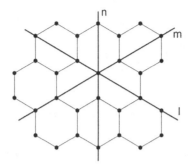

Fig. 10.3 For $G = \langle \rho_l, \rho_m, \rho_n \rangle \subset AGL(\mathbb{R}^2)$ the area inside the hexagon is a prefunda-mental domain for G

affine reflection, for which H_i is the mirror. For each $i \in I$ let A_i be one of the two open half-spaces of $V - H_i$ and $A = \cap_{i \in I} A_i$. Assume that $A \neq \emptyset$. Suppose that for $i \neq j$ $A_{ij} = A_i \cap A_j$ is a prefundamental domain for the subgroup $G_{ij} = \langle \rho_i, \rho_j \rangle$ of $AGL(V)$ generated by ρ_i and ρ_j. Then the following statements hold:

(1) *The subset A is a prefundamental domain for the subgroup G of $AGL(V)$ generated by $\{\rho_i\}_{i \in I}$.*

(2) *$(G, \{\rho_i \mid i \in I\})$ is a Coxeter system of type $M = (m_{ij})_{i,j \in I}$, where $m_{ij} = ord(\rho_i \rho_j)$.*

Proof. Let $M = (m_{ij})_{i,j \in I}$ be the Coxeter matrix, where m_{ij} is the order of $\rho_i \rho_j$. Let $(W(M), S = \{s_i\}_{i \in I})$ be the Coxeter group of type M. Let $G \subset AGL(V)$ be the subgroup generated by $\{\rho_i\}_{i \in I}$. Let $\gamma : W(M) \to G$ be the group homomorphism defined by $\gamma(s_i) = \rho_i$. Note that the group action of $W(M)$ on $A(V)$ is obtained from γ, denote by $wX = \gamma(w)X$ for $w \in W(M)$ and $X \subset A(V)$. Now we claim that for each $k \in \mathbb{N}$ the following holds:

(P_k) For all $i \in I$ and $w \in W(M)$ such that $l(w) \leq k$, either $wA \subset A_i$ or $wA \subset s_i A_i$. If $wA \subset s_i A_i$, then $l(s_i w) = l(w) - 1$.

First, we shall show that our statements (1) and (2) follow from the statement (P_k). For the statement (1) suppose that $A \cap wA \neq \emptyset$ for some $w \in W(M)$. Since $A = \cap_{i \in I} A_i$ it follows that $A_i \cap wA \neq \emptyset$ for all $i \in I$. By definition of A_i, $A_i \cap s_i A_i = A_i \cap \gamma(s_i) A_i = A_i \cap \rho_i A_i = \emptyset$ for each i. From the statement $(P_{l(w)})$ it follows that $wA \subset A_i$ for all $i \in I$ and hence $wA \subset A$. By the assumption $w^{-1}A \cap A \neq \emptyset$ and we obtain that $w^{-1}A \subset A$,

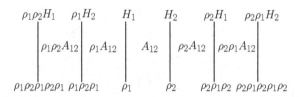

Fig. 10.4   In the case of $H_1 \parallel H_2$

that is, $A \subset wA$. Since $s_i wA = s_i A \subset s_i A_i$, from the statement $(P_{l(s_i w)})$ it follows that $l(w) = l(s_i s_i w) = l(s_i w) - 1$, that is, $l(s_i w) = l(w) + 1$ for all $i \in I$. Since $s_i^2 = 1$, we obtain that w is the identity, in other words, $A \cap wA \neq \emptyset$ if and only if $w = 1$. We proved that the statement (1) holds for the group action of $W(M)$ on $A(V)$ and it follows that the statement (1) holds for the group action of G on $A(V)$. For the statement (2) we may assume that $\gamma(w) = id$. Since $A \cap \gamma(w)A = A \neq \emptyset$, from the proof of the statement (1) it follows that w is the identity in $W(M)$ and hence $ker(\gamma) = \{1\}$. Therefore $W(M)$ and G are isomorphic and the statement (2) is proved.

Now we shall prove that the statement (P_k) is true in several steps.

(Step 1.) Let us consider the case $I = \{1, 2\}$. To prove that the statement (P_k) is true in this case, we distinguish the cases $H_1 \parallel H_2$ and $H_1 \nparallel H_2$. In the case of $H_1 \parallel H_2$ since A_{12} is a prefundamental domain for G_{12} we can see that $A = A_{12} = A_1 \cap A_2$ appears between H_1 and H_2. In particular, $\rho_2 A \subset A_1$, $\rho_1 A \subset A_2$, $\rho_1 \rho_2 A \subset \rho_1 A_1$, $\rho_2 \rho_1 A \subset \rho_2 A_2$ and so on, see Fig. 10.4. Then it is easy to see that for each $k \in \mathbb{N}$, $w \in W(M)$ with $l(w) \leq k$, the domain wA is contained in either A_i or $s_i A_i$ for $i = 1, 2$. Note that if $wA \subset s_i A_i$ for $i = 1, 2$, then $w = s_i w'$. It follows that $l(s_i w) = l(w') = l(w) - 1$ and (P_k) holds for every $k \in \mathbb{N}$. In the case of $H_1 \nparallel H_2$, similarly we can show that (P_k) holds for every $k \in \mathbb{N}$, see Fig. 10.5. See also Fig. 10.3.

In other words, we proved that for all $t \in \{i, j\}$ and $w \in G_{ij}$ such that $l(w) \leq k$, the domain wA is contained in either A_t or $s_t A_t$. If wA is contained in $s_t A_t$, then $l(s_t w) = l(w) - 1$.

Now we prove that (P_k) holds for any I by induction on k. It is obvious that (P_0) holds.

(Step 2.) We claim that if (P_k) holds for $k \geq 1$, then for each $w \in W$ such that $l(w) \leq k$ and $i \in I$, $l(s_i w) < l(w)$ if and only if $wA \subset s_i A_i$. If $l(s_i w) < l(w)$, then $l(w) = l(s_i (s_i w)) = l(s_i w) + 1$. From (P_{k-1}) it follows that $s_i wA \subset A_i$. If $wA \subset s_i A_i$, then from (P_k) follows that $l(s_i w) = $

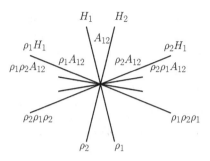

Fig. 10.5   In the case of $H_1 \nparallel H_2$

$l(w) - 1 < l(w)$ and our claim is proved. Similarly, we can show that $l(s_i w) > l(w)$ if and only if $wA \subset A_i$.

(Step 3.) Suppose that (P_k) holds for some $k \geq 0$. Let $w \in W$ such that $l(w) = k + 1$. Then there exist $i \in I$ and $w' \in W$ such that $w = s_i w'$ and $l(w') = k$. Since $w'A \subset A_i$ or $w'A \subset s_i A_i$, $wA = s_i w'A \subset s_i A_i$ or $wA = s_i w'A \subset s_i(s_i A_i) = A_i$. Suppose that $j \in I, j \neq i$. From Lemma 10.3 it follows that there exist $u \in \langle s_i, s_j \rangle$ and left $\langle s_i, s_j \rangle$-reduced $v \in W(M)$ such that $w' = uv$ and $l(w') = l(u) + l(v)$. Since v is left $\langle s_i, s_j \rangle$-reduced, $l(s_i v) > l(v)$ and $l(s_j v) > l(v)$. By Step 2, we obtain that $vA \subset A_{ij}$ and $w'A \subset uA_{ij}$. Then $wA = s_i w'A \subset s_i u A_{ij}$. Note that $l_{ij}(s_i u) \leq l_{ij}(u) + 1 \leq l(w') + 1 = k + 1$, where $l_{ij}(u)$ is the length of u in $\langle s_i, s_j \rangle$. By Step 1, we obtain that for $s_i u \in \langle s_i, s_j \rangle$ $s_i u A_{ij} \subset A_j$ or $s_i u A_{ij} \subset s_j A_j$, and hence $wA \subset s_i u A_{ij} \subset A_j$ or $wA \subset s_i u A_{ij} \subset s_j A_j$. Moreover if $wA \subset s_i u A_{ij} \subset s_j A_j$, then we obtain that

$$l(s_j w) = l(s_j s_i u v) \leq l(s_j s_i u) + l(v)$$
$$\leq l_{ij}(s_j s_i u) + l(w') - l_{ij}(u)$$
$$\leq l_{ij}(s_i u) - 1 + k - l_{ij}(u) \leq k.$$

Since $l(w) = k + 1$, we can see that $l(s_j w) = k = l(w) - 1$ and (P_{k+1}) holds. By induction we proved that (P_k) holds for all $k \in \mathbb{N} \cup \{0\}$. $\qquad\square$

Corollary 10.2. *Assume that the assumptions on Theorem 10.2 hold. Let $\gamma : W(M) \to G$ be the isomorphism in the proof of Theorem 10.2. Then for all $i, j \in I$ and $w \in W(M)$, the followings hold:*

(1) *Either $\gamma(w)A \subset A_i$ and $l(s_i w) = l(w) + 1$, or $\gamma(w)A \subset \gamma(s_i)A_i$ and $l(s_i w) = l(w) - 1$*

(2) *If $i \neq j$, there exists $w_{ij} \in \langle s_i, s_j \rangle$ such that $\gamma(w)A \subset \gamma(w_{ij})A_{ij}$ and $l(w) = l(w_{ij}^{-1} w) + l(w_{ij})$.*

Proof. The statement (1) can be obtained from the statement (P_k) in the proof of Theorem 10.2. For the statement (2), by Lemma 10.2 there exist $w_{ij} \in \langle s_i, s_j \rangle$ and $\langle s_i, s_j \rangle$-reduced element v such that $w = w_{ij}v$ and $l(w) = l(v) + l(w_{ij}) = l(w_{ij}^{-1}w) + l(w_{ij})$. Since l_{ij} and l coincide on $\langle s_i, s_j \rangle$, the statement (2) follows from the statement (1). $\qquad\square$

Definition 10.13. If $\rho \colon G \to GL(V)$ is a linear representation on a vector space V, then the *contragredient representation* ρ^* is defined by
$$\rho_g^* f = (v \mapsto f(\rho_g^{-1} v))$$
for all $f \in V^*$, $g \in G$.

Lemma 10.4. $Ker\rho = Ker\rho^*$. *In particular, ρ is faithful if and only if ρ^*.*

Proof. Let $g \in Ker\rho$. Since $\rho_g = Id$, we obtain that $\rho_g^* f = (v \mapsto f(\rho_g^{-1} v)) = (v \mapsto f(v)) = f$, and hence $g \in Ker\rho^*$. The converse also can be proved analogously. $\qquad\square$

For $v \in V, v \neq 0$, set $A_v = \{x \in V^* \mid x(v) > 0\}$. This is an open half-space in V^*.

Theorem 10.3. *Let $\rho \colon W(M) \to GL(V)$ be the reflection representation. Then, in V^*, the half-spaces $A_i = \{x \in V^* \mid x(e_i) > 0\}$ and the linear transformations $\rho_i^* \in GL(V^*)$ satisfy the following conditions.*

(1) *For each $i \in \{1, 2, \ldots, n\}$ the transformation ρ_i^* is a reflection in V^*, and A_i and $\rho_i^* A_i$ are the half-spaces separated by its mirror.*
(2) *For $i, j \in \{1, 2, \ldots, n\}$ such that $i \neq j$, $A_{ij} = A_i \cap A_j$ is a prefundamental domain for the subgroup G_{ij} of $AGL(V^*)$ generated by ρ_i^* and ρ_j^*.*
(3) *The intersection $\bigcap_{i \in \{1,2,\ldots,n\}} A_i \neq \emptyset$ is a prefundamental domain for the subgroup G acting on V^*.*
(4) *ρ is faithful.*

Proof. For the statement (1) we consider ρ_i^* for $i \in \{1, 2, \ldots, n\}$. For $f \in V^*$ we obtain that
$$\begin{aligned}
\rho_i^* \rho_i^* f(\cdot) &= \rho_i^* (f(\cdot) - f(e_i)B(\cdot, e_i)) \\
&= f(\cdot) - f(e_i)B(\cdot, e_i) - (f(\cdot) - f(e_i)B(\cdot, e_i))(e_i)B(\cdot, e_i) \\
&= f(\cdot) - f(e_i)B(\cdot, e_i) - (f(e_i) - f(e_i)B(e_i, e_i))B(\cdot, e_i) \\
&= f(\cdot) - f(e_i)B(\cdot, e_i) + f(e_i)B(\cdot, e_i) \\
&= f(\cdot),
\end{aligned}$$

and hence ρ_i^* is a reflection. It is easy to see that A_i and $\rho_i^* A_i$ are the half-spaces separated by its mirror $\{x \in V^* \mid x(e_i) = 0\}$.

For the statement (2) let $i, j \in \{1, 2, \ldots, n\}$ such that $i \neq j$. Note that $G_{ij} = \rho^*(\langle s_i, s_j \rangle)$, where $\langle s_i, s_j \rangle$ is the subgroup of $W(M)$ generated by $\{s_i, s_j\}$. We will show that if $A_{ij} \cap \rho_w^* A_{ij} \neq \emptyset$ for $w \in \langle s_i, s_j \rangle$, then $w = 1$. Let U be the linear subspace of V spanned by e_i, e_j. Let us define a surjective map π from V^* to U^* by

$$\pi(f) = f|_U.$$

Then it is easy to see that G_{ij} of $AGL(V^*)$ is preserved with respect to π since $\rho_i^*|_U = \rho_i^*$ and $\rho_j^*|_U = \rho_j^*$. Put $K_v = \{x \in U^* \mid x(v) > 0\}$ for $v \in U$. Then $K_v = \pi A_v$. Put $K_{ij} = K_{e_i} \cap K_{e_j}$. Then we obtain $K_{ij} \cap \sigma_w^* K_{ij} \neq \emptyset$ implies $w = 1$. Suppose that $A_{ij} \cap \rho_w^* A_{ij} \neq \emptyset$. Then for $f \in A_{ij} \cap \rho_w^* A_{ij}$ $f|_U \in \pi A_{ij} = K_{ij}$ and $f|_U \in \pi \rho_w^* A_{ij} = \sigma_w^* K_{ij}$. From the above observation it follows that $w = 1$.

For the statement (3), we consider the intersection $\bigcap_{i \in \{1, 2, \ldots, n\}} A_i$. It is a non-empty set because it contains linear map f such that $f(e_i) = 1$. Recall that from Theorem 10.1 it follows that the restriction of ρ to the subgroup $\langle s_i, s_j \rangle$ of w is faithful and $\rho_i^* \rho_j^*$ has the order m_{ij}. From Theorem 10.2 we obtain that $A = \bigcap_{i \in \{1, 2, \ldots, n\}} A_i$ is a prefundamental domain of ρ^* and $(\rho^*(W(M)), \{\rho_i^* \mid i \in \{1, 2, \ldots, n\}\})$ is a Coxeter system of type M. Therefore ρ^* is faithful, and hence ρ is faithful by Lemma 10.4. $\quad\square$

10.2 Groups G_n^2 and Coxeter groups $C(n, 2)$

Let us consider the group homomorphism l from G_n^2 to S_n given by $l(a_{ij}) = (i, j)$. It is easy to see that there exists a homomorphism m from $C(n, 2)$ to S_n given by $m(b_{ij}) = (i, j)$. We call an element β *pure* if $l(\beta) = 1$. Denote the normal subgroup $ker(l)$ by PG_n^2, which is similar to the pure braid group of the braid group.

Let us start to construct a mapping from G_n^2 to $C(n, 2)$. Let $w = a_{i_1, j_1} a_{i_2, j_2} \ldots a_{i_k, j_k}$ be a word representation from G_n^2. For $s = 0, 1, \ldots, k$ we denote the product of the first s letters of w by w_s presenting w_0 the empty word. For $t = 0, 1, \ldots, k-1$, we define the permutation $\sigma_t = l(w_s)^{-1}$ with $\sigma_0 = id$. In other words, σ_0 is the identical permutation, and $\sigma_t = (i_t, j_t)(i_{t-1}, j_{t-1}) \ldots (i_1, j_1)$. Let us define

$$\bar{w} = b_{\sigma_0(i_1), \sigma_0(j_1)} b_{\sigma_1(i_2), \sigma_1(j_2)} b_{\sigma_2(i_3), \sigma_2(j_3)} \ldots b_{\sigma_{k-1}(i_k), \sigma_{k-1}(j_k)} \in C(n, 2).$$

Example 10.5. For example, let $w = a_{12} a_{13} a_{23} \in G_3^2$. By the above construction we obtain that $w_1 = a_{12}$, $w_2 = a_{13} a_{12}$ and $w_3 = $

$w = a_{12}a_{13}a_{23}$. Then $\sigma_1 = (12), \sigma_2 = (12)(13)$. We obtain $\bar{w} = b_{\sigma_0(1),\sigma_0(2)}b_{\sigma_1(1),\sigma_1(3)}b_{\sigma_2(2),\sigma_2(3)} = b_{12}b_{23}b_{12}$, see Fig. 10.6. Note that $w = a_{12}a_{13}a_{23} = a_{23}a_{13}a_{12} = w'$ in G_3^2 and $\bar{w}' = b_{23}b_{12}b_{23}$. Hence $\bar{w} = \bar{w}'$.

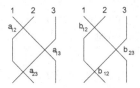

Fig. 10.6   Rewriting word $w = a_{12}a_{13}a_{23}$ from G_3^2

Theorem 10.4. *If two words w and w' generate two equal elements of G_n^2, then the words $\bar{w}$ and $\bar{w}'$ generate equal elements of $C(n,2)$. Moreover, $l(w) = m(\bar{w})^{-1}$.*

Proof. It suffices to check the following three cases:

(1) w' is obtained from w by an addition or a removal of a pair of equal subsequent letters.
(2) w' is obtained from w by applying the relation $a_{ij}a_{kl} = a_{kl}a_{ij}$ for distinct i, j, k, l.
(3) w' is obtained from w by applying the relation $a_{ij}a_{ik}a_{jk} = a_{jk}a_{ik}a_{ij}$ for distinct i, j, k.

In the first case, at some step s we have two letters $a_{ij}a_{ij}$ and a permutation σ_s. When rewriting the word w we obtain that $b_{\sigma_{s-1}(i)\sigma_{s-1}(j)}b_{\sigma_s(i)\sigma_s(j)}$, which corresponds to $a_{ij}a_{ij}$. Since $\sigma_s(i) = \sigma_{s-1}(j)$, $\sigma_s(j) = \sigma_{s-1}(i)$ and $\sigma_{s-1} = \sigma_{s+1}$, $b_{\sigma_{s-1}(i)\sigma_{s-1}(j)}b_{\sigma_s(i)\sigma_s(j)} = b_{\sigma_{s-1}(i)\sigma_{s-1}(j)}b_{\sigma_{s-1}(i)\sigma_{s-1}(j)} = 1$ and hence $\bar{w}'$ is obtained from $\bar{w}$ by an addition or a removal of a pair of equal subsequent letters $b_{st}b_{st}$.

In the second case, if in w at s-th and $(s + 1)$-th steps we have two commuting letters $a_{ij}a_{kl}$, then the corresponding letters in w' appear at s-th and $(s + 1)$-th steps and will commute as well. Since the values of the permutation from k and l does not depend on the values of the permutation from i and j, $\bar{w} = \bar{w}'$ when the corresponding transposition is applied.

Let us consider the third case. Assume that the letters a_{ij}, a_{ik}, a_{jk} are positioned at s-th, $(s + 1)$-th and $(s + 2)$-th steps. By simple observation we obtain that $\sigma_s(i) = \sigma_{s-1}(j)$ and $\sigma_s(j) = \sigma_{s-1}(i)$. Hence it follows that

$$\sigma_{s+1}(j) = \sigma_s(j) = \sigma_{s-1}(i),$$

$$\sigma_{s+1}(k) = \sigma_s(i) = \sigma_{s-1}(j),$$

and

$$\sigma_{s+1}(i) = \sigma_s(k) = \sigma_{s-1}(k).$$

Then the corresponding three letters to $a_{ij}a_{ik}a_{jk}$ in $\bar{w}$ is

$$b_{\sigma_{s-1}(i)\sigma_{s-1}(j)}b_{\sigma_{s-1}(j)\sigma_{s-1}(k)}b_{\sigma_{s-1}(i)\sigma_{s-1}(j)}.$$

When we consider three letters $a_{jk}a_{ik}a_{ij}$ in the word w'

$$\sigma'_{s+1}(k) = \sigma'_s(i) = \sigma'_{s-1}(i),$$

$$\sigma'_{s+1}(i) = \sigma'_s(k) = \sigma'_{s-1}(j),$$

and

$$\sigma'_{s+1}(j) = \sigma'_s(j) = \sigma'_{s-1}(k).$$

Note that $\sigma'_{s-1} = \sigma_{s-1}$. Hence the corresponding three letters to $a_{jk}a_{ik}a_{ij}$ in $\bar{w}'$ is $b_{\sigma_{s-1}(j)\sigma_{s-1}(k)}b_{\sigma_{s-1}(i)\sigma_{s-1}(j)}b_{\sigma_{s-1}(j)\sigma_{s-1}(k)}$. Since the relation $b_{ij}b_{ik}b_{ij} = b_{ik}b_{ij}b_{ik}$ holds in $C(n,2)$, we obtain the equality $\bar{w} = \bar{w}'$. $\qquad\square$

Chapter 11

Realisation of Spaces with G_n^k Action. The braid groups for higher-dimensional spaces

The aim of the subsequent chapters is to construct spaces where G_n^k (or a finite index subgroup of it) acts faithfully.

On this way, we shall define the braid groups for higher dimensional spaces (Part 4). This gives rise to a definition for *braid groups* for arbitrary Euclidean and real projective spaces — the objects for study of its own. Note that the braid groups for manifolds other than $\mathbb{R}^n$, $\mathbb{R}P^n$ will be defined in Part 4.

We continue by constructing invariants of arbitrary topological spaces using the G_n^k groups.

11.1 Realisation of the groups G_{k+1}^k

In the present section we tackle the problem of the realisation of the groups G_n^k for the special case $n = k + 1$ (the general case is handled in the subsequent section).

More precisely, we shall prove the following

Theorem 11.1. *There is a subgroup $\tilde{G}_{k+1}^k$ of the group G_{k+1}^k of index 2^{k-1} which is isomorphic to $\pi_1(\tilde{C}'_{k+1}(\mathbb{R}P^{k-1}))$.*

The space $\tilde{C}'_n$ and the group $\tilde{G}_{k+1}^k$ will be defined later in this chapter. The simplest case of the above Theorem is

Theorem 11.2. *The group $\tilde{G}_4^3$ (which is a finite index subgroup) is isomorphic to $\pi_1(FBr_4(\mathbb{R}P^2))$, the 4-strand braid group on $\mathbb{R}P^2$ with two point fixed.*

This theorem gives a solution to the word problem in the group G_4^3.

11.1.1   *Preliminary definitions*

We begin with the definition of spaces $C'_n(\mathbb{R}P^{k-1})$ and maps from the corresponding fundamental groups to the groups G_n^k.

Let us fix a pair of natural numbers $n > k$. A point in $C'_n(\mathbb{R}P^{k-1})$ is an ordered set of n pairwise distinct points in $\mathbb{R}P^{k-1}$ such that any $(k-1)$ of them are in general position. Thus, for instance, if $k = 3$, then the only condition is that these points are pairwise distinct. For $k = 4$ for points $x_1, \ldots, x_n$ in $\mathbb{R}P^3$ we impose the condition that no three of them belong to the same line (though some four are allowed to belong to the same plane), and for $k = 5$ a point in $C'_n(\mathbb{R}P^4)$ is a set of ordered n points in $\mathbb{R}P^4$, with no four of them belonging to the same 2-plane.

Let us use the projective coordinates $(a_1 : a_2 : \cdots : a_k)$ in $\mathbb{R}P^{k-1}$ and let us fix the following $k-1$ points in general position, $y_1, y_2, \ldots, y_{k-1} \in \mathbb{R}P^{k-1}$, where $a_i(y_j) = \delta_i^j$. Let us define the subspace $\tilde{C}'_n(\mathbb{R}P^{k-1})$ taking those n-tuples of points $x_1, \ldots, x_n \in \mathbb{R}P^{k-1}$ for which $x_i = y_i$ for $i = 1, \ldots, k-1$.

We say that a point $x \in C'_n(\mathbb{R}P^{k-1})$ is *singular*, if the set of points $x = (x_1, \ldots, x_n)$, corresponding to x, contains some subset of k points lying on the same $(k-2)$-plane.

Let us fix two non-singular points $x, x' \in C'_n(\mathbb{R}P^{k-1})$. We shall consider smooth paths $\gamma_{x,x'} : [0,1] \to C'_n(\mathbb{R}P^{k-1})$. For each such path there are values t for which $\gamma_{x,x'}(t)$ is not in the general position (some k points belong to the same $(k-2)$-plane). We call these values $t \in [0,1]$ *singular*.

On the path γ, we say that a moment t of passing through a singular point x, corresponding to the set $x_{i_1}, \ldots, x_{i_k}$, is *transverse* (or stable) if for any sufficiently small perturbation $\tilde{\gamma}$ of the path γ, in the neighbourhood of the moment t there exists exactly one time moment t' corresponding to some set of points $x_{j_1}, \ldots, x_{j_k}$ not in general position.

Definition 11.1. We say that a path is *good and transverse* if the following holds:

(1) The set of singular values t is finite;
(2) For every singular value $t = t_l$ corresponding to n points representing $\gamma_{x,x'}(t_l)$, there exists only one subset of k points belonging to a $(k-2)$-plane;
(3) Each singular value is *transverse*.

Definition 11.2. We say that the path without singular values is *void*.

We shall call such paths *braids*. We say that a braid whose ends x, x' coincide with respect to the order, is *pure*. We say that two braids γ, γ' with endpoints x, x' are *isotopic* if there exists a continuous family $\gamma_{x,x'}^s, s \in [0,1]$ of smooth paths with fixed ends such that $\gamma_{x,x'}^0 = \gamma, \gamma_{x,x'}^1 = \gamma'$. By a small perturbation, any path can be made good and stable (if endpoints are generic, we may also require that the endpoints remain fixed).

Definition 11.3. A path from x to x' is called a *braid* if the points representing x are the same as those representing x' (possibly, in different orders); if x coincides with x' with respect to order, then such a braid is called *pure*.

There is an obvious concatenation structure on the set of braids: for paths $\gamma_{x,x'}$ and $\gamma'_{x',x''}$, the concatenation is defined as a path $\gamma''_{x,x''}$ such that $\gamma''(t) = \gamma(2t)$ for $t \in [0, \frac{1}{2}]$ and $\gamma''(t) = \gamma'(2t-1)$ for $t \in [\frac{1}{2}, 1]$; this path can be smoothed in the neighbourhood of $t = \frac{1}{2}$; the result of such smoothing is well defined up to isotopy.

Thus, the sets of braids and pure braids (for fixed x) admit a group structure. This latter group is certainly isomorphic to the fundamental group $\pi_1(C'_n(\mathbb{R}P^{k-1}))$. The former group is isomorphic to the fundamental group of the quotient space by the action of the permutation group.

11.1.2   The realisability of G_{k+1}^k

The main idea of the proof of Theorem 11.1 is to associate with every word in G_{k+1}^k a braid in $\tilde{C}'_{k+1}(\mathbb{R}P^{k-1})$.

Let us start with the main construction from [Manturov, 2015d].

With each good and stable braid from $PB_n(\mathbb{R}P^{k-1})$ we associate an element of the group G_n^k as follows. We enumerate all singular values of our path $0 < t_1 < \cdots < t_l < 1$ (we assume than 0 and 1 are not singular). For each singular value t_p we have a set m_p of k indices corresponding to the numbers of points which are not in general position. With this value we associate the letter a_{m_p}. With the whole path γ (braid) we associate the product $f(\gamma) = a_{m_1} \ldots a_{m_l}$.

Theorem 11.3. [Manturov, 2015d] *The map f takes isotopic braids to equal elements of the group G_n^k. For pure braids, the map f is a homomorphism $f : \pi_1(C'_n(\mathbb{R}P^{k-1})) \to G_n^k$.*

Now we claim that

Every word from G_{k+1}^k can be realised by a path of the above form.

Note that if we replace $\mathbb{R}P^{k-1}$ with $\mathbb{R}^{k-1}$, the corresponding statement will fail. Namely, starting with the configuration of four points, $x_i, i = 1, 2, 3, 4$, where x_1, x_2, x_3 form a triangle and x_4 lies inside this triangle, we see that any path starting from this configuration will lead to a word starting from $a_{124}, a_{134}, a_{234}$ but not from a_{123}. In some sense the point 4 is "locked" and the points are not in the same position.

Fig. 11.1   The "locked" position for the move a_{123}

The following well known theorem (see, e.g., [Berrick, Cohen, Wong and Wu, 2005]) plays a crucial role in the construction.

Theorem 11.4. *For any two sets of $k + 1$ points in general position in $\mathbb{R}P^{k-1}$, $(x_1, \ldots, x_{k+1})$ and $(y_1, \ldots, y_{k+1})$ there is an action of $PGL(k, \mathbb{R})$ taking all x_i to y_i.*

For us this will mean that there is no difference between all possible "non-degenerate starting positions" for $k + 1$ points in $\mathbb{R}P^{k-1}$.

We shall consider paths in $\tilde{C}'_{k+1}(\mathbb{R}P^{k-1})$ starting and ending in two points from this set (possibly, the same).

We shall denote homogeneous coordinates in $\mathbb{R}P^{k-1}$ by $(a_1 : \cdots : a_k)$ in contrast to points (which we denote by $(x_1, \ldots, x_{k+1})$).

11.1.3   *Constructing a braid from a word in G_{k+1}^k*

Our main goal is to construct a braid from a word. To this end, we need a base point for the braid. For the sake of convenience, we shall use not one, but rather 2^{k-1} reference points. For the first k points $y_1 = (1 : 0 : \cdots : 0), \ldots, y_k = (0 : \cdots : 0 : 1)$ fixed, we will have 2^{k-1} possibilities for the choice of the last point. Namely, let us consider all possible strings of length k of ± 1 with the last coordinate $+1$:

$$(1 : 1 : \cdots : 1 : 1), (1 : \cdots : 1 : -1 : 1), \ldots, (-1 : -1 : \cdots : -1 : 1)$$

with $a_k = +1$. We shall denote these points by y_s where s records the first $(k-1)$ coordinates of the point.

Now, for each string s of length $k - 1$ of ± 1, we set $z_s = (y_1, y_2, \ldots, y_k, y_s)$.

The following lemma is evident.

Lemma 11.1. *For every point $z \in \mathbb{R}P^{k-1}$ with projective coordinates $(a_1(z) : \cdots : a_{k-1}(z) : 1)$, let $\tilde{z} = (sign(a_1(z)) : sign(a_2(z)) : \cdots : sign(a_{k-1}(z)) : 1)$. Then there is a path between $(y_1, \ldots, y_k, z)$ and $(y_1, \ldots, y_k, \tilde{z})$ in $\tilde{C}'_{k+1}(\mathbb{R}P^k)$ with the first points $y_1, \ldots, y_k$ fixed, and the corresponding path in $\tilde{C}'_{k+1}$ is void.*

Proof. Indeed, it suffices just to connect z to $\tilde{z}$ by the shortest geodesic. $\qquad\square$

From this we easily deduce the following

Lemma 11.2. *Every point $y \in \tilde{C}'_{k+1}(\mathbb{R}P^{k-1})$ can be connected by a void path to $(y_1, \ldots, y_k, y_s)$ for some s.*

Proof. Indeed, the void path can be constructed in two steps. At the first step, we construct a path which moves both y_k and y_{k+1}, so that y_k becomes $(0 : \cdots : 0 : 1)$, and at the second step, we use Lemma 11.1. To realise the first step, we just use linear maps which keep the hyperplane $a_k = 0$ fixed. $\qquad\square$

The lemma below shows that the path mentioned in Lemma 11.2 is unique up to homotopy.

Lemma 11.3. *Let γ be a closed path in $\tilde{C}'_{k+1}(\mathbb{R}P^{k-1})$ such that the word $f(\gamma)$ is empty. Then γ is homotopic to the trivial braid.*

Proof. In $\tilde{C}'_{k+1}$, we deal with the motion of points, where all but x_k, x_{k+1} are fixed.

Consider the projective hyperplane $\mathcal{P}_1$ passing through $x_1, \ldots, x_{k-1}$ given by the equation $a_k = 0$.

We know that none of the points x_k, x_{k+1} is allowed to belong to $\mathcal{P}_1$. Hence, we may fix the last coordinate $a_k(x_k) = a_k(x_{k+1}) = 1$.

Now, we may pass to the affine coordinates of these two points (still to be denoted by $a_1, \ldots, a_k$).

Now, the condition $\forall i = 1, \ldots, k - 1 : a_j(x_k) \neq a_j(x_{k+1})$ follows from the fact that the points $x_1, \ldots, \hat{x}_j, \ldots, x_{k+1}$ are generic.

This means that $\forall i = 1, \ldots, k-1$ the sign of $a_i(x_k) - a_i(x_{k+1})$ remains fixed.

Now, the motion of points x_k, x_{k+1} is determined by their coordinates $a_1, \ldots, a_j$, and since their signs are fixed, the configuration space for this motions is simply connected.

This means that the loop γ is described by a loop in a two-dimensional simply connected space. $\qquad\square$

Our next strategy is as follows. Let us start from $(y_1, \ldots, y_k, y_{1,\ldots,1})$. Having a word in G_{k+1}^k, we shall associate with this word a path in $\tilde{C}'_{k+1}(\mathbb{R}P^{k-1})$. After each letter, we shall get to $(y_1, \ldots, y_k, y_s)$ for some s.

After making the final step, one can calculate the coordinate of the $(k+1)$-th points. They will be governed by Lemma 11.6 (see ahead). As we shall see later, those words we have to check for the solution of the word problem in G_{k+1}^k, will lead us to closed paths, i.e., pure braids.

Let us be more detailed.

Lemma 11.4. *Let y be a non-singular set of points in $\mathbb{R}P^{k-1}$. We shall treat y as an unordered set. Then for every (unordered) set of k numbers $i_1, i_2, \ldots, i_k \in [k+1]$, there exists a path $y_{i_1 \ldots i_k}(t) = y(t)$ in $C'_{k+1}(\mathbb{R}P^{k-1})$, having $y(0) = y(1) = y$ (as unordered sets) as the starting point and the final point and with only one singular moment corresponding to the numbers $i_1, \ldots, i_k$ which encode the points not in general position; moreover, we may assume that at this moment all points except i_1, are fixed during the path.*

Moreover, the set of paths possessing this property is connected: any other path $\tilde{y}(t)$, possessing the above properties, is homotopic to $y(t)$ in this class.

Proof. Indeed, for the first statement of Lemma, it suffices to construct a path for some initial position of points and then apply Theorem 11.4.

For the second statement, let us take two different paths γ_1 and γ_2 satisfying the conditions of the Lemma. By a small perturbation, we may assume that for both of them, $t = \frac{1}{2}$ is a singular moment with the same position of y_{i_1}.

Now, we can contract the loop formed by $\gamma_1|_{t\in[\frac{1}{2},1]}$ and the inverse of $\gamma_2|_{t\in[\frac{1}{2},1]}$ by using Lemma 11.3 as this is a small perturbation of a void braid. We are left with $\gamma_1|_{t\in[0,\frac{1}{2}]}$ and the inverse of $\gamma_2|_{t\in[0,\frac{1}{2}]}$ which is contractible by Lemma 11.3 again. $\qquad\square$

Remark 11.1. Note that in the above lemma, we deal with the space $C'_{k+1}(\mathbb{R}P^{k-1})$, not with $\tilde{C}'_{k+1}(\mathbb{R}P^{k-1})$. On the other hand, we may always choose $i_1 \in \{k, k+1\}$; hence, the path in question can be chosen in $\tilde{C}'_{k+1}(\mathbb{R}P^{k-1})$.

Now, for every subset $m \subset [n], Card(m) = k+1$ we can create a path p_m starting from any of the base points listed above and ending at the corresponding basepoints.

Now, we construct our path step-by step by applying Lemma 11.4 and returning to some of base points by using Lemma 11.2.

From [Manturov, 2015d], we can easily get the following

Lemma 11.5. *Let $i_1, \ldots, i_{k+1}$ be some permutation of $1, \ldots, k+1$. Then the concatenation of paths $p_{i_1 i_2 \ldots i_k} p_{i_1 \ldots i_{k-1} i_k} \cdots p_{i_2 i_3 \ldots i_k}$ is homotopic to the concatenation of paths in the inverse order*

$$p_{i_2 i_3 \ldots i_{k+1}} \cdots p_{i_1 i_3 i_4 \ldots i_{k+1}} p_{i_1 i_2 \ldots i_k}.$$

Proof. Indeed, in [Manturov, 2015d], some homotopy corresponding to the above mentioned relation corresponding to *some* permutation is discussed. However, since all basepoints are similar to each other as discussed above, we can transform the homotopy from [Manturov, 2015d] to the homotopy for any permutation. □

Lemma 11.6. *For the path starting from the point $(y_1, \ldots, y_k, y_s)$ constructed as in Lemma 11.4 for the set of indices $[k+1] \setminus j$, we get to the point $(y_1, \ldots, y_k, y_{s'})$ such that:*

(1) if $j = 1, \ldots, k$, then s' differs from s only in coordinate a_j;
(2) if $j = k+1$, all coordinates of s' differ from those coordinates of s by sign.

Denote the map from words in G_{k+1}^k to paths between basepoints by g.

By construction, we see that for every word w we have $f(g(w)) = w \in G_{k+1}^k$.

Now, we define the group $\tilde{G}_{k+1}^k$ as the subgroup of G_{k+1}^k which is taken by g to *pure braids*, i.e., to those paths with coinciding initial and final points. From Lemma 11.6, we see that this is a subgroup of index $(k-1)$: there are exactly $(k-1)$ coordinates.

Let us pass to the proof of Theorem 11.2. Our next goal is to see that equal words can originate only from homotopic paths.

To this end, we shall first show that the map f from Theorem 11.3 is an isomorphism for $n = k + 1$. To perform this goal, we should construct the inverse map $g : \tilde{G}^k_{k+1} \to \pi_1(\tilde{C}'_{k+1}(\mathbb{R}P^{k-1}))$.

Note that for $k = 3$ we deal with the pure braids $PB_4(\mathbb{R}P^2)$.

Let us fix a point $x \in C'_4(\mathbb{R}P^2)$. With each generator $a_m, m \subset [n], Card(m) = k$ we associate a path $g(m) = y_m(t)$, described in Lemma 11.4. This path is not a braid: we can return to any of the 2^{k-1} base points. However, once we take the concatenation of paths corresponding to $\tilde{G}^k_{k+1}$, we get a braid.

By definition of the map f, we have $f(g(a_m)) = a_m$. Thus, we have chosen that the map f is a surjection.

Now, let us prove that the kernel of the map f is trivial. Indeed, assume there is a pure braid γ such that $f(\gamma) = 1 \in G^k_{k+1}$. We assume that γ is good and stable. If this path has l critical points, then we have the word corresponding to it $a_{m_1} \ldots a_{m_l} \in G^k_{k+1}$.

Let us perform the transformation $f(\gamma) \to 1$ by applying the relations of G^k_{k+1} to it and transforming the path γ respectively. For each relation of the sort $a_m a_m = 1$ for a generator a_m of the group G^k_{k+1}, we see that the path γ contains two segments whose concatenation is homotopic to the trivial loop (as follows from the second statement of Lemma 11.4).

Whenever we have a relation of length $2k + 2$ in the group G^k_{k+1}, we use the Lemma 11.5 to perform the homotopy of the loops.

Thus, we have proved that if the word $f(\gamma)$ corresponding to a braid $\gamma \in G^k_{k+1}$ is equal to 1 in G^k_{k+1} then the braid γ is isotopic to a braid γ' such that the word corresponding to it is empty. Now, by Lemma 11.3, this braid is homotopic to the trivial braid.

11.1.4   *The group H_k and the algebraic lemma*

The aim of the present section is to reduce the word problem in G^k_{k+1} to the word problem in a certain subgroup of it, denoted by H_k.

Let us rename all generators of G^k_{k+1} lexicographically:

$$b_1 = a_{1,2,\ldots,k}, \ldots, b_{k+1} = a_{2,3,\ldots,k+1}.$$

Let H_k be the subgroup of G^k_{k+1} consisting of all elements $x \in G^k_{k+1}$ that can be represented by words with no occurrences of the last letter b_{k+1}.

Our task is to understand whether a word in G_{k+1}^k represents an element in H_k. That question shall be called the *generalised word problem for H_k in G_{k+1}^k*. In general, let us introduce the following definition.

Definition 11.4. Let G be a group and $H \subset G$ be its subgroup. The problem of algorithmic determination whether a given word $w \in G$ belongs to the subgroup H is called the *generalised word problem for H in G*.

To solve this problem, we recall the map defined in [Manturov and Nikonov, 2015]. Consider the group $F_{k-1} = \mathbb{Z}_2^{*2^{k-1}} = \langle c_m | c_m^2 = 1 \rangle$, where all generators c_m are indexed by $(k-1)$-element strings of 0 and 1 with only relations being that the square of each generator is equal to 1. We shall construct a map[1] from G_{k+1}^k to F_{k-1} as follows.

Take a word w in generators of G_{k+1}^k and list all occurrences of the last letter $b_{k+1} = a_{2,\dots,k+1}$ in this word. With any such occurrence we first associate the string of indices $0, 1$ of length k. The j-th index is the number of letters b_j preceding this occurrence of b_{k+1} modulo 2. Thus, we get a string of length k for each occurrence.

Let us consider "opposite pairs" of strings $(x_1, \dots, x_k) \sim (x_1 + 1, \dots, x_k + 1)$ as equal. Now, we may think that the last (k-th) element of our string is always 0, so, we can restrict ourselves with $(x_1, \dots, x_{k-1}, 0)$. Such a string of length $k-1$ is called the *index* of the occurrence of b_{k+1}.

Having this done, we associate with each occurrence of b_{k+1} having index m the generator c_m of F_{k-1}. With the word w, we associate the word $f(w)$ equal to the product of all generators c_m in order.

In [Manturov and Nikonov, 2015], the following Lemma is proved:

Lemma 11.7. *The map $f : G_{k+1}^k \to F_{k-1}$ is well defined.*

Now, let us prove the following crucial

Lemma 11.8. *If $f(w) = 1$, then $w \in H_k$.*

In other words, the free group F_{k-1} yields the only obstruction for an element from G_{k+1}^k to have a presentation with no occurrence of the last letter.

Proof. Let w be a word such that $f(w) = 1$. If $f(w)$ is empty, then there is nothing to prove. Otherwise w contains two "adjacent" occurrences of the same index. This means that $w = Ab_{k+1}Bb_{k+1}C$, where A and C are

---

[1]This map becomes a homomorphism when restricted to a finite index subgroup.

some words, and B contains no occurrences of b_{k+1} and the number of occurrences of $b_1, b_2, \ldots, b_k$ in B are of the same parity.

Our aim is to show that w is equal to a word with smaller number of b_{k+1} in G_{k+1}^k. Then we will be able to induct on the number of b_{k+1} until we get a word without b_{k+1}.

Thus, it suffices for us to show that $b_{k+1} B b_{k+1}$ is equal to a word from H_k. We shall induct on the length of B. Without loss of generality, we may assume that B is reduced, i.e., it does not contain adjacent $b_j b_j$.

Let us read the word B from the very beginning $B = b_{i_1} b_{i_2} \ldots$. If all elements $i_1, i_2, \ldots$ are distinct, then, having in mind that the number of occurrences of all generators in B should be of the same parity, we conclude that $b_{k+1} B = B^{-1} b_{k+1}$, hence $b_{k+1} B b_{k+1} = B^{-1} b_{k+1} b_{k+1} = B^{-1}$ is a word without occurrences of b_{k+1}.

Now assume $i_1 = i_p$ (the situation when the first repetition is for $i_j = i_p, 1 < j < p$ is handled in the same way). Then we have $b_{k+1} B = b_{k+1} b_{i_1} \ldots b_{i_{p-1}} b_{i_1} B'$. Now we collect all indices distinct from $i_1, \ldots, i_{p-1}, k+1$ and write the word P containing exactly one generator for each of these indices (the order does not matter). Then the word $W = P b_{k+1} b_{i_1} \ldots b_{i_{p-1}}$ contains any letter exactly once and we can invert the word W as follows: $W^{-1} = b_{i_{p-1}} \ldots b_{i_1} b_{k+1} P^{-1}$. Thus, $b_{k+1} B = P^{-1} (P b_{k+1} b_{i_1} \ldots b_{i_{p-1}}) b_{i_1} B' = P^{-1} b_{i_{p-1}} \ldots b_{i_1} b_{k+1} P^{-1} b_{i_1} B'$.

We know that the letters in P (hence, those in P^{-1}) do not contain b_{i_1}. Thus, the word $P^{-1} b_{i_1}$ consists of distinct letters. Now we perform the same trick: create the word $Q = b_{i_2} b_{i_3} \ldots b_{i_{p-1}}$ consisting of remaining letters from $\{1, \ldots, k\}$, we get:

$$P^{-1} b_{i_{p-1}} \ldots b_{i_1} b_{k+1} P^{-1} b_{i_1} B'$$

$$= P^{-1} b_{i_{p-1}} \ldots b_{i_1} Q Q^{-1} b_{k+1} P^{-1} b_{i_1} B'$$

$$= P^{-1} b_{i_{p-1}} \ldots b_{i_1} Q b_{i_1} P b_{k+1} Q B'.$$

Thus, we have moved b_{k+1} to the right and between the two occurrences of the letter b_{k+1}, we replaced $b_{i_1} \ldots, b_{i_{p-1}} b_{i_1}$ with just $b_{i_2} \ldots b_{i_{p-1}}$, thus, we have shortened the distance between the two adjacent occurrences of b_{k+1}.

Arguing as above, we will finally cancel these two letters b_{k+1} and perform the induction step. $\qquad \square$

Theorem 11.5. *Generalised word problem for H_k in G_{k+1}^k is solvable.*

Proof. Indeed, having a word w in generators of G_{k+1}^k, we can look at the image of this word by the map f. If it is not equal to 1, then, from [Manturov and Nikonov, 2015], it follows that w is non-trivial, otherwise we can construct a word $\tilde{w}$ in H_k equal to w in G_{k+1}^k. $\quad\square$

Remark 11.2. A shorter proof of Theorem 11.5 based on the same ideas was communicated to the authors by A.A. Klyachko. We take the subgroup K_k of G_{k+1}^k generated by products $B_\sigma = b_{\sigma_1} \ldots b_{\sigma_k}$ for all permutations σ of k indices. This group K_k contains the commutator of H_k and is a normal subgroup in G_{k+1}^k.

Moreover, the quotient group G_{k+1}^k/K_k is naturally isomorphic to the free product $(H_k/K_k) * \langle b_{k+1}\rangle$. Hence, the problem whether an element of G_{k+1}^k/K_k belongs to H_k/K_k is trivially solved, which solves the initial problem because of the normality of K_k in G_{k+1}^k.

Certainly, to be able to solve the word problem in H_k, one needs to know a presentation for H_k. It is natural to take $b_1, \ldots, b_k$ for generators of H_k. Obviously, they satisfy the relations $b_j^2 = 1$ for every j.

To understand the remaining relations for different k, which shall be of the form $[(b_i b_j)^2, (b_k b_l)^2] = 1$ for all distinct 4-tuples of indices i, j, k, l, we shall need geometrical arguments.

We have completely constructed the isomorphism between the (finite index subgroup of) the group G_{k+1}^k and a fundamental group of some configuration space.

This completely solves the word problem in G_4^3 for braid groups in projective spaces are very well studied. The same can be said about the conjugacy problem in $\tilde{G}_4^3$.

Besides, we have seen that the word problem for the case of general G_{k+1}^k can be reduced to the case of H_k.

It would be very interesting to compare the approach of G_n^k with various generalizations of braid groups, e.g., Manin–Schechtmann groups [Manin and Schechtmann, 1990; Kapranov and Voevodsky, 1994].

11.2 Realisation of G_n^k for $n \neq k+1$. Partial flag varieties

In the present section, we shall tackle the question, *how to represent all words from G_n^k by paths in some configuration space if $n > k+1$.*

The answer will be: for whatever k, we shall deal with configurations of n points in $\mathbb{R}P^{n-2}$, and the condition will be: k points are not in the general position.

An obvious objection by the reader should be: *this condition is not of codimension* 1.

Certainly, this is true, and to handle the problem, let us be more detailed.

11.2.1 *A simple partial case*

We begin with a simple example: the group G_4^2.

By $\tilde{C}(\mathbb{R}P^2, 4)$ (for shortness, $\tilde{C}_4^2$) we denote the following configuration space. A point of $\tilde{C}_4^2$ is a collection of four points $z_1, z_2, z_3, z_4 \in \mathbb{R}P^2$ (not necessarily pairwise distinct) together with six selected (projective) lines l_{ij}, passing through z_i, z_j. Here all l_{ij} belong to $\mathbb{R}P^1$.

Note that there is a natural "line forgetting map" $f : \tilde{C}_4^2 \to (\mathbb{R}P^2)^4$ which takes (z_i, l_{ij}) to (z_i).

Note that the restriction of this map to the subspace where the four points are pairwise distinct is a bijection. If there is exactly one pair of coinciding points, say, $z_1 = z_2$, then this map is a fibration with fibre $\mathbb{R}P^1$.

Now, let us make a restriction to the above configuration space: by $\tilde{C}'(\mathbb{R}P^2, 4)$ (for shortness, $\tilde{C}'_4^2$) we denote those points $(z_i, \ldots, l_{ij}, \ldots)$ of $\tilde{C}_4^2$ where no three distinct points z_i, z_j, z_k are collinear.

A point in $\tilde{C}'(\mathbb{R}P^2, 4)$ with $z_i = z_j$ for some $i \neq j$ will be called *non-generic*. Note that the set of all non-generic points in $\tilde{C}'(\mathbb{R}P^2, 4)$ has codimension 1.

Fix two generic points $p, p^* \in \tilde{C}'(\mathbb{R}P^2, 4)$. A *generic path* from p to p^* is a smooth path in the space $\tilde{C}'^k_n$ connecting the points p and p^* and passing through a finite number of non-generic points.

With any generic path γ from p to p^* we associate a word in letters $a_{12}, a_{13}, a_{14}, a_{23}, a_{24}, a_{34}$ as follows: when passing through a point where z_i coincides with z_j, we write down a_{ij}.

Thus we get a word $w(\gamma)$ in a_{ij} and the corresponding element of G_4^2.

Theorem 11.6. *The above element of G_4^2 depends only on the homotopy class of the path γ.*

In particular, if $p = p^*$, we get a map from $\pi_1(\tilde{C}'_4^2)$ to G_4^2.

Certainly, we have already had a map from homotopy groups of "some" configuration spaces to G_n^k.

The main advantage here is that there is an "inverse" operation: having a word w in letters of G_4^2, we can restore a certain path γ_w which gives rise to such a word.

Let w be such a word. We put some four generic points z_1, z_2, z_3, z_4 in $\mathbb{R}P^2$.

The point p will be $(z_1, z_2, z_3, z_4, l_{12}, l_{13}, l_{14}, l_{23}, l_{24}, l_{34})$ (note that l_{ij} are automatically restored from z_i).

Take the first letter $w_1 = a_{ij}$ of w.

In the path γ corresponding to a_{ij} all points z_k except z_1 will be fixed. Connect z_1 with z_2 by the shortest geodesic; let z_1' be a point such that $z_1 z_1'$ is a geodesic segment containing z_2. In the path γ the point z_1 will go through z_2 to z_1'; l_{12} will stay the same as an element of $\mathbb{R}P^1$; the motion of other $l_{1j}, j \neq 2$ is uniquely restored by the motion of z_j.

After that we get to a generic point $(z_1', z_2, z_3, z_4, l_{12}', \dots)$.

Take the second letter $w_2 = a_{kl}$ of the word w and proceed as above.

By construction, we get the following

Statement 11.1. $w(\gamma_w) = w$.

11.2.2 General construction

Fix $n > k + 1$. We will follow the same pattern as used above in the case of G_4^2.

Consider the configuration space $\tilde{C}_n^k$. A point of the space $\tilde{C}_n^k$ is a collection of n points $z_1, \dots, z_n \in \mathbb{R}P^{n-2}$ such that any $k-1$ of them are in general position, together with a collection of $(k-1)$-planes $P_{i_1 \dots i_k}$ passing through the points $z_{i_1}, \dots, z_{i_k}$ for each k–set $(i_1, \dots, i_k)$, $i_j \in \{1, \dots, n\}$. Furthermore, we define a restriction of this space: by $\tilde{C'}_n^k$ we denote the set of such points $(z_i, \dots, P_{i_1 \dots i_k}, \dots)$ of $\tilde{C}_n^k$ that, if some $k+1$ points $z_{i_1}, \dots, z_{i_{k+1}}$ are not in general position, then some k among them are not in general position.

A point z in $\tilde{C'}_n^k$ such that some k points $z_{i_1}, \dots, z_{i_k}$ are not in general position will be called *non-generic*. In the same manner as in the case of the group G_4^2 one can define a *generic path* in the space $\tilde{C'}_n^k$ connecting two generic points p, p^* and associate with it a word in letters a_{m_i} where m_i denote the k-subsets of the set $\{1, \dots, n\}$ and a letter $a_m, m = (i_1, \dots, i_k)$, is written when passing through a point where $z_{i_1}, \dots, z_{i_k}$ are not in general position. Thus we obtain a word $w(\gamma)$ in letters a_{m_i} and the corresponding element of G_n^k.

Theorem 11.7. *The above element of G_n^k depends only on the homotopy class of the path γ.*

In particular, if $p = p^*$, then we get a map from $\pi_1(\tilde{C}'_n^k)$ to the group G_n^k.

This theorem may be proved by checking the possible transformations of the words, induced by homotopies of the path γ. For example, if we homotop the path by going through a non-generic point z ("puncture a plane", in essence) and then return, going through the same plane (at a non-generic point z'), not necessary coinciding with the point z but such that the same set of $z_{i_1}, \ldots, z_{i_k}$ are not in general position, we obtain a subword $a_m a_m$ which is trivial due to the relations in the group G_n^k and thus may be removed.

Just as in Section 11.2.1, Theorem 11.6, the advantage here is that we have an "inverse" operation: given a word in letters of G_n^k we can restore a path γ which gives rise to such a word. To be more precise, the following holds:

Proposition 11.1. *For every word $w \in G_n^k$ there exists a path γ_w in the space $\tilde{C}'_n^k$ such that $w(\gamma_w) = w$. Moreover, for a given word the path γ_w may be constructed algorithmically.*

11.3 The G_n^k-complex

The aim of the present section is to construct a complex whose fundamental group is commensurable with the group G_n^k.

We start with the toy model G_3^2. The group has $\binom{3}{2} = 3$ generators a_{12}, a_{13}, a_{23} which are all involutions, and all the relations (except $a_{ij}^2 = 1$) can be represented by words of length six: $a_{ij}a_{ik}a_{jk}a_{ij}a_{ik}a_{jk} = 1$.

Let $\{0, 1\}^3$ be the 0-skeleton of the 3-dimensional cube in the space with coordinates corresponding to $(12), (13), (23)$.

We add the 1-skeleton Σ_1 of the 3-dimensional cube with the above mentioned vertices.

The skeleton can be considered as the graph for the group

$$\mathbb{Z}_2 * \mathbb{Z}_2 * \mathbb{Z}_2 = \langle a_{12}, a_{13}, a_{23} | a_{ij}^2 = 1 \rangle.$$

Here we associate the unit element with $(0, 0, 0)$.

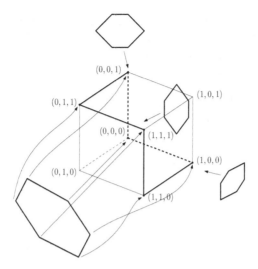

Fig. 11.2 Black edges of the cube form the boundary of the hexagon corresponding to the word $a_{12}a_{13}a_{23}a_{12}a_{13}a_{23}$

We shall deal with paths and loops starting from the base point. Note that elements of the group

$$\mathbb{Z}_2 * \mathbb{Z}_2 * \mathbb{Z}_2.$$

correspond not to homotopy classes of paths starting from the fixed base point rather than to homotopy classes of loops. There are eight classes of elements (depending on the parity of the number of occurrences of a_{12}, a_{13}, a_{23}). They correspond to the eight vertices of the cube and four classes of elements of the above mentioned group.

In order to add the relations, we should paste 2-cells (hexagons). Each cell misses exactly two opposite vertices, say, $a_{12}a_{13}a_{23}a_{12}a_{13}a_{23}$ starting from $(0,0,0)$ goes through $(1,0,0),(1,1,0),(1,1,1),(0,1,1),(0,0,1)$ and misses $(0,1,0)$ and $(1,0,1)$, see Fig. 11.2. Thus, every vertex is incident to exactly three hexagons.

Thus, we get a 2-complex to be denoted by Σ_3^2.

Statement 11.2. Elements of the group G_3^2 are in bijection with homotopy types of paths in Σ_3^2 starting from $(0,0,0)$.

It is easy to see that the universal covering over this complex is the plane with its hexagonal tiling (see Section 10.1.1).

Now, we are ready to do the same for the group G_n^k.

Our discrete cube will be $\binom{n}{k}$-dimensional; its coordinates will correspond to k-tuples of indices from $\{1, \ldots, n\}$. We paste $2k$-gons corresponding to relations of the tetrahedron type and quadrilaterals corresponding to commutativity relations.

Statement 11.3. Elements of the group G_n^k are in bijection with homotopy types of paths in Σ_n^k starting from $(0, 0, 0)$.

Chapter 12

Word and Conjugacy Problems in G_{k+1}^k Groups

In the present chapter we study the word and conjugacy problems in the groups G_{k+1}^k. Notice that, as was remarked in Section 11.1.4, the word problem in the group G_4^3 is solved by Theorem 11.5. So in the present chapter we concentrate on the conjugacy problem for G_4^3 and the word problem for G_5^4.

This chapter is based on the joint work with A.B. Karpov [Fedoseev, Karpov, and Manturov, 2020].

12.1 Conjugacy problem in G_4^3

We begin with the conjugacy problem for the group G_4^3.

12.1.1 *The existence of the algorithmic solution of the conjugacy problem in the group G_4^3*

Let us denote the free groups on generators $a_1, \ldots, a_n$ by $F(a_1, \ldots, a_n)$.

Theorem 12.1. *The group G_4^3 with presentation*

$$\langle a, b, c, d \mid a^2 = b^2 = c^2 = d^2 = (abcd)^2 = (bcad)^2 = (cabd)^2 = 1 \rangle,$$

has another presentation

$$\langle a, b, c \mid a^2 = b^2 = c^2 = 1 \rangle \underset{C}{*} (F(x, y, z) \rtimes \langle d \mid d^2 = 1 \rangle)$$

where

$$C = \langle abc, bca, cab \rangle = F(x, y, z), \ x = abc, \ y = bca, \ z = cab,$$

and

$$dxd = x^{-1}, dyd = y^{-1}, dzd = z^{-1}.$$

Proof. We will define two groups:

$$A = \langle a, b, c \,|\, a^2 = b^2 = c^2 = 1 \rangle$$

and

$$B = F(x, y, z) \rtimes \langle d \,|\, d^2 = 1 \rangle.$$

The presentation of the group $A \underset{C}{*} B$ with A and B defined above is:

$$\langle a, b, c, x, y, z, d \,|\, a^2 = b^2 = c^2 = 1, d^2 = 1,$$

$$dxd = x^{-1}, dyd = y^{-1}, dzd = z^{-1}, x = abc, y = bca, z = cab \rangle.$$

Replacing x, y, z with $(abc), (bca), (cab)$ in the relations $dxd = x^{-1}, dyd = y^{-1}, dzd = z^{-1}$ we obtain the relations

$$dabcd = cba, dbcad = acb, dcabd = bac.$$

Now we can eliminate x, y, z and come to the presentation:

$$\langle a, b, c, d | a^2 = b^2 = c^2 = 1, d^2 = 1, dabcd = cba, dbcad = acb, dcabd = bac \rangle.$$

It is easy to see that this presentation is equivalent to the standard presentation of G_4^3. $\qquad\square$

The book [Lyndon and Schupp, 2001] contains the following theorem:

Theorem 12.2. *Let $P = D \underset{R}{*} E$, and let $u = c_1 \ldots c_n$ be a cyclically reduced element of P where $n \geq 2$. Then every cyclically reduced conjugate of u can be obtained by cyclically permuting $c_1 \ldots c_n$ and then conjugating the result by an element of the amalgamated part R.*

The main result of the present section is the following theorem:

Theorem 12.3. *There is an algorithm with input consisting of two words from $G_4^3 \setminus C$ which returns 1, if these words are conjugated in G_4^3, and returns 0, if not.*

It turns out that the conjugacy problem in the group G_4^3 is closely related to the so-called *twisted conjugacy problem*:

Definition 12.1. Let ϕ be an automorphism of a group F. It is said that *the ϕ-twisted conjugacy problem is solvable in the group F* if for any elements $u, v \in F$, we can algorithmically decide whether there exists such g in F that $\phi(g^{-1})ug = v$. For example, the id-twisted conjugacy problem is the standard conjugacy problem in F.

Moreover, it is said that *the twisted conjugacy problem is solvable in the group F* if the ϕ-twisted conjugacy problem is solvable for any $\phi \in Aut(F)$.

Remark 12.1. The generalised word problem (see Definition 11.4) for C in G_4^3 is solvable. Also it is proved that $C \lhd G_4^3$.

Proof of Theorem 12.3. Consider two words $v, w \in G_4^3 \setminus C$.

1. First let us suppose that one of those words belongs neither to the subgroup A, nor to the subgroup B. Then we can notice by Theorem 12.2 that if they are conjugated, then some cyclic permutation of the first word will belong to the coset wC of the second one. Indeed, a conjugation by an element of C cannot change the coset. Thus all suitable cyclic permutations of the first word should belong the coset wC or, equivalently, these permutations and the second word should have the same image in the quotient group G_4^3/C under the natural homomorphism. This quotient group is isomorphic to $(\mathbb{Z}_2 \oplus \mathbb{Z}_2) * \mathbb{Z}_2$, since $A/C \cong \mathbb{Z}_2 \oplus \mathbb{Z}_2$ and $B/C \cong \mathbb{Z}_2$. The word problem is obviously solvable in the group $(\mathbb{Z}_2 \oplus \mathbb{Z}_2) * \mathbb{Z}_2$.

Now let us have two words from one C-coset: v and w. We need to determine whether they are conjugated in the group G_4^3.

Notice that if h is in C, then

$$hwh^{-1} = (ww^{-1})hwh^{-1} = wh^w h^{-1} = w\varphi_w(h)h^{-1},$$

where $\varphi_w \colon C \to C$ is an exterior automorphism (since w does not belong to C) sending h to h^w. C is a normal subgroup $C \lhd G_4^3$, so φ is indeed an automorphism. Thus if $hwh^{-1} = v$, then $w\varphi_w(h)h^{-1} = v$, so $w^{-1}v = \varphi_w(h)h^{-1}$.

Therefore we have reduced the conjugacy problem of the words v and w in the group G_4^3 to the following question: whether there exists an element h of the subgroup C such that $w^{-1}v = \varphi_w(h)h^{-1}$. In other words, we need to solve a twisted conjugacy problem in the subgroup C.

Notice also that the group C is a free group of finite rank (in fact, it is a group of rank 3). It is shown in [Bogopolsky, Martino, Maslakova and Ventura, 2006] that the twisted conjugacy problem is solvable in finitely generated free groups. Therefore, in this case the conjugacy problem in the group G_4^3 is solvable.

2. Now let us consider the case, when at least one of the words v, w lies either in the subgroup A or the subgroup B. Without loss of generality let v be a word from $A \setminus C$. Denote the cyclically reduced word for v by $\tilde{v}$. A conjugate to the word $\tilde{v}$ is of the form $b_n^{-1}a_n^{-1} \ldots b_1^{-1}a_1^{-1}\tilde{v}a_1 b_1 \ldots a_n b_n$, where $a_i \in A, b_i \in B$. Suppose that this word is cyclically reduced and $b_1 \neq 1$.

Consider the word $\hat{v} = a_1^{-1}\tilde{v}a_1$. If $\hat{v}$ is an element of C, then $\tilde{v}$ is an element of C as well, since C is a normal subgroup of G_4^3. But it contradicts the assumption that $v \in A \setminus C$. Therefore $\hat{v} \notin C$. It means that we can cyclically reduce the word $b_n^{-1} a_n^{-1} \ldots b_1^{-1} a_1^{-1} \tilde{v} a_1 b_1 \ldots a_n b_n$ obtaining the word $\hat{v} = a_1^{-1} \tilde{v} a_1 \in A \setminus C$. In other words, cyclically reduced elements conjugated to v can only be elements of A and conjugating elements have to belong to A as well.

Using the same arguments we deduce that cyclically reduced elements conjugated to an element of B can only be elements of B and conjugating elements have to belong to B as well. It allows us to reduce the conjugacy problem for words from A and B in G_4^3 to the conjugacy problem in A and B correspondingly. It is obvious that the conjugacy problem in the subgroups A and B is solvable.   □

We proved that the conjugacy problem is solvable in the group G_4^3. We will go a bit further though, presenting in the next section the explicit algorithm of the conjugacy problem solution.

12.1.2 *Algorithm of solving the conjugacy problem in G_4^3*

First, we present an algorithm solving the twisted conjugacy problem in the subgroup C.

The input of the algorithm is the set $\{u, v, \phi\}$, where u, v are words from C, and ϕ is an automorphism of C.

We already have a free basis for C and, adding a new letter z, we get a free basis for $C' = C * \langle z \rangle$. We will define some objects to help ourselves.

Definition 12.2. We say that the *extension of ϕ*, which is denoted by $\phi' \in Aut(C')$, is defined by the formula $\phi'(z) = uzu^{-1}$, $\phi'(c) = \phi(c)$ for $c \in C$.

Let us denote by γ_g the inner automorphism of C' defined by the formula $\gamma_g(h) = g^{-1}hg$ for any $h \in C'$. Furthermore, we introduce the notation $\gamma_g(\phi)$ for the composition of γ_g and the extension ϕ' by the formula

$$\gamma_g(\phi)(h) = \gamma_g(\phi'(h)).$$

$Fix(\gamma_v(\phi))$ denotes the set of fixed points of that mapping.

Lemma 12.1. *The words u and v are ϕ-twisted conjugated if and only if $Fix(\gamma_v(\phi))$ contains an element of the form $g^{-1}zg$ for some $g \in C$ (and, in this case, g is a valid twisted conjugating element).*

Proof. In fact, suppose that $v = (\phi(g))^{-1}ug$ for some $g \in F$. A simple calculation shows that $g^{-1}zg$ is then fixed by $\gamma_v(\phi)$. Conversely, if $g^{-1}zg$ is fixed by $\gamma_v(\phi)$ for some $g \in F$, then $gv^{-1}(\phi(g))^{-1}u$ commutes with z. And this implies $gv^{-1}(\phi(g))^{-1}u = 1$, since this word contains no occurrences of z. Hence, $v = (\phi(g))^{-1}ug$ so u and v are ϕ-twisted conjugated (with g being the twisted conjugating element). $\square$

We can algorithmically find a generating set for $Fix(\gamma_v(\phi))$; we can also decide if this subgroup contains an element of the form $g^{-1}zg$ for some $g \in F$. One can, for example, look at the corresponding (finite) core-graph for $Fix(\gamma_v(\phi))$ (algorithmically computable from a set of generators) and see if there is some loop labelled z at some vertex connected to the base-point by a path not using the letter z. And, if this is the case, then the label of such a path provides the g, i.e. the required twisted conjugating element.

Now we have all the necessary instruments to describe an algorithm of solving conjugacy problem for words in G_4^3.

The input of algorithm is the set $\{w, v\}$, where w, v are words from G_4^3. If either of those words lies in the subgroup A or B, then the algorithmic solution is obvious. Now let the words $w, v \in G_4^3 \setminus (A \cup B)$. In that case the algorithm is as follows.

(1) Write down all cyclic permutations of w.
(2) For any permutation $\widehat{w}$ check if it is in the C−coset vC, using the algorithm of solving the word problem in $G_4^3/C \cong (\mathbb{Z}_2 \oplus \mathbb{Z}_2) * \mathbb{Z}_2$. If no, this permutation does not fit. If yes, go to step 3.
(3) For permutation $\widehat{w}$ we define the automorphism $\varphi_{\widehat{w}} \in Aut(C)$, defined this way: $\varphi_{\widehat{w}} : h \to h^{\widehat{w}}$.
(4) Define the group $F = C * \langle z \rangle$, define the automorphism $\psi \in Aut(F)$ this way: its restriction to C is $\varphi_{\widehat{w}}$ and $\psi(z) = z$.
(5) For $\varphi_{\widehat{w}v^{-1}} \circ \psi \in Aut(F)$ construct a finite basis of the group $Fix(\varphi_{\widehat{w}v^{-1}} \circ \psi)$.
(6) Having this finite basis of $Fix(\varphi_{\widehat{w}v^{-1}} \circ \psi)$ construct its core-graph.
(7) Check if there is a vertex with loop labeled z connected to the marked vertex with a path without z-labels.
(8) If there is such a vertex, then there exists a word h in C with $w^{-1}v = \varphi_{\widehat{w}}(h)eh^{-1}$ where e is the empty word, so the empty word and $w^{-1}v$ are $\varphi_{\widehat{w}}$-conjugated. In this way $\widehat{w}$ and v are also conjugated.
(9) Otherwise, there is no such word, and w and v are not conjugated.

12.2 The word problem for G_5^4

Consider the group G_5^4 and its subgroup H_4 (see Section 11.1.4). As was remarked before in Section 11.1.4, if the word problem is solvable in the subgroup H_k, then it is solvable in the group G_{k+1}^k as well. Therefore we need to study the group H_4 to understand the solvability of the word problem in the group G_5^4.

12.2.1 *Presentation of the group H_4*

The following lemma is central in the proof of algorithmic solvability of the word problem in the group G_5^4.

Lemma 12.2. *The subgroup $H_4 \subset G_5^4$ has a presentation*
$$\langle a, b, c, d \, | \, a^2 = b^2 = c^2 = d^2 = 1, \, [(ab)^2, (cd)^2] = 1,$$
$$[(ac)^2, (bd)^2] = 1, \, [(ad)^2, (bc)^2] = 1 \rangle.$$

Proof. Consider five points in $\mathbb{R}P^3$. Consider a motion of 5 points and forbid the situation when three points lie on the same straight line.

Now, since we study the group H_4, we may consider a motion of the following type: four point (say, $1, 2, 3, 4$) are in general position and fixed, and the fifth point 5 moves around, see Fig. 12.1. Generators of the group H_4 correspond to the moments when some four points lie in the same plane. Therefore they are in one-to-one correspondence with triples of points defining the plane where in that moment lies the moving point 5. For example, the triple (123) corresponds to the moment, when the points $1, 2, 3, 5$ lie in the same plane. We may say that the point 5 moves in the manifold $M = \mathbb{R}P^3 \setminus \{\text{tetrahedron } (1234)\}$. Here saying "tetrahedron" we mean the configuration of $\mathbb{R}P^1$.

Consider two loops γ_1, γ_2 depicted in Fig. 12.1, upper part. The fundamental group of the space M is formed by such paths looping around the edges of the tetrahedron.

Now let us allow the points $1, 2, 3, 4$ to move. There is a finite number of singular moments when the four points lie on a plane and the tetrahedron transforms into a quadrilateral with crossing diagonals, see Fig. 12.1, middle part. Notice that the tetrahedron lies in a projective space, not in Euclidian $\mathbb{R}^3$. At that point the forbidden set, which we cut from the manifold in which the point 5 moves, changes and becomes singular, denote the resulting manifold by M_{sing}.

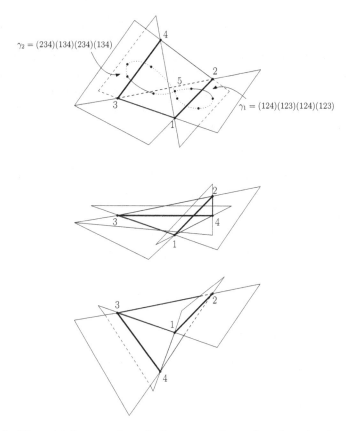

Fig. 12.1 The point 5 moves along the loops γ_1 and γ_2; when the point 4 moves, the tetrahedron becomes inverted

If the points continue moving, then the tetrahedron becomes "inverted", see Fig. 12.1, lower part. Thus two copies of the manifold M become glued along the singular moment. This gluing may be imagined in the following way. Let us take the manifold M and deform it into the manifold M_{sing}. We obtain a submanifold in $M \times [0, 1]$ such that the manifold M_{sing} is embedded into its boundary. Then we perform the same operation with the second copy of the manifold M and glue the resulting manifolds along the submanifold M_{sing}.

Now we can describe the space in which the point 5 moves: it is a manifold $\widehat{M}$ obtained by gluing together copies of the manifold M along all singular moments of the movement of the point 5 as described above. Fig. 12.1 depicts two copies of manifold M (upper and lower parts of the figure) and the manifold M_{sing} along which they are glued (middle part of the figure).

Each singular moment produces a relation in the group H_4. Before the singular moment t we had two loops: γ_1 and γ_2. After the singular moment t we obtain two new loops γ_1', γ_2', which go around the same edges of the tetrahedron as the loops γ_1, γ_2 (the edges (12) and (34) in the example in Fig. 12.1). We have the following equalities:

$$\gamma_1' = \gamma_1, \quad \gamma_2' = \gamma_1^{-1} \gamma_2 \gamma_1.$$

We consider the fundamental group of the manifold $\widehat{M}$. To obtain it, at each singular moment t we glue two fundamental groups of the copies of the manifold M and obtain the equality $\gamma_2' = \gamma_2$. Therefore the loops commute and we get the relation

$$(124)(123)(124)(123)(234)(134)(234)(134) =$$

$$(234)(134)(234)(134)(124)(123)(124)(123).$$

Introducing the notation $a = (123), b = (124), c = (134), d = (234)$ we formulate the relation

$$(baba)(dcdc) = (dcdc)(baba).$$

The relations of the form $a^2 = b^2 = c^2 = d^2 = 1$ are evident.
Taking it into account, we obtain the relations of the form

$$(abab)(cdcd) = (cdcd)(abab).$$

Considering the loops around other pairs of opposite edges, we obtain the rest of the relations. Since there are no more singular configurations of the points movement, there are no other relations. The lemma is proved.

□

12.2.2 The Howie diagrams

To study the group H_4 and the word problem in it we will need the so-called *Howie diagrams* — a geometric way of representing group elements, in some aspects resembling the van Kampen diagrams (see, for example, [Howie, 1983], where Howie diagrams are called *relative diagrams*). Just like in the case of van Kampen diagrams (see Section 1.1), one can consider disc diagrams, spherical diagrams, etc. In the present section we will only need disc Howie diagrams.

To be more precise, the Howie diagrams are defined in the following way. First we define a *relative group presentation*.

Definition 12.3. A *relative group presentation* is an expression of the form $\mathcal{P} = \langle G, x \mid R \rangle$, where G is a group, x is a set disjoint from G, and R is a set of cyclically reduced words in the free product $G * \langle x \rangle$.

This notion was introduced by Bogley and Pride, see [Bogley and Pride, 1992]. Now let us fix a group G and let the set x be one-element. Consider a map M on the plane and let the edges of M (its 1-cells) be oriented. We will treat the faces (2-cells) of the map as polygons, and for each face F saying a *corner of the face* we mean an intersection of a sufficiently small neighbourhood of a vertex of the face with the face F itself.

Let us decorate every edge of the map with the letter x. Furthermore, we decorate every corner of every face of the map with a word $w \in G$. Now we can define *face-labels* and *vertex-labels* of the map M.

Definition 12.4. Let M be a map as described above. A *vertex-label* of a vertex v is a word $\nu \in G$ obtained by reading all the corner labels while walking around the vertex v in the clockwise direction.

A *face-label* of a face F of the map M is a word $\nu \in G * \langle x \rangle$ obtained by reading all corner labels and edge labels while walking around the boundary of the face F in the counterclockwise direction. As usual, if the orientation of an edge is compatible with the counterclockwise direction convention, then we shall read x, otherwise we read x^{-1}.

Now we have all the tools to define the Howie diagrams.

Definition 12.5. Let $K = \langle G, x \mid R \rangle$ be a relative group presentation with the set x being one-element. A decorated map M as described above is called a *Howie diagram* over that relative presentation, if the following conditions are satisfied:

(1) for each interior vertex of the map its vertex-label is trivial in the group G;
(2) for every interior 2-cell F of the map its face-label w should lie in the set R of the relations of the relative presentation $\langle G, x \mid R \rangle$.

Remark 12.2. Notice that in [Howie, 1983] the relative diagrams are considered on the sphere and for that reason the corresponding maps have one vertex distinguished. Here we work with planar diagrams and may treat all vertices equally.

The Howie diagrams are useful because the following analogue of the van Kampen lemma holds for them.

Lemma 12.3. *Let H be a group, t be a letter, and G be the group given by a relative presentation*

$$G = \langle H, t \mid R \rangle.$$

Let furthermore the natural mapping $\iota: H \to G$ be injective. Then for any word $w \in H * \langle t \rangle$ its image $\iota(w)$ is trivial in G if and only if there exists a Howie diagram with single exterior cell over this presentation such that its exterior cell face-label is the word w.

12.2.3 The solution to the word problem in H_4

Now we shall use the Howie diagrams techniques and, in particular, Lemma 12.3 to tackle the word problem in H_4. First, consider the subgroup $H_3 \subset H_4$ generated by the letters $\{a, b, c\}$. It is isomorphic to $\langle a \rangle_2 * \langle b \rangle_2 * \langle c \rangle_2$. The group H_4 may be given by the following relative presentation:

$$H_4 = \langle H_3, d \mid R \rangle,$$

where the set of relations R is naturally obtained from the presentation of H_4 given in Lemma 12.2:

$$R = \{d^2 = 1, \ [(ab)^2, (cd)^2] = 1, \ [(ac)^2, (bd)^2] = 1, \ [(ad)^2, (bc)^2] = 1\},$$

where the square brackets denote the usual commutator. By cyclic permutations we can transform those relation into such a form, that the letter d is at the end of every relation:

(1) $d^2 = 1$,
(2) $(cababc)d(c)d(baba)d(c)d = 1$,
(3) $(bacacb)d(b)d(caca)d(b)d = 1$,
(4) $(acbcba)d(a)d(bcbc)d(a)d = 1$.

The inclusion mapping $\iota: H_3 \to H_4$ is injective, so Lemma 12.3 holds for this relative presentation. Therefore, a word w is trivial in the group H_4 if and only if there exists a Howie diagram with single exterior cell (that is, a connected Howie diagram) with the word w appearing as the face-label of the exterior face of the diagram. Let us denote this diagram by D.

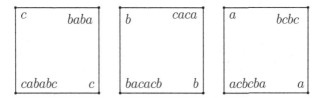

Fig. 12.2   Three types of cells of the diagram D'

Now we transform the diagram D in the following manner. First, we reduce all the bigons corresponding to the relations $d^2 = 1$ replacing each bigon with vertices v_1, v_2 with an edge connecting the same vertices and oriented arbitrarily. After the removal of bigons, we "forget" the orientation of the edges. This operation is legal due to the relation $d^2 = 1$. Finally, we forget the label on the edges of the diagram, since all of them are equal to d. This operation does not change the vertex-labels, but transforms the face-labels. We will come back to that change later.

Thus we obtain an unoriented diagram D' without bigonal cells. Furthermore, every relation of the set R (except for the $d^2 = 1$ relations, which are no longer present in the obtained unoriented diagram) includes exactly four subwords from the group H_3. That means that every interior face of the diagram D' has four vertices. Note that the diagram D' may not be homeomorphic to a disc. Nevertheless, it is connected and homeomorphic to a tree, some vertices of which are "blown up" into a submap homeomorphic to a disc.

Each cell of the map D' corresponds to a relation of the set R. Moreover, there are no bigonal cells. It means that the diagram has only cells depicted in Fig. 12.2. We shall call them the *types* of the cells of the diagram.

Whenever two cells are glued together by two or more edges, at least one vertex becomes interior. By definition of the Howie diagram, the vertex-label of that vertex must be trivial in the group H_3. Consider the leftmost part of Fig. 12.2 which corresponds to the second relation of the set R. There are three different labels corresponding to a cell of that type: $c, baba$ and $cababc$. Among their pairwise concatenations there is only one word, trivial in H_3: the word c^2. Therefore, there are only two ways to glue together two cells of this type by at least two edges. Those situations are depicted in Fig. 12.3. Analogous operations may be performed for two other types of cells. At the same time, no two cells of different types may be glued by two or more edges, because no trivial in H_3 vertex-label may be obtained this way.

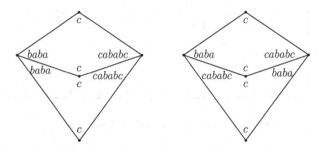

Fig. 12.3   Possible ways to glue together two cells of the first type by two or more edges

Now let us consider an interior vertex of the diagram D'.

Statement 12.1. For every interior vertex v of the diagram D' there are either two or at least four cells of the diagram incident to that vertex.

Proof. The case of a vertex incident to exactly two cells was studied above. There are six possible glueings of two cells producing such vertices (two possible ways for each of three types of cells). Now let us proof that all other interior vertices are incident to at least four cells.

   Suppose there exists an interior vertex v incident to exactly three cells. Then the vertex-label l of the vertex v is of the form $l = l_1 l_2 l_3$, where each l_i is from the set $W = \{a, b, c, baba, caca, bcbc, cababc, bacacb, acbcba\}$. The label l must be trivial in the group H_3. Without the loss of generality we may say that it means that $l_1^{-1} = l_2 l_3$. It is easy to check that for all possible pairs of words of this set their product does not lie in the set W. Therefore the label l cannot be trivial and there is no such vertex v.   □

   Let us apply one more transformation to the diagram D'. For each pair of cells glued together by two edges we replace them with one cell, deleting the two edges by which the cells are glued and replacing the labels in the corners of the cells with their products, see Fig. 12.4. We call this operation a *contraction* of cells.

   After this transformation is applied to all suitable pairs of cells, we obtain a decorated map, denote it by M. Notice that now the face-labels are not necessary words from the set of relations R and therefore the map may be not a Howie diagram. On the other hand, it is a map such that its every interior vertex is incident to at least four cells and every two cells are glued together by at most one edge. In other words, the map M is a $(4, 4)$-map in terms of the small cancellation theory.

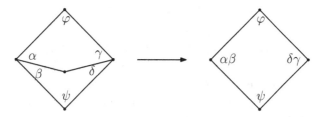

Fig. 12.4   Transforming two cells glued by two edges into one

Every cell of the map M is obtained by contracting a chain of the cells of one type of the diagram D'. We extend the notation and say that a cell F obtained by contraction of a chain of cells of type i is of type i. Let us describe what kind of labels the cells of the map M may have in their corners. Suppose that a cell F is of the first type (other types are treated analogously). Then two opposite corners of the new cell contain the letter c and the other two corners contain some words from the group $\langle cababc \rangle * \langle baba \rangle$. Moreover, if we know the label u in a corner of this cell, then we can uniquely determine the label u' in the opposite corner. Indeed, if $u = c$, then $u' = c$ as well. If $u \in \langle cababc \rangle * \langle baba \rangle$, then the opposite label may be obtained by replacing all occurrences of the word $baba$ in the label u with $cababc$ and vice versa.

To prove the main statement of the present subsection we shall use the following theorem (see [Lyndon and Schupp, 2001]):

Theorem 12.4. *Let M be a connected $(4, 4)$-map. Then the number of 2-cells of the map does not exceed*

$$\frac{1}{4} \left(\sum_{F \subset M} [4 - i(F)] \right)^2 ,$$

where the sum is taken over all 2-cells of the map and $i(F)$ denotes the number of 2-cells incident to the cell F (i.e. the cells with a common edge with F).

Notice that every interior cell of the map M has at least 4 adjacent cells and thus does not yield a positive summand in that sum. Let us denote the cells with common edges with the outer boundary of the map by D_i and their number by N. Each edge of the outer boundary of the map corresponds to a letter d in the word w. Therefore the number N cannot be greater than the number of letters d in the word w. Thus the total number of the cells of the map M cannot exceed $\frac{1}{4}(4|w|_d)^2 = 4|w|_d^2$, where $|w|_d$ denotes the number of letters d in the word w.

We have demonstrated that if a word $w \in H_4$ is trivial, then there exists a connected $(4,4)$-map M decorated as described above such that the face-label of its exterior cell is the word w. As follows from Theorem 12.4 the total number maps (without labels) that can be potentially transformed by labelling into the map M with the word w on its outer boundary is finite. Now we shall prove the following statement:

Lemma 12.4. *Let w be a word in H_4 and let M be a $(4,4)$-map homeomorphic to a disc with the number of cells not exceeding $4|w|_d^2$. Then there exists an algorithm determining whether there is such a labelling of the corners of the cells of M that the vertex-labels of all interior vertices of the maps were trivial and the face-label of the exterior cell of the map M was exactly w.*

Proof. The proof shall be by induction on the number of cells of the map. The general strategy is as follows: we shall change the word w and the map M by finding relations inside the word w and removing them in such a way, that the number of cells in M decreases.

Let us call two edges of a map *neighbouring* if they have a common vertex. The map M is finite, hence it has at least one cell with two neighbouring edges incident to the exterior cell of the map (such edges should also be called *exterior*). Let us take such a cell, denote it by F, and consider the label l in the corner between the two neighbouring exterior edges. Let us consider the possible values of the label l. Without loss of generality we should suppose that the cell F is of the first type.

1. l could be a word from the group $\langle cababc \rangle * \langle baba \rangle$. Notice that since l is the label in the corner between two exterior edges, the word w has a subword of the form dld. Therefore there are finitely many possible values of l. For each of them we can uniquely determine labels in all other corners of the cell F. Denote the label in the opposite corner by l'. Now we can change the word w by replacing the subword dld with the word $cd(l')^{-1}dc$ and get a new map M' with fewer cells (effectively we contract the cell F with the exterior cell of the map M in the same manner as we contracted interior cells earlier).

2. If $l = c$, then in the neighbouring corner the cell F has a label $u \in \langle cababc \rangle * \langle baba \rangle$. The word u uniquely determines the label u' in the

opposite corner. To perform the contraction of the cell F with the exterior cell we need to prove that there are finitely many possible values of u.

The corner with the label u is incident to a vertex which we shall denote by v. Only a finite number of cells are incident to the vertex v, let us denote their number by K (this number may easily be bounded considering the total number of cells of the map M). Notice that v is an exterior vertex, because it is an endpoint of an exterior edge. Therefore, the vertex-label of v is a subword of the word w and hence it cannot be longer than the word w. Now suppose that the length of the label u is greater than $|w| + K$.

The labels in the corners of the cells incident to the vertex v may either equal a, b or c, or lie in one of the groups $\langle cababc \rangle * \langle baba \rangle, \langle bacacb \rangle * \langle caca \rangle, \langle acbbca \rangle * \langle bcbc \rangle$ depending on the type of the corresponding cell. It is easy to see that with the addition of each label the length of u may decrease at most by one. That means that $|l(v)| \geq |u| - (K-1) > |w| + K - K + 1 = |w| + 1$. That is impossible, therefore the length of u cannot be greater than $|w| + K$.

We proved that there is only finite number of possible values of u. For each of them we can uniquely determine the value of u' and contract the cell F with the exterior cell, obtaining a map with fewer cells.

There are no other possibilities for the values of the label l. It means that if we take a cell with two neighbouring exterior edges, then we may obtain finitely many maps with fewer cells. By induction we either find a decomposition of the word w into a product of generators or prove that there is no required Howie diagram. $\quad\square$

Now we are ready to prove the main theorem of this section.

Theorem 12.5. *The word problem is algorithmically solvable in the group H_4.*

Proof. Consider a word $w \in H_4$. As we have shown, it is trivial in H_4 if and only if there exists a $(4, 4)$-map with labels in the corners of its cells such that the vertex-label of each interior vertex of the map is trivial in the group and the face-label of the exterior cell is the word w.

As we noticed before, this map is a tree with a disc submap at each vertex. The total number of cells in the map is not greater than $4|w|_d^2$, the number of the edges of the tree is bounded by the same number. Therefore the number of such trees is finite.

Now for each disc submap we may perform the reduction algorithm from Lemma 12.4. Reducing each disc submap in this manner we come down to a tree and some new word $\tilde{w}$. Letters d break the word $\tilde{w}$ into a finite sequence of subwords $w_1, \ldots, w_n$. To check whether the map admits a labelling giving the word $\tilde{w}$, we need to choose a starting vertex (that can be done in a finite number of ways) and place the labels $w_1, \ldots, w_n$ into the vertices of the tree walking around it starting from the chosen vertex (when we reach a leaf of the tree, we shall go back along the same edge). That gives us a finite algorithm of determining whether the necessary labelling exists.

Hence, we have constructed a finite constructive algorithm which takes a word w and returns "true" if there exists a labelled map with the word w as the face-label of its outer cell, or "false" otherwise. Therefore the word problem in the group H_4 is solvable. $\square$

As was remarked in Section 11.1.4, the solvability of the word problem in H_k yields the solvability of the word problem in G_{k+1}^k. Thus we obtain the following theorem.

Theorem 12.6. *The word problem is algorithmically solvable in the group* G_5^4.

Chapter 13

The Groups G_n^k and Invariants of Manifolds

In the main principle formulated in Chapter 7, we spoke about the dynamics of n *particles* in a certain topological space.

In all examples considered in the previous chapters, all particles were *points* in certain topological spaces which led us to various *configuration spaces*. In the present chapter, we shall deal with *submanifolds* instead of *points* playing roles of particles, hence, with *moduli spaces* of such submanifolds.

One of the goals of the present chapter is *to construct invariants of arbitrary manifolds*. The groups G_n^k allow one to study *braids* of certain spaces, however, the configuration spaces of points for the G_n^k method are quite restrictive. If we deal with submanifolds rather than just points, then we have much more degrees of freedom.

We shall start with the example of projective duality: looking at hypersurfaces in projective planes, we can understand what sort of submanifolds we require.

13.1 Projective duality

Consider the main examples from Chapters 7 and 11.

Assume we deal with a collection of n points in $\mathbb{R}P^{k-1}$ such that any $k-2$ of them are in general position. Our property of codimension 1 will be:

$$k-1 \text{ points are not generic.}$$

For example, when dealing with points in $\mathbb{R}P^1$, this means that some two points coincide and leads to a map from the configuration space to G_n^2, in the case of $\mathbb{R}P^2$ we deal with braids, and we get a map to G_n^3, for points in $\mathbb{R}P^3$ without collinear triples, we get a map to G_n^4, and so on.

Now, we just pass to the dual space. Namely, in the projective space of dimension $k - 1$, a point $(a_0 : \cdots : a_{k-1})$ corresponds to the hyperplane. Here we require that *any $k - 1$ planes are generic, that is, the intersection of any $k - 1$ planes is a point.* This is the definition of our restricted moduli space $M'(\mathbb{R}P^{k-1}, n)$. Certainly, the projective duality leads to a homeomorphism $i: M'(\mathbb{R}P^{k-1}, n) \to C'(\mathbb{R}P^{k-1}, n)$ which takes the hyperplane with coordinates $(a_0 : \cdots : a_{k-1})$ to the point with the same coordinates.

Our *good k-condition* is:

some k hyperplanes share a point.

This condition is of codimension 1, it precisely corresponds to the condition "*some k points are not generic in $C'(\mathbb{R}P^{k-1}, n)$*".

Just as in Section 11.1, we take all good and transverse paths in $M'(\mathbb{R}P^{k-1})$ with fixed ends and take those paths to words in G_n^k.

In this case the isomorphism $i: M'(\mathbb{R}P^{k-1}, n) \to C'(\mathbb{R}P^{k-1}, n)$ allows one to follow Theorem 7.2. Namely, we get

Theorem 13.1. *Having fixed a generic point in $M'(\mathbb{R}P^{k-1})$, we have a well-defined map from $\pi_1(M'(\mathbb{R}P^{k-1}), *)$ to elements of G_n^k.*

If we analyse the case of $n = 5, k = 3$, then we see the exact meaning of the tetrahedron equation (7.1).

Namely, for $k = 3$ we look for hyperplanes (i.e., 2-dimensional planes), such that each three of them intersect in a point. In this case, the letter a_{ijkl} corresponds exactly to a collection of four hyperplanes which pass through a point.

If we look at what happens in the neighbourhood of such a point, then we shall see "inversion of a tetrahedon". Namely, we consider four planes $P_1 = \{x = 0\}, P_2 = \{y = 0\}, P_3 = \{z = 0\}, P_4 = \{x + y + z + \varepsilon = 0\}$ in the affine space $\mathbb{R}P^3$.

If we slightly perturb it to get $\varepsilon > 0$ or $\varepsilon < 0$, then we get two tetrahedra.

Now, consider some fifth plane which is not parallel to any of the four above planes. Then it passes through each of them in some order.

If in the picture $\varepsilon < 0$ we have the order

$$(1, 2, 3), (1, 2, 4), (1, 3, 4), (2, 3, 4),$$

then in the picture $\varepsilon > 0$ we have the opposite order

$$(2, 3, 4), (1, 3, 4), (1, 2, 4), (1, 2, 3).$$

This leads us exactly to the Zamolodchikov relation

$$a_{1234}a_{1235}a_{1245}a_{1235}a_{2345} = a_{2345}a_{1345}a_{1245}a_{1235}a_{1234}.$$

13.2   Embedded hypersurfaces

The aim of the present chapter is to "copy" the above pattern to the general case of a moduli space $M(S, \Sigma_1, \ldots, \Sigma_n)$, where S is a certain manifold of dimension k, and Σ_j are submanifolds of codimension 1.

Copying the condition on page 272 we require that under the above embedding, any $k - 1$ submanifolds $\Sigma_{i_1}, \ldots, \Sigma_{i_{k-1}}$ should have exactly one intersection.

Then, using the main Theorem 7.2, we get the following

Theorem 13.2. *Let M^k be a closed manifold of dimension k and let $\Sigma_1, \ldots, \Sigma_n$ be a fixed collection of manifolds of dimension $k - 1$.*

Consider the moduli space $M(M^k, \Sigma_1, \ldots, \Sigma_n)$ of fixed collection of manifolds ($M(M, \Sigma_i)$, for short). Then for each connected component $\#_j M(M, \Sigma_i)$ there is a well defined map from the fundamental group of this connected component to the group G_n^{k+1}.

Hence we get a collection of subgroups in G_n^{k+1} which can serve as an invariant of the entire manifold M.

13.2.1   *Examples*

If we deal with case when $M = S^{k-1}$ and all Σ_i are homeomorphic to S^{k-2}, $k > 3$ then we get precisely the subgroup in G_n^k corresponding to "realisable" elements, i.e., "braids" [Manturov, 2015d].

Example 13.1. If $M = \mathbb{R}P^{k-1}, \Sigma_1 = \cdots = \Sigma_k = \mathbb{R}P^{k-2}$, then we get exactly the image of the braid group of the projective space in G_n^k.

Example 13.2. If $k = 3$ (i.e., the dimension of the above total manifold is 2), then the above construction works for any manifold except the sphere.

Namely, for the torus we have $M = T^2$ and $\Sigma_1 = \cdots = \Sigma_n = S^1$. Thanks to the gradient flow of Chapter 3 we can think of the submanifolds Σ_i as simple geodesics. Let $(u_i, v_i) \in \mathbb{Z}^2$ be the coordinates of Σ_i in $H_1(M, \mathbb{Z})$ with the basis Σ_1, Σ_2, i.e. $\Sigma_i = u_i \Sigma_1 + v_i \Sigma_2$ in $H_1(M, \mathbb{Z})$. The condition that any two submanifolds Σ_i have exactly one intersection point implies that $|u_i| = 1, |v_i| = 1$ for any $i \geq 3$. For the same reason any two submanifolds Σ_i, Σ_j cannot have the same homological coordinates and we cannot have simultaneously two submanifolds with coordinates $(\pm 1, \pm 1)$. Thus, $n \leq 3$ and up to isomorphisms the homological classes of the submanifolds in the chosen basis are $\Sigma_1 = (1, 0), \Sigma_2 = (0, 1), \Sigma_3 = (1, 1)$, see Fig. 13.1.

Let $n = 3$. The fundamental group of the moduli space $(T^2, \Sigma_1, \Sigma_2, \Sigma_3)$ is generated by loops γ_i, $i = 1, 2, 3$, where γ_i is a circular move of Σ_i along a transversal direction. All the generators $\gamma_1, \gamma_2, \gamma_3$ are mapped to the generator a_{123} of the group G_3^3.

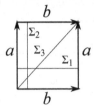

Fig. 13.1   Submanifolds in the torus

For the Klein bottle K^2 we have the following situation:

(1) the orientation-reversing cycles among the submanifolds Σ_i must belong to the same homology class;
(2) there can be at most one orientation-preserving cycle among Σ_i.

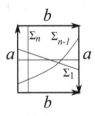

Fig. 13.2   Submanifolds in the Klein bottle

Let $\Sigma_1, \ldots, \Sigma_{n-1}$ be orientation-reverting cycles and let Σ_n be an orientation-preserving cycle, see Fig. 13.2. Let γ be the loop in the fundamental group of the moduli space which moves the cycle Σ_n twice along an orientation-reversing cycle, e.g. Σ_1. We should move the cycle along Σ_1 two times because the final orientation of the Σ_n must coincide with the initial one. The image of γ in G_n^3 is equal to $\left(\prod_{k=1}^{(n-1)(n-2)/2} a_{i_k j_k n} \right)^2$

where $\Sigma_{i_k} \cap \Sigma_{j_k}$ is the k-th intersection point the submanifold Σ_n encounters while moving along the orientation-reversing cycle. For example, for the configuration of the orientation-reversing cycles given in Fig. 13.3, the image of γ equals to $\left(\prod_{i=1}^{n-2} \prod_{j=i+1}^{n-1} a_{ijn} \right)^2$. For other configurations of the orientation-reversing cycles the result will differ by conjugation.

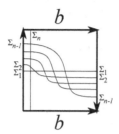

Fig. 13.3   A configuration of submanifolds in the Klein bottle

However, a careful reader may note that under some circumstances, the space $M'(M, \Sigma)$ is empty. This happens, for example, in the case when M is a sphere. Then any two closed manifolds Σ_1, Σ_2 in this sphere of codimension one have evenly many intersection which contradicts the main condition.

We shall tackle this case in the next subsection.

13.3   Immersed hypersurfaces

The most important case which is not covered by the main Theorem 13.2 is the case when the ambient manifold M^k is the sphere. In this case, the moduli space described above is *void*.

How to overcome this difficulty and to construct meaningful invariants?

Let us start with the 2-sphere and a collection of 1-circles in it.

When performing an isotopy, the number of intersection points between two circles changes, moreover, generically, *two circles never have exactly one intersection point*. This means that we have to be more restrictive.

Namely, we may think that our circles are *immersed in our 2-surface*. For any collection of $k - 1$ circles (in our case, 2 circles) we take exactly one true intersection point.

Note that in the case when the ambient manifold is the sphere of any dimension S^n, and submanifolds are just spheres S^{n-1} of one dimension less, the above principle does not work at all: $H_j(S^n) = 0$ for $j < n$, hence, in general position, n spheres S^{n-1} have zero intersection algebraically, which means that generically the number of intersections is *even*.

Below, we suggest two solutions to the above problem. The first one is to replace the group Γ_n^3 with the group G_n^3 from Section 9.6.

We already know how to modify groups Γ_n^k for the case of more than one intersection point and we apply this for the different situation.

The other approach deals with the situation when we *immerse* several S^{k-1} into a S^k and for each $k-1$-tuple of S^{k-1} we fix exactly one intersection point and keep track of that one.

We are sure that this approach (with minor or major modifications) works for arbitrary manifolds and submanifolds, but here we restrict ourselves with the case of 2-dimensional manifolds and curves in them.

13.4   Circles in 2-manifolds and the group G_n^3

Fix a positive integer n and let us consider an oriented sphere S^2. We shall consider the space of all embeddings of n oriented 1-dimensional circles $S_1, \ldots, S_n$ to S^2 such that every pair of two circles S_i, S_j has exactly two common points.

Denote this moduli space by $M_{imm}(S^2, n)$.

We claim that there is a well-defined map from $\pi_1(M_{imm}(S^2, n))$ to G_n^3.

Indeed, generically we do not have any triple intersections. When considering a path in the moduli space, we meet codimension 1 situations: three circles have an intersection.

As for the case of line on the plane, we have two configurations of codimension 2:

(1) two triple points (note that points may belong to the same triple of circles);
(2) a quadruple point.

Fig. 13.4   A configuration of codimension 2 that gives relation $a_{ikj}a_{ijl} = a_{ijl}a_{ikj}$

The first case corresponds to commutativity relations $a_{ijk}a_{i'j'k'} = a_{i'j'k'}a_{ijk}$ for any triples of indices, see for example Fig. 13.4. The second case gives a tetrahedron relation

$$a_{ijk}a_{ijl}a_{ikl}a_{jkl} = a_{jkl}a_{ikl}a_{ijl}a_{ijk}.$$

These relations with the relation $a_{ijk}^2 = 1$ define a presentation for the group $G = \mathbb{Z}_2^{\oplus \frac{1}{3}n(n-1)(n-2)}$. Thus, we get a map from $\pi_1(M_{imm}(S^2, n))$ to G. It is easy to see that the image of the map is trivial.

Indeed, given indices i, j, k, there are three possible configurations of the circles S_i, S_j, S_k on the sphere, see Fig. 13.5. In the corresponding transformation graph all the cycles have even number of generators a_{ijk} and even number of generators a_{jik}.

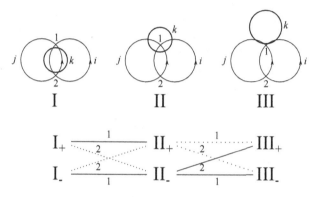

Fig. 13.5 Configurations of circles S_i, S_j, S_k and the transformation graph. The signs of the configurations correspond to the orientation of the circle S_k. The label of an edge is the vertex the circle S_k goes over. The solid edges correspond to the generator a_{ijk}, the dotted edges correspond to a_{jik}.

Since the configurations of codimension 1 and 2 are local, the result can be generalised to the moduli space consisting of embeddings of circles into a surface Σ such that any two circles have two intersection points in Σ.

13.5 Immersed curves in M^2

In this section we shall again restrict ourselves to the two-dimensional case, namely, we consider the moduli space of immersed curves in M^2. The situation with immersed hypersurfaces will cause much more difficulties.

Fix a 2-surface Σ and a natural number n; our moduli space will consist of n closed curves $\gamma_1, \ldots, \gamma_n$ smoothly immersed in Σ together with a choice of one transverse intersection for each pair of curves: $P_{ij} \in \gamma_i \cap \gamma_j$. The total number of intersection points between any two curves may be an arbitrary positive number.

We denote this space by $P(\Sigma, n)$.

Theorem 13.3. *There is a well defined map from $\pi_1(P(\Sigma, n))$ to G_n^3.*

Proof. We are looking for those moments when fixed intersection points for some triple of points coincide: $P_{ij} = P_{ik} = P_{ik}$ and associate the generator a_{ijk} of the group G_n^3 to that.

The rest of the proof follows from the main principle from Chapter 7. $\square$

Remark 13.1. In the above setup we can drop the manifold Σ and deal with curves by themselves. In order to fix transverse intersection, we should deal with diagrams of *n-component free links*, such that every two components have exactly one intersection. We see that any equivalence of diagrams in this class, presented by a sequence of Reidemeister moves, produces a word in G_n^3 (Fig. 13.6).

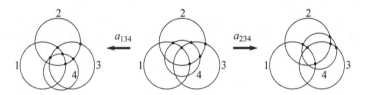

Fig. 13.6   Transformations of free links and the corresponding words in G_n^3. The crossings of the link are marked with dots.

13.6   A map from knots to 2-knots

There is a nice theory of free braids (G_n^2-braids) which naturally closes up to give rise to free knots. Moreover pure classical braids, as we already studied in Section 9.1, are related to words in G_n^3. It gives rise to a natural question: does there exist a link counterpart in G_n^3? One may wonder to consider some "braids" encoded by one-dimensional lines with trisecants and close them up as we did it in the case of G_n^2. In other words, is it possible to get a one-dimensional formalism of, say, monoids representing "G_n^3-knots" where we consider new generators corresponding to maximal and minimal points along with trisecant generators a_{ijk}? Naturally, in this approach, lines should correspond to secants on the horizontal line and triple intersection should correspond to the situation, when three points belong to the same horizontal line.

We would like to construct a one-dimensional formalism with a_{ijk} corresponding to triple points and some other letters corresponding to local maxima and local minima.

Let us start with an unsuccessful attempt. When trying to close G_n^3 to get "knots", we meet the following problem. Consider the situation when we have two maxima, see Fig. 13.7.

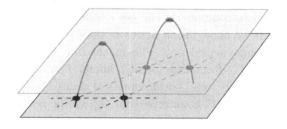

Fig. 13.7   Two maximal points on the same horizontal plane

Then, when looking at the total configuration space, we face the following problem. When knot in $\mathbb{R}^3$ passes through the point with two maxima on the same level, its "knot G_n^3-counterpart" undergoes a fusion, see Fig. 13.8.

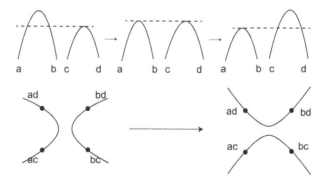

Fig. 13.8   Fusion with two maxima on the same level

Hence, the expected map from classical knot to "one-dimensional closure of G_n^3-braid" will have (as its target) not "G_n^3-like graphs" modulo

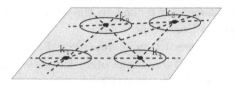

Fig. 13.9   Four points of K on the same horizontal plane

some "G_n^3-moves" but rather graphs modulo G_n^k moves and fusion moves (Fig. 13.8) when two components can get reconstructed in one component and vice versa. If we allow fusion moves in their full generality, the graph can change in an unexpected way. This means that we do not have a genuine 1-dimensional G_n^3-formalism for such knots. The remedy turns out to be a map from knots to "free" surface knots of genus 1, see Definition 4.10. Hence, the genuine "knot" counterpart for G_n^3 will be not one-dimensional but rather two-dimensional.

Now we construct a free surface knot of genus 1 from a given knot in $\mathbb{R}^3$. Let K be an oriented knot in $\mathbb{R}^3$. Assume that a height function given for K with respect to a fixed vector V. Let us denote by $\mathbb{R}^3 = \mathbb{R}^2 \times \mathbb{R}_V$, where $\mathbb{R}_V$ is the real line according to V. We call a plane, which is orthogonal to V, a *horizontal plane*, and lines on a horizontal plane *horizontal lines*. We shall be especially interested in the height function for K; whenever mentioning "minimum" or "maximum" we assume extreme with respect to this function.

Definition 13.1. We say that $K \subset \mathbb{R}^3$ is *generic* if the following conditions hold:

(1) No four points k_1, k_2, k_3, k_4 of $K \subset \mathbb{R}^3$ are on a horizontal quadrisecant;
(2) For each horizontal plane there are no local extrema or only one local extremum;
(3) K has finitely many horizontal trisecants;
(4) If k_1 and k_2 are on the horizontal line for some horizontal plane $\mathbb{R}^2 \times \{s\}$, then there exist open neighborhoods I_{k_1} of k_1 and I_{k_2} of k_2 in K and $\epsilon > 0$ such that $I_{k_1} \cap I_{k_2} \cap (\mathbb{R}^2 \times \{t\})$ on some horizontal line for every $t \in (s - \epsilon, s + \epsilon)$.

Now we assume that the knot K is generic. Let $S(K) = T^2 = K \times S^1$ be the torus in $\mathbb{R}^3$. Let us assume that the orientation of torus agrees to the orientation of K. Without loss of generality we may assume that there are no self-intersections of $S(K)$. Now, we make an abstract surface knot $D(K) = (S(K), D)$ as follows.

Let k_1 and k_2 be two points of K on some horizontal plane $\mathbb{R}^2 \times \{s\}$. Then there exist four points of $S(K)$ on the horizontal line $L_{k_1 k_2}$, which is oriented from k_1 to k_2. Let k_1^s, k_1^e, k_2^s and k_2^e be the points on the line $L_{k_1 k_2}$, which are placed as the indicated order according to the orientation of $L_{k_1 k_2}$, see Fig. 13.10.

If both of local orientations of K around k_1 and k_2 decrease (or increase) according to V, then we call the points k_1^{-1} and k_2^{-1} (k_1^{+1} and k_2^{+1}) *equivalent*. If one of local orientations of K around k_1 and k_2 decreases according to V and the other increases, then we call the points k_1^{+1} and k_2^{-1} (k_1^{-1} and k_2^{+1}) are *equivalent*. For example, in Fig. 13.10, since both of local orientations of K around k_1 and k_2 increase according to V, k_1^{+1} and k_2^{+1} are equivalent, at the same time, k_1^{-1} and k_2^{-1} are equivalent. Notice that this relation does not depend on the orientation of $L_{k_1 k_2}$, but depends on the orientation of K.

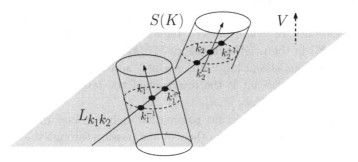

Fig. 13.10   Four points of $S(K)$ on the horizontal line $L_{k_1 k_2}$

Let D be the set of all p of $S(K)$ such that there exists a point p' in $S(K)$ such that p and p' are equivalent. Let D_c be the set of all local maximum and local minimum points of $S(K)$ according to the vector V. Then D must be the set of curves on the torus $S(K)$ which are paired by the equivalence defined in the previous paragraph. We assume that for each curve in D the orientation is given according to the orientation of K. In the end we obtain a surface diagram $D(K) = (S(K), D)$ without the "over/under" information. See Fig. 13.11.

By $\alpha(K)$ we denote the free surface knot represented by $D(K)$. Now, we are ready to formulate the main theorem.

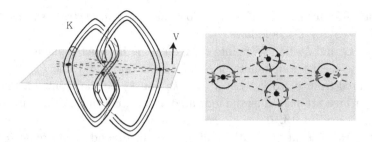

Fig. 13.11 $D(K)$ from the trefoil knot in $\mathbb{R}^3$

Theorem 13.4. *If knots K and K' are isotopic in $\mathbb{R}^3$, then the abstract surface knots $\alpha(K)$ and $\alpha(K')$ are equivalent.*

Proof. The proof essentially is a case-by-case consideration of all situations where K fails to be generic.

If we pass through a horizontal quadrisecant passing through some four points $k_1, k_2, k_3, k_4 \in K$ with the same z-coordinate, then we get two "antipodal" seventh Roseman moves. Indeed we can see four sheets of the surface $D(K)$ in the neighborhoods of $k_1^{\alpha_1}, k_2^{\alpha_2}, k_3^{\alpha_3}, k_4^{\alpha_4}$, which are equivalent. These four sheets have four secants, and the effect of the passing through the quadrisecant is the inversion of the tetrahedron. The same happens for $k_1^{-\alpha_1}, k_2^{-\alpha_2}, k_3^{-\alpha_3}, k_4^{-\alpha_4}$.

If we pass through a secant passing through two maxima k_1 and k_2, then two maxima contact each other. When passing through the secant we can see the second Roseman moves. $\square$

PART 4
Manifolds of Triangulations

Chapter 14

Introduction

The present part is the culmination of the whole book: here we construct "braid-group like" invariants of arbitrary manifolds.

Returning back to 1954, J.W. Milnor in his first paper on link group [Milnor, 1954] formulated the idea that manifolds which share lots of well known invariant (homotopy type, etc.) may have different *link groups*.

We shall look at braids on a manifold as loops in some configuration space related to the manifold. Following G_n^k-ideology, we mark singular configurations, while moving along a loop in the configuration space, to get a word — an element of a "G_n^k-like" group Γ_n^k. Singular configurations will arise here from the moments of transformation of triangulations spanned by the configuration points; in this case the good property of codimension one is

"k points of the configuration lie on a sphere of dimension $k - 3$ and there are no points inside the sphere".

In this part, we shall consider an arbitrary closed d-manifold M^d and for n large enough we shall construct the braid group $B_n(M^d) = \pi_1(C_n(M^d))$, where $C_n(M^d)$ is the space of generic (to some extent) configurations of points in M^d that is dense enough to span a triangulation of M. This braid group will admit a natural map to the group Γ_n^{d+2}. Here the number n will be estimated in terms of minimal number of triangles of a triangulation of M.

Part 4 is organised as follows. In Chapter 14.1 we consider points configurations and their triangulations in the plane and two-dimensional surfaces. Chapter 15 describes triangulation transformations in higher-dimensional spaces. Thus, we characterise the local structure of the manifold of triangulations of the manifold M^n. We propose there two approaches to the manifold of triangulations: a geometrical one which deals with Delaunay triangulations of M^n, and a purely combinatorial language [Nabutovsky, 1996a] which speaks in terms of PL-structure of M^n.

14.1 The manifold of triangulations

In the present section, for a manifold M^d and a natural number $n \gg d$, we shall define two non-compact manifolds of triangulations. They will be both of dimension nd. The geometric one $M_g^{nd}(M, n)$ will depend on a Riemannian metric g on M, and the topological one $M^{nd}(M, n)$ will depend only on the topological type of M.

The crucial property of these configuration spaces is the existence of "natural" maps from them to some special groups Γ_n^k.

For two-dimensional manifolds, these groups are Γ_n^4, for three-dimensional manifolds, the groups are Γ_n^5.

Fix a smooth manifold M of dimension d with a Riemannian metric g on it. Let $n \gg d$ be a large natural number. Consider the set of all Delaunay triangulations of M with n vertices $x_1, \ldots, x_n$ (if such triangulations exist). Such Delaunay triangulations are indexed by sets of vertices $x_1, \ldots, x_n$, hence, the set of such triangulation forms a subset of the configuration space $C_n(M)$. It is clear that the above subset will be an open (not necessarily connected) manifold of dimension dn: if some n pairwise distinct points $(x_1, \ldots, x_n)$ form a Delaunay triangulation then the same is true for any collection of points $(x_1', \ldots, x_n')$ in the neighbourhood of $(x_1, \ldots, x_n)$. We call a set of points $x_1, \ldots, x_n$ *admissible* if such a Delaunay triangulation exists.

Hence, we get a non-compact (open) manifold M_g^{dn}. This manifold has a natural structure. Namely, we call two collections $(x_1, \ldots, x_n)$ and $(x_1', \ldots, x_n')$ *adjacent* if there is a path $(x_{1,t}, \ldots, x_{n,t})$ for $t \in [0, 1]$ such that $(x_{1,0}, \ldots, x_{n,0}) = (x_1, \ldots, x_n)$ and $(x_{1,1}, \ldots, x_{n,1}) = (x_1', \ldots, x_n')$ and there exists exactly one $t_0 \in [0, 1]$ such that when passing through $t = t_0$ the Delaunay triangulation $(x_{1,t}, \ldots, x_{n,t})$ undergoes a flip (also called *Pachner move* or *bistellar move*). Generally, the Pachner moves, which do not change the number of points, may be not enough. In that case, we shall have many components and the invariant (with values in Γ_n^k) will be a set of groups. In dimension 3 the Pachner moves are sufficient [Matveev, 1984].

The flip (Pachner move) corresponds to a position when some $d + 2$

points $(x_{i_1,t_0}, \ldots, x_{i_{d+2},t_0})$ belong to the same $(d-1)$-sphere S^{d-1} such that no other point x_j lies inside the ball B^d bounded by this sphere.

Note that the above definition heavily depends on the metrics of the manifold M. For example, if we take M to be the torus glued from a square 1×10, the combinatorial structure of the triangulation manifold will heavily differ from the combinatorial structure for the case of the triangulation manifold for the case of the torus glued from a square 1×1.

The goal of the next section will be to construct the *topological manifold of triangulations*.

The idea is pretty simple. We try to catch all simplicial decompositions which may eventually happen for the manifold of triangulations for whatever metrics, and glue them together to get an open (non-compact) manifold.

Every triangulation of manifold has a natural stratification. Namely, every point of M_g^{dn} is given by a collection of $(x_1, \ldots, x_n)$. Such a collection is *generic* if there is no sphere S^{d-1} such that some $x_{i_1}, \ldots, x_{i_{d+2}}$ belong to this sphere with no other points among x_k belonging to this sphere.

We say that a point $(x_1, \ldots, x_n)$ is *of codimension* 1, if there exists exactly one sphere with exactly $d+2$ points on it and no points inside it; analogously, *codimension* 2 *strata* correspond to either one sphere with $d+3$ points or two spheres containing $d+2$ points each.

In dimension 2 this corresponds either one point of valency five in Voronoï tiling or two points of valency four in Voronoï tiling.

Hence, we have constructed a stratified (open) manifold M_g^{dn}. We call it the *metrical triangulation manifold*.

Note that the manifold M_g^{dn} may be not connected, i.e. there can exist non-equivalent triangulations. On the other hand, if one considers the spines of some manifold, they all can be transformed into each other by Matveev–Piergallini moves [Matveev, 1984].

Denote the connected components of M_g^{dn} by $(M_g^{dn})_1, \ldots, (M_g^{dn})_p$.

The *geometrical n-strand braid groups* $B(M_g, n)_j$ of M_g are the fundamental groups

$$\{\pi_1((M_g^{dn})_j)\}, \quad j = 1, \ldots, p.$$

By a *generalised cell* of such a stratification we mean a connected component of the set of generic points of M_g^{dn}.

We say that two generalised cells C_1 and C_2 are *adjacent* it there exist two points, say, $x = (x_1, \ldots, x_n)$ in C_1 and $x' = (x'_1, \ldots, x'_n)$ in C_2 and a path $x_t = (x_1(t), \ldots, x_n(t))$, such that $x_i(0) = x_i$ and $x_i(1) = x_i'$ such that all points on this path are generic except for exactly one point, say, corresponding to $t = t_0$, which belongs to the stratum of codimension 1.

Now, let us pass to the definition of the topological braid groups. To this end, we shall construct the 2-frame of the *topological manifold of triangulation*.

Consider a manifold M^n. We consider all possible metrics g_α as above. They naturally lead to $M_{g_\alpha}^{dn}$.

We say that two generic strata of $M_{g_\alpha}^{dn}, M_{g_\beta}^{dn}$ are *equivalent* if there is a homeomorphism of $M_{g_\alpha}^{dn} \to M_{g_\beta}^{dn}$ taking one stratum to the other.

A 0-cell of the manifold of triangulations is an equivalence class of generic strata.

Analogously, we define 1-cells of the manifold of triangulations as equivalence classes of pairs of adjacent vertices for different metrics $M_{g_\alpha}^{dn}, M_{g_\beta}^{dn}$ to the pair of adjacent vertices equivalent to the initial ones.

In a similar manner, we define 2-cells as equivalence classes of discs for metrics $M_{g_\alpha}^{dn}$ such that:

(1) vertices of any disc are points in 0-strata;
(2) edges of the disc connect vertices from adjacent 0-strata; each edge intersects codimension 1 set exactly in one point;
(3) the cycle is spanned by a disc which intersects codimension 2 set exactly at one point;
(4) the equivalence relation is defined by homeomorphism taking disc to disc, edge to an equivalent edge and vertex to an equivalent vertex and respects the stratification.

Thus, we get the 2-frame of the manifold M_{top}^{dn} which might be disconnected.

Now, we define the *topological braid groups* of $B(M, n)_j, j = 1, \ldots, p$, as fundamental groups $\{\pi_1((M_{top}^{dn})_j)\}, j = 1, \ldots, p$.

Chapter 15

The Two-dimensional Case

In Chapter 7 we defined a family of groups G_n^k for two positive integers $n > k$, and formulated the following principle:

If dynamical systems describing a motion of n particles admit some good codimension one property governed by exactly k particles, then these dynamical systems have a topological invariant valued in G_n^k.

The main examples coming from G_n^k-theory are homeomorphisms from the n-strand pure braid group to the groups G_n^3 and G_n^4, introduced in Chapter 8. If we consider a motion of n pairwise distinct points on the plane and choose the property "some three points are collinear", then we get a homomorphism from the pure n-strand braid group PB_n to the group G_n^3. If we choose the property "some four points belong to the same circle or line", we shall get a group homomorphism from PB_n to G_n^4.

In other words, in our examples we look for "walls" in the configuration space $C_n(\mathbb{R}^2)$ where some three points are collinear (or some four points are on the same circle or line). This condition can be well defined for $C_n(\mathbb{R}P^2)$.

But what would be the "good" codimension one property if we try to study similar configuration spaces or braids for some other topological spaces? First, our conditions will heavily depend on the metric: the property "three points are collinear" is metrical. On the other hand, even having some good metric chosen, we meet other obstacles because we need to know what is a "line". For example, there is no unique geodesic passing through two points in the general case. When finding all possible geodesics and trying to write a word corresponding to them, we shall see that "the word will contain infinitely many letters" in the case of irrational cable.

The detour for this problem will be as follows: we shall consider the condition "locally" and instead of "global configurations in spaces". We shall consider only Voronoï tiling or Delaunay triangulations. From this point of view, we deal with the space of triangulations with a fixed number of vertices, where any two adjacent triangulations are related by a Pachner move [Nabutovsky, 1996b], which is closely related to the group Γ_n^4 (see Definition 15.1).

On the other hand, triangulations of spaces and Pachner moves are also related to Yang–Baxter maps (see [Dynnikov, 2002]). Moreover, in [Cho, Yoon and Zickert, 2018; Hikami and Inoue, 2015] a boundary-parabolic $PSL(2, \mathbb{C})$-representation of $\pi_1(S^3 \backslash K)$ for a hyperbolic knot K is studied by using cluster algebras and *flips* — Pachner moves for 2-dimensional triangulations. Since the group Γ_n^4 is closely related to triangulations of spaces and Pachner moves, it can be expected to obtain invariants by means of the group Γ_n^4 not only for braids, but also for knots, which are obtained by closing braids.

Now we recall the definition of restricted configuration spaces $C_n'(\mathbb{R}^{k-1})$, introduced in Section 11.1: a point of $C_n'(\mathbb{R}^{k-1})$ is an ordered set of n distinct points in $\mathbb{R}^{k-1}$, such that every $(k-1)$ points of it are in general position. In particular, for $k = 3$, the only condition is that no two points among the given n points coincide and the fundamental group $\pi_1(C_n'(\mathbb{R}^{3-1}))$ is precisely the Artin pure braid group. For $k = 4$, for points $x_1, \ldots, x_n$ in three-space we require that no three points are collinear (though some four points can belong to the same plane). We call elements in $\pi_1(C_n'(\mathbb{R}^{4-1}))$ *braids on n strands for $\mathbb{R}^3$*. We call elements in $\pi_1(C_n'(\mathbb{R}P^{4-1}))$ *braids on n strands for $\mathbb{R}P^3$*. In Section 11.1 the following statement is proved:

Proposition 15.1. *There exists a group homomorphism f_n^k from*

$$\pi_1(C_n'(\mathbb{R}P^{k-1}))$$

to G_n^k.

As before, for a path $\gamma \in \pi_1(C_n'(\mathbb{R}P^{k-1}))$ the mapping $f_n^k(\gamma)$ is defined by writing a_m in a sequence, when exactly one k-tuple $(x_i)_{i \in m}$ of points belongs to a $(k-2)$-plane, where m is the set of indices for these k points $\{x_i\}_{i \in m} \subset \mathbb{R}P^{k-1}$ on the $(k-2)$-plane.

15.1 The group Γ_n^4 definition. A group homomorphism from PB_n to Γ_n^4

Definition 15.1. The group Γ_n^4 is the group generated by

$$\{d_{(ijkl)} \mid \{i, j, k, l\} \subset \bar{n}, |\{i, j, k, l\}| = 4\}$$

subject to the following relations:

(1) $d^2_{(ijkl)} = 1$ for $(i, j, k, l) \subset \bar{n}$,
(2) $d_{(ijkl)}d_{(stuv)} = d_{(stuv)}d_{(ijkl)}$, for $|\{i, j, k, l\} \cap \{s, t, u, v\}| < 3$,
(3) $d_{(ijkl)}d_{(ijlm)}d_{(jklm)}d_{(ijkm)}d_{(iklm)} = 1$ for distinct i, j, k, l, p,
(4) $d_{(ijkl)} = d_{(kjil)} = d_{(ilkj)} = d_{(klij)} = d_{(jkli)} = d_{(jilk)} = d_{(lkji)} = d_{(lijk)}$
   for distinct i, j, k, l.

The group Γ^4_n is naturally related to triangulations of 2-surfaces and the Pachner moves for the two dimensional case, called "flip", see Fig. 15.1. More precisely, a generator $d_{(ijkl)}$ of Γ^4_n corresponds to a flip in the sequence constituting the Pentagon relation as described in Fig. 15.5, the most important relation for the group Γ^4_n.

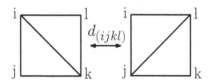

Fig. 15.1   Flip on a rectangle $ijkl$

More specifically, the relation $d_{(ijkl)}d_{(ijlm)}d_{(jklm)}d_{(ijkm)}d_{(iklm)} = 1$ corresponds to the sequence of flips applied to a triangulation of a pentagon.

In this section we construct a group homomorphism f_n from PB_n to Γ^4_n. The topological background for that is very easy: we consider codimension 1 "walls" which correspond to generators (flips) and codimension 2 singularities correspond to relations (of the group Γ). Having this, we construct a map on the level of generators and prove its correctness.

15.1.1   *Geometric description of the mapping from PB_n to Γ^4_n.*

Let us consider b_{ij} as the dynamical system described in Section 8.2, page 176. The homomorphism f_n from PB_n to Γ^4_n can be defined as follows: for the above dynamical system for each generator b_{ij}, let us enumerate the moments $0 < t_1 < t_2 < \cdots < t_l < 1$ such that at the moment t_k four points belong to one circle without points inside the circle. At the moment t_k, if P_s, P_t, P_u, P_v are positioned on one circle as the indicated order and there are no points inside the circle, then we define $\delta_k = d_{(stuv)}$. With the pure braid b_{ij} we associate the product $f_n(b_{ij}) = \delta_1 \delta_2 \ldots \delta_l$.

Remark 15.1. From the choice of base points $\{P_1, \ldots, P_n\}$ it follows that points $\{P_1, \ldots, P_{j-1}\} \cup \{P_{p+1}, \ldots P_{q-1}\}$ are placed inside the circle passing through points P_j, P_p, P_q for $j < p < q$. That is, inside the circle passing through P_j, P_p, P_q for $j < p < q$ there are $j - 1 + q - p - 1$ points from $\{P_1, \ldots, P_n\}$.

15.1.2 Algebraic description of the mapping from PB_n to Γ_n^4.

Let us describe the mapping $f_n : PB_n \to \Gamma_n^4$ as follows: Let us denote

$$
d_{\{p,q,(r,s)_s\}} = \begin{cases}
d_{(pqrs)} & \text{if } p < q < s, \\
d_{(prsq)} & \text{if } p < s < q, \\
d_{(rspq)} & \text{if } s < p < q, \\
d_{(qprs)} & \text{if } q < p < s, \\
d_{(qrsp)} & \text{if } q < s < p, \\
d_{(rsqp)} & \text{if } s < q < p.
\end{cases}
$$

Remark 15.2. Note that the generator $d_{\{p,q,(r,s)_s\}}$ corresponds to four points P_p, P_q, P_r, P_s such that they are placed on a circle according to the order of p, q, s and the point P_r is placed close to P_s for the orientation P_r to P_s to be the counterclockwise orientation, see Fig. 15.2. The subscript s of $(r, s)_s$ means that the point P_s does not move, but the point P_r moves turning around the point P_s after this moment. In other words, the generator $d_{\{p,q,(r,s)_s\}}$ corresponds to the moment when the point P_r is "moving" closely to the point P_s, turning around P_s. We would like to highlight that $d_{\{p,q,(r,s)_s\}} \neq d_{\{p,q,(s,r)_s\}}$.

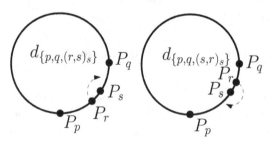

Fig. 15.2   For $p < s < q$, $d_{\{p,q,(r,s)_s\}} = d_{(prsq)}$, but $d_{\{p,q,(s,r)_s\}} = d_{(psrq)}$.

Let us denote $k \in \{p, q, r\}$ such that $min\{p, q, r\} < k < max\{p, q, r\}$ by $mid\{p, q, r\}$. Let us define $\gamma_{\{p,q,(i,j)_j\}}$ as follows: If $min\{p, q, j\} < i <$

$mid\{p, q, j\}$ or $i > max\{p, q, j\}$, then

$$\gamma_{\{p,q,(i,j)_j\}} =$$
$$\begin{cases} d_{\{p,q,(i,j)_j\}}, & \text{if } min\{p,q,j\} - mid\{p,q,j\} + max\{p,q,j\} - 2 = 0, \\ 1, & \text{if } min\{p,q,j\} - mid\{p,q,j\} + max\{p,q,j\} - 2 \neq 0. \end{cases}$$
$$(15.1)$$

If $i < min\{p, q, j\}$ or $mid\{p, q, j\} < i < max\{p, q, j\}$, then

$$\gamma_{\{p,q,(i,j)_j\}} =$$
$$\begin{cases} d_{\{p,q,(i,j)_j\}}, & \text{if } min\{p,q,j\} - mid\{p,q,j\} + max\{p,q,j\} - 2 = 1, \\ 1, & \text{if } min\{p,q,j\} - mid\{p,q,j\} + max\{p,q,j\} - 2 \neq 1. \end{cases}$$
$$(15.2)$$

As asserted in Remark 15.1 $min\{p, q, j\} - mid\{p, q, j\} + max\{p, q, j\} - 2$ is the number of points inside the circle passing points P_j, P_p, P_q. Formulas (15.1) and (15.2) are describing that if four points P_i, P_j, P_p, P_q belong to one circle without points inside the circle, then we write a generator $d_{\{p,q,(i,j)_j\}}$, otherwise, the identity 1.

Let $b_{ij} \in PB_n$, $1 \leq i < j \leq n$, be a generator. Consider the elements

$$\Delta^I_{i,(i,j)} = \prod_{p=2}^{j-1} \prod_{q=1}^{p-1} \gamma_{\{p,q,(i,j)_j\}},$$

$$\Delta^{II}_{i,(i,j)} = \prod_{p=1}^{j-1} \prod_{q=1}^{n-j} \gamma_{\{(j-p),(j+q),(i,j)_j\}},$$

$$\Delta^{III}_{i,(i,j)} = \prod_{p=1}^{n-j-1} \prod_{q=0}^{p-1} \gamma_{\{(n-p),(n-q),(i,j)_j\}},$$

$$\Delta_{i,(i,j)} = \Delta^{II}_{i,(i,j)} \Delta^I_{i,(i,j)} \Delta^{III}_{i,(i,j)}.$$

Now we define $f_n : PB_n \to \Gamma_n^4$ by

$$f_n(b_{ij}) = \Delta_{i,(i,(i+1))} \cdots \Delta_{i,(i,(j-1))} \Delta_{i,(i,j)} \Delta_{i,(j,i)} \Delta^{-1}_{i,((j-1),i)} \cdots \Delta^{-1}_{i,((i+1),i)},$$

for $1 \leq i < j \leq n$.

Theorem 15.1. *The map* $f_n : PB_n \to \Gamma_n^4$, *defined above, is a well defined homomorphism.*

Proof. When we consider isotopies between two pure braids, it suffices to take into account only singularities of codimension at most two. Singularities of codimension one give rise to generators, and relations come from singularities of codimension two. Now we list the cases of singularities of codimension two explicitly.

(1) One point moving on the plane is tangent to the circle, which passes through three points, see Fig. 15.3. This corresponds to the relation $d^2_{(ijkl)} = 1$.

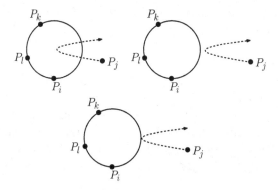

Fig. 15.3   A point P_j moves, being tangent to the circle, which passes through P_i, P_k and P_l

(2) There are two sets A and B of four points each, which are on the same circles such that $|A \cap B| \le 2$, see Fig. 15.4 with $A = \{ijkl\}, B = \{stuv\}$. This corresponds to the relation $d_{(ijkl)} d_{(stuv)} = d_{(stuv)} d_{(ijkl)}$.

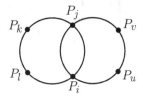

Fig. 15.4   Two sets A and B of four points on the circles such that $|A \cap B| = 2$

(3) There are five points $\{P_i, P_j, P_k, P_l, P_m\}$ on the same circle. We obtain the sequence of five subsets of $\{P_i, P_j, P_k, P_l, P_m\}$ with four points on the same circle, which corresponds to the flips on the pentagon, see

Fig. 15.5. This corresponds to the relation

$$d_{(ijkl)}d_{(ijlm)}d_{(jklm)}d_{(ijkm)}d_{(iklm)} = 1.$$

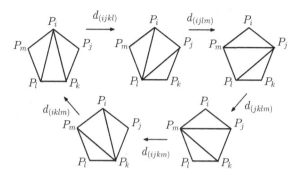

Fig. 15.5 A sequence of five subsets of $\{P_i, P_j, P_k, P_l, P_m\}$, corresponding to flips on the pentagon

$\square$

15.2 A group homomorphism from PB_n to $\Gamma_n^4 \times \Gamma_n^4$

In this section we construct the group homomorphism f_n^2 from PB_n to $\Gamma_n^4 \times \Gamma_n^4$ for a positive even integer $n = 2k$. Roughly speaking, f_n^2 will be defined by reading generators of $\Gamma_n^4 \times \Gamma_n^4$, which correspond to four points on the same circle, but we distinguish between them with respect to the number of points inside the circle.

15.2.1 *Geometric description of the mapping from PB_n to $\Gamma_n^4 \times \Gamma_n^4$.*

Let us consider the dynamical system for the generator b_{ij}, which is described in Section 8.2, page 176. Let us enumerate $0 < t_1 < t_2 < \cdots < t_l < 1$ such that at the moment t_k four points are positioned on one circle (or on the line). Note that by the assumption of the dynamical system for a generator b_{ij}, there are no four points on the circle at the beginning.

Let us assume that at the moment t_k the points P_s, P_t, P_u, P_v are positioned on one circle in the indicated order. If there are even number of points inside the circle, where the four points P_s, P_t, P_u, P_v are placed, then t_k corresponds to $\delta_k = (d_{(stuv)}, 1) \in \Gamma_n^4 \times \Gamma_n^4$. Otherwise, t_k corresponds to $\delta_k = (1, d_{(stuv)}) \in \Gamma_n^4 \times \Gamma_n^4$.

If P_s, P_t, P_u, P_v are positioned on one line l at the moment t_k in the indicated order, then there are two connected components of $\mathbb{R}^2 \backslash l$. If one (hence, any) of connected components of $\mathbb{R}^2 \backslash l$ contains an even number of points, then t_k corresponds to $\delta_k = (d_{(stuv)}, 1) \in \Gamma_n^4 \times \Gamma_n^4$. Otherwise, t_k corresponds to $\delta_k = (1, d_{(stuv)}) \in \Gamma_n^4 \times \Gamma_n^4$. With the pure braid b_{ij} we associate the product $f_n^2(b_{ij}) = \delta_1 \delta_2 \ldots \delta_l$.

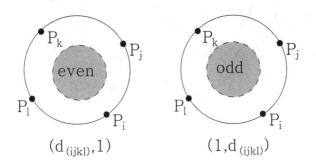

Fig. 15.6   Geometric description for $f_{2,n}$

Remark 15.3. Since we consider the case of $n = 2k$, the numbers of points inside and outside the circle, where four points are positioned, are same modulo 2. Analogously, the numbers of points contained in each connected component of $\mathbb{R}^2 \backslash l$ are same modulo 2.

15.2.2   *Algebraic description of the mapping from PB_n to $\Gamma_n^4 \times \Gamma_n^4$.*

Let us define $\delta_{\{p,q,(i,j)_j\}}$ as follows. If $min\{p,q,j\} < i < mid\{p,q,j\}$ or $i > max\{p,q,j\}$, then

$$\delta_{\{p,q,(i,j)_j\}} = \begin{cases} (d_{\{p,q,(i,j)_j\}}, 1), & \text{if } j + p + q \equiv 0 \text{ mod } 2, \\ (1, d_{\{p,q,(i,j)_j\}}), & \text{if } j + p + q \equiv 1 \text{ mod } 2. \end{cases}$$

If $i < min\{p,q,j\}$ or $mid\{p,q,j\} < i < max\{p,q,j\}$, then

$$\delta_{\{p,q,(i,j)_j\}} = \begin{cases} (d_{\{p,q,(i,j)_j\}}, 1), & \text{if } j + p + q \equiv 1 \text{ mod } 2, \\ (1, d_{\{p,q,(i,j)_j\}}), & \text{if } j + p + q \equiv 0 \text{ mod } 2. \end{cases}$$

Let $b_{ij} \in PB_n$, $1 \le i < j \le n$ be a generator. Consider the elements

$$D_{i,(i,j)}^I = \prod_{p=2}^{j-1} \prod_{q=1}^{p-1} \delta_{\{p,q,(i,j)_j\}}, \qquad (15.3)$$

$$D_{i,(i,j)}^{II} = \prod_{p=1}^{j-1} \prod_{q=1}^{n-j} \delta_{\{(j-p),(j+q),(i,j)_j\}}, \qquad (15.4)$$

$$D_{i,(i,j)}^{III} = \prod_{p=1}^{n-j-1} \prod_{q=0}^{p-1} \delta_{\{(n-p),(n-q),(i,j)_j\}}, \qquad (15.5)$$

$$D_{i,(i,j)} = D_{i,(i,j)}^{II} D_{i,(i,j)}^I D_{i,(i,j)}^{III}. \qquad (15.6)$$

Now we define $f_n^2 : PB_n \to \Gamma_n^4 \times \Gamma_n^4$ by

$$f_n^2(b_{ij}) = D_{i,(i,(i+1))} \dots D_{i,(i,(j-1))} D_{i,(i,j)} D_{i,(j,i)} D_{i,((j-1),i)}^{-1} \dots D_{i,((i+1),i)}^{-1}$$

for $1 \le i < j \le n$.

Theorem 15.2. *The map $f_n^2 : PB_n \to \Gamma_n^4 \times \Gamma_n^4$, which is defined as above, is a well defined homomorphism.*

Proof. The proof for this statement is analogous to the proof for Theorem 15.3. □

15.3 A group homomorphism from PB_n to $\Gamma_n^4 \times \cdots \times \Gamma_n^4$

For a positive integer $n > 5$, the homomorphism $f_n^2 : PB_n \to \Gamma_n^4$ can be extended to a mapping from PB_n to the product of $([\frac{n-4}{2}] + 1)$ copies of Γ_n^4 as follows. We denote the product of $([\frac{n-4}{2}] + 1)$ copies of the group Γ_n^4 by

$$\Gamma_{n,0}^4 \times \Gamma_{n,1}^4 \times \cdots \times \Gamma_{n,[\frac{n-4}{2}]}^4.$$

The idea is that we can not only distinguish between "evenly many points inside the circle" or "oddly many points inside the circle", but also just count this number of points modulo $n - 4$.

Let $r = n - 4$. For the dynamical system corresponding to the generator b_{ij}, which is described in Section 2, let us enumerate $0 < t_1 < t_2 < \cdots < t_l < 1$ such that at the moment t_k four points P_s, P_t, P_u, P_v are positioned on one circle or one line. In our construction, lines are considered as circles passing through "the infinity point". The orders of points P_s, P_t, P_u, P_v on one line are not permutations, but circular permutations of points P_s, P_t, P_u, P_v.

The idea of the construction of the map $f_n^r : PB_n \to \Gamma_{n,0}^4 \times \Gamma_{n,1}^4 \times \cdots \times \Gamma_{n,[\frac{r}{2}]}^4$ is the following. When four points lie on one circle C, there are $n - 4$ points inside and outside the circle on the plane. We want to count the number of points inside the circle, but we cannot distinguish, where is the inside or the outside of the circle, since we consider the plane without an orientation. So, we consider the numbers of points on both of connected components of $\mathbb{R}\backslash C$. If there are α points contained in one of connected components of $\mathbb{R}\backslash C$, then there are $n - 4 - \alpha$ points contained on the other connected component and $n - 4 - \alpha \equiv -\alpha \bmod r = n - 4$. Now, when four points lie on one circle C and each of components contain α and $-\alpha \pmod r$ points where four points lie on one circle C and each of components contain α and $-\alpha \pmod r$ points where $\alpha \le [\frac{r}{2}]$, the moment corresponds to a generator from α-th copy of Γ_n^4. Now we define the map $f_n^r : PB_n \to \Gamma_{n,0}^4 \times \Gamma_{n,1}^4 \times \cdots \times \Gamma_{n,[\frac{r}{2}]}^4$ more precisely.

If P_s, P_t, P_u, P_v are positioned on one line L or one circle C at the moment t_k in the indicated order, then there are two connected components of $\mathbb{R}^2\backslash L$ (or $\mathbb{R}^2\backslash C$). If the numbers of points contained in connected components of $\mathbb{R}^2\backslash L$ (or $\mathbb{R}^2\backslash C$) are α and $n - 4 - \alpha \bmod r$ for some $\alpha \le [\frac{r}{2}]$, then t_k corresponds to $\delta_k = (1, \ldots, 1, d_{(stuv)}, 1, \ldots, 1) \in \{1\} \times \ldots \{1\} \times \Gamma_{n,\alpha}^4 \times \{1\} \times \ldots \{1\} \subset \Gamma_{n,0}^4 \times \Gamma_{n,1}^4 \times \cdots \times \Gamma_{n,[\frac{r}{2}]}^4$.

With the pure braid b_{ij} we associate the product $f_n^r(b_{ij}) = \delta_1 \delta_2 \ldots \delta_l$.

Algebraically this construction can be presented as follows:

$$D_{i,(i,j)}^I = \prod_{p=2}^{j-1} \prod_{q=1}^{p-1} \delta_{\{p,q,(i,j)_j\}}^r,$$

$$D_{i,(i,j)}^{II} = \prod_{p=1}^{j-1} \prod_{q=1}^{n-j} \delta_{\{(j-p),(j+q),(i,j)_j\}}^r,$$

$$D_{i,(i,j)}^{III} = \prod_{p=1}^{n-j-1} \prod_{q=0}^{p-1} \delta_{\{(n-p),(n-q),(i,j)_j\}}^r,$$

where

$$\delta_{\{p,q,(i,j)_j\}}^r = (1, \ldots, d_{\{p,q,(i,j)_j\}}, 1, \ldots, 1)$$
$$\in \{1\} \times \ldots \{1\} \times \Gamma_{n,\alpha}^4 \times \{1\} \times \ldots \{1\}$$
$$\subset \Gamma_{n,0}^4 \times \Gamma_{n,1}^4 \times \cdots \times \Gamma_{n,[\frac{r}{2}]}^4,$$

if

$$\begin{cases} min\{j,p,q\} - mid\{j,p,q\} + max\{j,p,q\} - 2 \equiv \\ \alpha \ mod \ r, \quad \text{if } min\{p,q,j\} < i < mid\{p,q,r\} \text{ or } i > max\{p,q,j\}, \alpha \leq [\tfrac{r}{2}], \\ -\alpha \ mod \ r, \quad \text{if } min\{p,q,j\} < i < mid\{p,q,r\} \text{ or } i > max\{p,q,j\}, \alpha \geq [\tfrac{r}{2}], \\ \alpha - 1 \ mod \ r, \quad \text{if } i < min\{p,q,j\} \text{ or } mid\{p,q,j\} < i < max\{p,q,j\}, \alpha \leq [\tfrac{r}{2}], \\ -\alpha - 1 \ mod \ r, \quad \text{if } i < min\{p,q,j\} \text{ or } mid\{p,q,j\} < i < max\{p,q,j\}, \alpha \geq [\tfrac{r}{2}]. \end{cases}$$

$$D_{i,(i,j)} = D^{II}_{i,(i,j)} D^{I}_{i,(i,j)} D^{III}_{i,(i,j)}.$$

Now we define $f_n^r : PB_n \to \Gamma^4_{n,0} \times \Gamma^4_{n,1} \times \cdots \times \Gamma^4_{n,[\frac{r}{2}]}$ by

$$f_n^r(b_{ij}) = D_{i,(i,(i+1))} \ldots D_{i,(i,(j-1))} D_{i,(i,j)} D_{i,(i,j)} D^{-1}_{i,(i,(j-1))} \cdots D^{-1}_{i,(i,(i+1))},$$

for $1 \leq i < j \leq n$.

Theorem 15.3. *The map $f_n^r : PB_n \to \Gamma^4_{n,0} \times \Gamma^4_{n,1} \times \cdots \times \Gamma^4_{n,[\frac{r}{2}]}$ defined above is a well defined homomorphism.*

Proof. This statement might be proved similarly to the proof of Theorem 15.1. Let us list cases of singularities of codimension two explicitly. From now on let us consider four points on one circle. For the case of four points on one line the proof is analogous. Note that the image of f_n^r from four points on one circle depends on the number of points inside the circle. Without loss of generality, the number of points inside the circle is less than or equal to the number of points outside the circle.

(1) One point moving on the plane is tangent to the circle, which passes through three points, see the middle in Fig. 15.3. Note that the number of points inside the circle does not change when the point P_j moves. It is easy to see that the image, when one point passes through the circle twice (upper left in Fig. 15.3), is equal to

$$(1, \ldots, 1, d_{(ijkl)} d_{(ijkl)}, 1, \ldots, 1) \in \Gamma^4_{n,0} \times \cdots \times \Gamma^4_{n,\alpha} \times \cdots \times \Gamma^4_{n,[\frac{r}{2}]},$$

where the α is the number of points mod r inside the circle, passing through P_i, P_k, P_l. If the point does not pass through the circle (upper right in Fig. 15.3), then the image is

$$(1, \ldots, 1, \ldots, 1) \in \Gamma^4_{n,0} \times \cdots \times \Gamma^4_{n,\alpha} \times \cdots \times \Gamma^4_{n,[\frac{r}{2}]}.$$

The equality of those two images is obtained by the relation $d^2_{(ijkl)} = 1$.

(2) There are two sets $A = \{P_i, P_j, P_k, P_l\}$ and $B = \{P_s, P_t, P_u, P_v\}$ of four points, which are on the same circles such that $|A \cap B| \leq 2$, see Fig. 15.4. If the numbers of points inside the circles, which pass through points

$\{P_i, P_j, P_k, P_l\}$ and $\{P_s, P_t, P_u, P_v\}$ respectively, are the same mod r, then the images of them are

$$(1, \ldots, d_{(ijkl)}d_{(stuv)}, 1, \ldots, 1) \in \Gamma^4_{n,0} \times \cdots \times \Gamma^4_{n,\alpha} \times \cdots \times \Gamma^4_{n,[\frac{r}{2}]},$$

and

$$(1, \ldots, d_{(stuv)}d_{(ijkl)}, 1, \ldots, 1) \in \Gamma^4_{n,0} \times \cdots \times \Gamma^4_{n,\alpha} \times \cdots \times \Gamma^4_{n,[\frac{r}{2}]}.$$

The equality of them follows from the relation $d_{(ijkl)}d_{(stuv)} = d_{(stuv)}d_{(ijkl)}$.

If the number of points inside the circles, where the points $\{P_i, P_j, P_k, P_l\}$ and $\{P_s, P_t, P_u, P_v\}$ are positioned respectively, are different mod r, then the images of them are

$$(1, \ldots, d_{(ijkl)}, \ldots, 1, \ldots 1)(1, \ldots, 1, \ldots, d_{(stuv)}, \ldots 1)$$
$$\in \Gamma^4_{n,0} \times \cdots \times \Gamma^4_{n,\alpha} \times \cdots \times \Gamma^4_{n,\beta} \times \cdots \times \Gamma^4_{n,[\frac{r}{2}]},$$

and

$$(1, \ldots, 1, \ldots, d_{(stuv)}, \ldots 1)(1, \ldots, d_{(ijkl)}, \ldots, 1, \ldots 1)$$
$$\in \Gamma^4_{n,0} \times \cdots \times \Gamma^4_{n,\alpha} \times \cdots \times \Gamma^4_{n,\beta} \times \cdots \times \Gamma^4_{n,[\frac{r}{2}]},$$

where α and β depend on the numbers of points inside the circles, which pass through $\{P_i, P_j, P_k, P_l\}$ and $\{P_s, P_t, P_u, P_v\}$, respectively. It is easy to obtain the equality of them.

(3) There are five points $\{P_i, P_j, P_k, P_l, P_m\}$ on the same circle. We obtain 10 generators. Note that if there are α points inside a circle, where four points lie, then in the very next time, when four points among $\{P_i, P_j, P_k, P_l, P_m\}$ lie on the same circle, there are $\alpha + 1$ points inside the circle, but in the very next step, there are α points inside the circle again. More precisely, we obtain the following element:

$$(1, \ldots, d_{(ijkl)}d_{(ijlm)}d_{(jklm)}d_{(ijkm)}d_{(iklm)}, d_{(ijkm)}d_{(iklm)}d_{(ijkl)}d_{(ijlm)}$$
$$d_{(jklm)}, \ldots, 1) \in \Gamma^4_{n,0} \times \cdots \times \Gamma^4_{n,\alpha} \times \Gamma^4_{n,\alpha+1} \times \cdots \times \Gamma^4_{n,[\frac{r}{2}]}.$$

From the relation

$$d_{(ijkl)}d_{(ijlm)}d_{(jklm)}d_{(ijkm)}d_{(iklm)} = 1$$
$$= d_{(ijkm)}d_{(iklm)}d_{(ijkl)}d_{(ijlm)}d_{(jklm)}$$

of Γ^4_n we obtain that it is equal to

$$(1, \ldots, 1, 1, \ldots, 1) \in \Gamma^4_{n,0} \times \cdots \times \Gamma^4_{n,\alpha} \times \Gamma^4_{n,\alpha+1} \times \cdots \times \Gamma^4_{n,[\frac{r}{2}]},$$

and the proof is completed.

$\square$

15.4 Braids in $\mathbb{R}^3$ and groups Γ_n^4

In Chapter 11, we studied the notion of braids for $\mathbb{R}^3$ and $\mathbb{R}P^3$. *A braid for* $\mathbb{R}^3$ *(or* $\mathbb{R}P^3$*)* is a path in a configuration space $C_n'(\mathbb{R}^3)$ (or $C_n'(\mathbb{R}P^3)$) which meets some conditions. If the initial and final points of the path in $C_n'(\mathbb{R}^3)$ coincide, then the path is called *a pure braid for* $\mathbb{R}^3$ ($\mathbb{R}P^3$). In the present section, we shall construct a group homomorphism from pure braids on n strands in $\mathbb{R}^3$ to Γ_n^4.

We shall consider (good and transverse, see Definition 11.1) pure braids on n strands in $\mathbb{R}^3$ and construct a group homomorphism from pure braids on n strands in $\mathbb{R}^3$ to the group Γ_n^4. Each element of Γ_n^4 corresponds to a moment when four points lie on $(4-2)$-dimensional plane in $\mathbb{R}^{4-1}$, but in the case of Γ_n^4 "the order" of four points on $(4-2)$-dimensional plane is very important. This order was ignored when the group homomorphism from pure braids in $\mathbb{R}^3$ to Γ_n^4 was constructed. Now we formulate more precisely how the group homomorphism from pure braids on n strands in $\mathbb{R}^3$ to Γ_n^4 is constructed.

Let γ be a good and transverse pure braid on n strands with base point $x = (x_1, \ldots, x_n)$ in $\mathbb{R}^3$. We call $t \in [0,1]$ *a special singular moment of* γ if the following hold:

(1) At the moment t four points x_p, x_q, x_r, x_s are on the same plane Π_t.
(2) The four points x_p, x_q, x_r, x_s in this order form a convex quadrilateral on Π_t.
(3) All of $\{x_1, \ldots, x_n\} \backslash \{x_p, x_q, x_r, x_s\}$ are located in the same connected component of $\mathbb{R}^3 \backslash \Pi_t$.

A normal vector v_t of Π_t pointing to the connected component, in which the points $\{x_1, \ldots, x_n\} \backslash \{x_p, x_q, x_r, x_s\}$ are placed, is called *the pointing vector* at the special singular moment t.

Remark 15.4.

(1) The plane Π_t admits a unique orientation with respect to v_t.
(2) Naturally, the quadrilateral admits an orientation with respect to v_t, see Fig. 15.7.

Let us enumerate all special singular moments $0 < t_1 < \cdots < t_l < 1$ of the path γ. For each t_s by definition of good pure braids on n strands there are exactly four points $\{x_p, x_q, x_r, x_s\}$ on the plane Π_{t_s}. As indicated in Remark 15.4 the convex quadrilateral on the plane Π_{t_s} with four

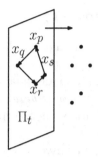

Fig. 15.7   A special singular moment, corresponding to $d_{(pqrs)}$

vertices $\{x_p, x_q, x_r, x_s\}$ admits an orientation with respect to v_{t_s}. If four points x_p, x_q, x_r, x_s are positioned as indicated order in accordance with the orientation with respect to v_{t_s}, then we associate the moment t_s to $d_{t_s} = d_{(pqrs)}$. Let us define the map $g \colon \pi_1(C'_n(\mathbb{R}^3)) \to \Gamma_n^4$ by the formula $g(\gamma) = d_{t_1} \ldots d_{t_l}$.

Theorem 15.4. *The map* $g \colon \pi_1(C'_n(\mathbb{R}^3)) \to \Gamma_n^4$ *is well-defined.*

Proof. We consider moments of isotopy between two paths, when the path at some moment in the isotopy between two paths is not good or not transverse. Let us list such cases explicitly.

(1) There are four points on a plane, which disappears after a small perturbation, see Fig. 15.8. This corresponds to the relation $d_{(pqrs)}^2 = 1$.

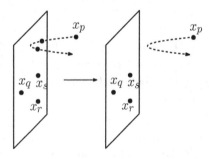

Fig. 15.8   Case 1: Four points on a plane, which disappears after a small perturbation

(2) At a moment there are two sets of four points m and m' with $|m \cap m'| < 3$, which are placed on planes at the same moment, see Fig. 15.9. This corresponds to the relation $d_{(ijkl)} d_{(stuv)} = d_{(stuv)} d_{(ijkl)}$.

(3) At a moment five points belong to the same plane plane. This is similar to the case of "five points on the circle" in the proof of Theorem 15.1.

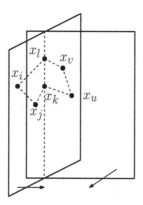

Fig. 15.9   Case 2: Two sets of four points $\{x_i, x_j, x_k, x_l\}$ and $\{x_l, x_k, x_u, x_v\}$ on planes at the same moment

This corresponds to the relation $d_{(ijkl)}d_{(ijlm)}d_{(jklm)}d_{(ijkm)}d_{(iklm)} = 1$ of Γ_n^4.

$\square$

Let $\{P_1(t), \ldots, P_n(t)\}_{t \in [0,1]}$ be n moving points in $\mathbb{R}^3$, corresponding to a path in $\pi_1(C_n'(\mathbb{R}^3))$. We may assume that the points $\{P_1(t), \ldots, P_n(t)\}_{t \in [0,1]}$ move inside a sphere with sufficiently large diameter. Let us fix four points $\{A, B, C, D\}$ on the sphere. This leads to a triangulation of the 3-ball with vertices

$$\{P_1(t), \ldots, P_n(t)\} \cup \{A, B, C, D\}$$

which can be obtained for each $t \in [0,1]$, see Fig. 15.10.

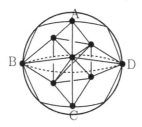

Fig. 15.10   Triangulation of 3-disc with $\{P_1(t), \ldots, P_n(t)\}$ inside the sphere and four points $\{A, B, C, D\}$ on the sphere

On the other hand, as described in Fig. 15.11, the moving of a vertex of the triangulation can be described by applying *the Pachner moves*

Fig. 15.11   Pachner move and moving of a vertex of the triangulation of 3-dimensional space

to the triangulation of a 3-dimensional space.   In other words, a path $\{P_0(t), \ldots, P_n(t)\}$ in $\pi_1(C_n'(\mathbb{R}^3))$ can be described by a finite sequence of "Pachner moves" applied to the triangulations of the sphere, see Fig. 15.12.

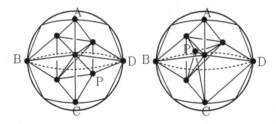

Fig. 15.12   Applying a Pachner move to the triangulation of a 3-dimensional space and the moving of a vertex

It can be expected that by using the triangulation of a sphere with $n+4$ points and the sequence of the Pachner moves, we obtain an invariant for (pure) braids.

15.5   Lines moving on the plane and the group Γ_n^k

Let $L = \{L_1, L_2, \ldots, L_n\}$ be n distinct lines on the plane. We shall consider lines $L(t) = \{L_1(t), \ldots, L_n(t)\}_{t \in [0,1]}$, moving on the plane, satisfying the following:

(1) $L(0) = L = L(1)$,
(2) $L_i(t) \neq L_j(t)$ for $i \neq j$ for every $t \in [0, 1]$,

and we call it *good moving* n *lines*.

Two sets of moving lines $L(t) = \{L_1(t), L_2(t), \ldots, L_n(t)\}$ and $L'(t) = \{L_1'(t), L_2'(t), \ldots, L_n'(t)\}$ are *equivalent*, if there is a continuous family of homeomorphisms $\{\gamma_s : \mathbb{R}^2 \to \mathbb{R}^2\}_{s \in [0,1]}$ such that

(1) for each $s \in [0,1]$, $\gamma_s(L_i(t))$ is a line,
(2) $\gamma_0(L(t)) = L(t)$ and $\gamma_0(L(t)) = L'(t)$.

We call the moment $t_0 \in [0,1]$ *singular* if there are four lines $L_i(t_0), L_j(t_0), L_k(t_0), L_l(t_0)$ at the moment t_0 such that they are tangent to one circle with an orientation, which is obtained from the orientation of plane, see Fig. 15.13.

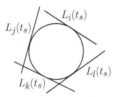

Fig. 15.13 Four lines $L_i(t_0), L_j(t_0), L_k(t_0), L_l(t_0)$, being tangent to a circle

If $L(t)$ has finitely many singular moments, then we call it *pleasant*.

15.5.1 A map from a group of good moving lines to G_n^4

Let $L(t)$ be a set of good and pleasant moving n lines. Let $t_1, \ldots, t_m$ be singular moments of $L(t)$. For each singular moment t_s if the four lines $L_i(t_s), L_j(t_s), L_k(t_s), L_l(t_s)$ are tangent to a circle, then we associate a generator $a_s = a_{ijkl}$ from G_n^4. Let $f^G(L(t)) = a_1 \ldots a_m \in G_n^4$.

Theorem 15.5. *If two sets $L(t)$ and $L'(t)$ of moving n lines are equivalent, then $f^G(L(t)) = f^G(L'(t))$ in G_n^4.*

15.5.2 A map from a group of good moving lines to Γ_n^4

Let $L(t)$ be a set of good and pleasant moving n lines. Let $t_1, \ldots, t_m$ be singular moments of $L(t)$. For each singular moment t_s if four lines $L_i(t_0), L_j(t_0), L_k(t_0), L_l(t_0)$ are tangent in the given order to a circle and the circle does not meet with other lines, then we associate the singular moment t_s with a generator $d_s = d_{(ijkl)}$ of Γ_n^4. But, at the moment t_s if four lines $L_i(t_0), L_j(t_0), L_k(t_0), L_l(t_0)$ are tangent to a circle but the circle meets with another line, then we associate the singular moment t_s with $d_s = 1$.

Let $f^\gamma(L(t)) = d_1 \ldots d_m \in \Gamma_n^4$.

Theorem 15.6. *If two sets $L(t)$ and $L'(t)$ of moving n lines are equivalent, then $f^\gamma(L(t)) = f^\gamma(L'(t))$ in Γ_n^4.*

Proof. When we consider continuous family of homeomorphisms between two sets of moving n lines, it suffices to consider only singularities of codimension at most two. Singularities of codimension one give rise to generators, and relations come from singularities of codimension two. Let us list up the cases of singularities of codimension two.

(1) A path of a line moving on the plane is tangent to the circle, which passes through three points. This corresponds to the relation $d^2_{(ijkl)} = 1$.

(2) There are two sets A and B of four lines, each of which are on the same circles such that $|A \cap B| \leq 2$. This corresponds to the relation $d_{(ijkl)}d_{(stuv)} = d_{(stuv)}d_{(ijkl)}$.

(3) There are five lines $\{L_i, L_j, L_k, L_l, L_m\}$ being tangent to the same circle. We obtain the sequence of five subsets of $\{L_i, L_j, L_k, L_l, L_m\}$ with four lines being tangent to the same circle. This corresponds to the relation $d_{(ijkl)}d_{(ijlm)}d_{(jklm)}d_{(ijkm)}d_{(iklm)} = 1$.

$\square$

Each moment, when four lines are tangent to a circle, which does not meet with another line, gives a relation, which consists of six angles between the four lines. More precisely, let $\{L_i, L_j, L_k, L_l\}$ be four lines, which are tangent to a circle in the indicated order. Let θ_{st} be the angle between lines L_s and L_t. Then the angles $\theta_{ij}, \theta_{jk}, \theta_{kl}, \theta_{il}, \theta_{ik}, \theta_{jl}$ satisfies

$$\cos\left(\frac{\theta_{ik}}{2}\right)\cos\left(\frac{\theta_{jl}}{2}\right) = \cos\left(\frac{\theta_{ij}}{2}\right)\cos\left(\frac{\theta_{kl}}{2}\right) + \cos\left(\frac{\theta_{jk}}{2}\right)\cos\left(\frac{\theta_{il}}{2}\right).$$

$$(15.7)$$

Remark 15.5. This relation actually is obtained from the Ptolemy relation, see Fig. 15.14. For example $\frac{l_{ij}}{2} = r\cos\left(\frac{\theta_{ij}}{2}\right)$, that is, $l_{ij} = 2r\cos\left(\frac{\theta_{ij}}{2}\right)$. Since the equation

$$l_{ik}l_{jl} = l_{ij}l_{kl} + l_{ik}l_{jl},$$

holds, the relation 15.8 can be obtained, see Fig. 15.14.

15.5.3 *A map from a group of good moving unit circles to* Γ_n^4

The moving lines can be replaced by moving unit circles $\{C_1, \ldots, C_n\}$ on the plane and every notion can be defined analogously. Singular moments

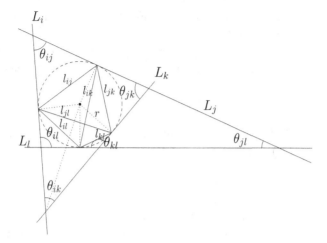

Fig. 15.14   The relation coming from tangent lines

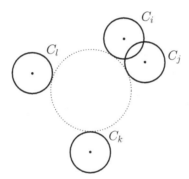

Fig. 15.15   Four circles $C_i(t_0), C_j(t_0), C_k(t_0), C_l(t_0)$, being tangent to a circle

of codimension one are the moments, when four unit circles C_i, C_j, C_k, C_l are tangent to a circle, see Fig. 15.15.

Remark 15.6. For moving circles, singular moments do correspond to moving lines and their tangent points to circle. Because if we have a point X on a circle C, then there uniquely exists a unit circle, which tangent to the circle C at the point X.

Let $C(t)$ be a set of good and pleasant moving n circles. Let $t_1, \ldots, t_m$ be singular moments of $C(t)$. For each singular moment t_s if the four lines $C_i(t_0), C_j(t_0), C_k(t_0), C_l(t_0)$ are tangent in the given order to a circle and the circle does not meet other circles, then we associate the singular moment t_s with a generator $d_s = d_{(ijkl)}$ of Γ_n^4. At the moment t_s if four circles $C_i(t_0), C_j(t_0), C_k(t_0), C_l(t_0)$ are tangent to a circle but the circle meets with another circle, then we associate the singular moment t_s with $d_s = 1$.

Let $f^C(L(t)) = d_1 \ldots d_m \in \Gamma_n^4$.

Theorem 15.7. *If two sets $L(t)$ and $L'(t)$ of moving n circles are equivalent, then $f^C(L(t)) = f^C(L'(t))$ in Γ_n^4.*

Each moment, when four circles are tangent to a circle, which does not meet with another circle, also gives rise to a relation. More precisely, let C_i, C_j, C_k, C_l be four lines, which are tangent to a circle C in the indicated order. Let P_i, P_j, P_k and P_l be centres of circles C_i, C_j, C_k and C_l. Let P be the center of the circle C. Let π_{st} be the angle between lines $\overline{P_s P_t}$ and $\overline{P_s P}$ (or $\overline{P_t P}$). Then the angles $\pi_{ij}, \pi_{jk}, \pi_{kl}, \pi_{il}, \pi_{ik}, \pi_{jl}$ satisfies

$$cosh(\pi_{ik}i)cosh(\pi_{jl}i) = cosh(\pi_{ij}i)cosh(\pi_{kl}i) + cosh(\pi_{jk}i)cosh(\pi_{il}i), \quad (15.8)$$

where $i^2 = -1$.

Remark 15.7. Note that $cosh(\pi i) = cos(\pi)$. Moreover, $2\pi_{ij} = \theta_{ij}$, where θ_{ij} is the angle between two tangent lines to C at points, where C_i and C_j are tangent to C, see Fig. 15.16.

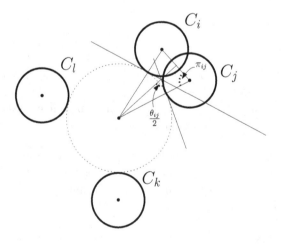

Fig. 15.16

15.6 A representation of braids via triangulations

The aim of the present section is to construct an (infinite-dimensional) representation of braid groups for arbitrary 2-surfaces valued in operators on the space of rational functions over a ring.

The main idea behind this representation is the operator which is associated with the flip.

Let Σ be a 2-surface with metric, and let N be a large natural number. We shall consider the space of all triangulations of Σ with N points.

Some of these triangulations originate from Delaunay triangulations with vertices in the given N points (for this to happen, some length constraints should hold).

When moving N points around, the Delaunay triangulation undergo some flips. With each flip, we associate a Ptolemy transformation:

$$y = \frac{ac + bd}{x}. \tag{15.9}$$

This equation describes the behaviour of variables when points undergo a flip.

In the sequel, by an *ordered triangulation* we mean a triangulation with enumerated vertices.

Consider the set of all triangulations of Σ with N enumerated vertices (hence, with $X = 3(N - 2g + 2)$ edges). We say that two triangulations are *adjacent* if they differ by a flip.

With any ordered triangulation having X edges, we associate X variables, one to each oriented edge. Unless otherwise is set, the edge is oriented from a vertex with a smaller number to a vertex with a larger number. If we associate a variable a to an edge, we associate $-a$ to the opposite edge.

For any two adjacent ordered triangulations T_1 and T_2, there is a bijection for all but one edges.

Now, consider the ring $\mathbb{Q}(x_1, \ldots, x_X)$ of all rational functions in X variables.

We shall relate variables for adjacent triangulations by a flip rule. If the triangulation T_2 is obtained from the triangulation T_1 by a flip within the quadrilateral a, b, c, d, then the diagonal y is equal to $\frac{ac+bd}{x}$. Later we shall see that this association is well defined because the denominators never equal zero.

Starting with a triangulation T, we can consider loops in the space of triangulations, i.e., sequences of triangulations $T = T_0 \to T_1 \to T_2 \to \cdots \to T_k = T$ where every two adjacent triangulations T_i, T_{i+1} differ by a flip.

Certainly, when we deal with loops, the labels of each T_i are expressible in terms of T_0, subsequently, for the triangulation $T_k = T$ we get a new set of labels.

Theorem 15.8. *If a loop $T = T_0 \to T_k = T$ in the space of triangulations originates from a braid which is homotopic to the identity, then the labels of T_k coincide with those of T_0.*

The main ingredient in the proof is the following Fig. 15.17.

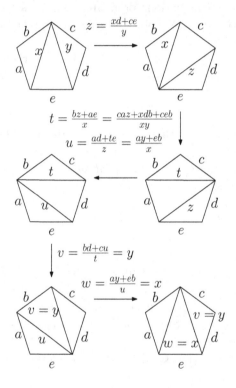

Fig. 15.17   The pentagon equation is satisfied

Let us consider the following

Example 15.1. Consider three points $1, 2, 3$ in the vertices of a triangle, one point 4 in the center of this triangle and one point 5 which goes around 4 in the neighbourhood of 4.

This gives rise to a sequence of triangulations given in Fig. 15.18.

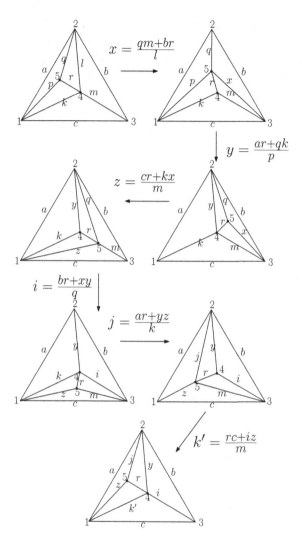

Fig. 15.18   Six triangulations corresponding to a rotation of point

Let us calculate the labels of edges corresponding to these pictures. For the initial one, we denote the labels by

$$a(12), b(23), c(13), k(14), l(24), m(34), p(15), q(25), r(45)$$

as shown above.

In all subsequent figures the points $1, 2, 3, 4$ as well as labels a, b, c, d, k, l, m remain unchanged.

Besides the initial triangulation T_0 and the final one T_6 it is worth taking glance at T_3. It looks similar to T_0, however, with one important difference: the small edge denoted by r is now oriented in the opposite direction.

This will be extremely important for the construction of knot invariants.

$$x = \frac{qm + br}{l};$$

$$y = \frac{ar + qk}{p};$$

$$z = \frac{cr + kx}{m} = \frac{crl + kqm + kbr}{ml};$$

$$i = \frac{br + xy}{q} = \frac{bprl + (ar + qk)(qm + br)}{pql};$$

$$j = \frac{ar + yz}{k} = \frac{arpml + (ar + qk)(crl + kqm + kbr)}{kpml};$$

$$o = \frac{rc + iz}{x} = \frac{k^2mq^2 + bklpr + cl^2pr + cklqr + akmqr + abkr^2 + aclr^2}{lmpq}.$$

From this we see that the variables corresponding to T_6 do not coincide with those corresponding to T_0, which says that the above invariant of this pure braid is non-trivial.

15.7 Decorated triangulations

The purpose of the present section is to introduce the notion of decorated triangulation and to construct a representation of braids group.

Definition 15.2. *A decorated triangulation* is a triangulation where all triangles are linearly ordered and in each triangle, a distinguished corner is fixed. The fixed corner is presented by a point on the corner.

An example of a decorated triangulation of the disk with 5 points is given in Fig. 15.19.

Definition 15.3. Let $\mathcal{C} = (\mathcal{C}, \otimes, \{P_{X,Y}\}_{X,Y \in Ob\mathcal{C}})$ be a symmetric monoidal category. A *basic algebraic system* (BAS) in $\mathcal{C}$ consists of an object $V \in Ob\mathcal{C}$ and two morphisms $R \in End(V)$ and $W \in End(V \otimes V)$ such that

Fig. 15.19   A decorated ideal triangulation of the disk

(1) $R_i^3 = id_V$,
(2) $W_{i,j}W_{j,k} = W_{j,k}W_{i,k}W_{i,j}$ in $End(V^{\otimes 3})$,
(3) $R_iR_jW_{j,i}R_i^{-1}W_{i,j} = P_{i,j}$ in $End(V^{\otimes 2})$, where $P_{i,j} := P_{V,V}$.

The morphisms $R_i : V \to V$ and $W_{i,j} : V \otimes V \to V \otimes V$ correspond to distinguished corner changes and the flip, respectively, see Fig. 15.20.

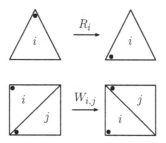

Fig. 15.20   Distinguished corner changes and the flip

Moreover the conditions (2) and (3) from Definition 15.3 can be described by the pentagon relation and changes of the order of triangles, see Fig. 15.21 and Fig. 15.22.

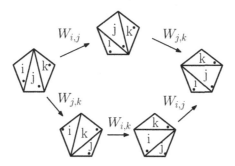

Fig. 15.21   Geometric description of the condition (2) from Definition 15.3

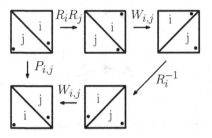

Fig. 15.22   Geometric description of the condition (3) from Definition 15.3

Example 15.2. Let **Set** be the category of sets with monoidal structure given by the Cartesian product, and the transpositions of the components. Let $V = \mathbb{R}_+^2$ and define $R : V \to V$ and $W : V \times V \to V \times V$ by

$$R(x_1, x_2) = (\frac{1}{x_1}, \frac{x_2}{x_1}), W((x_1, x_2), (y_1, y_2)) = \left(\frac{x_2 y_1}{x_1 y_2 + x_2}, \frac{y_2}{x_1 y_2 + x_2} \right).$$
(15.10)

Then this is a BAS.

Let us consider the set of all ordered triangulation of Σ with N enumerated vertices as the previous section. For each triangulation τ of Σ we can obtain a decorated triangulation $\tilde{\tau}$ by fixing a corner for each triangle. Recall that in the previous section edges of τ are considered as variables. Now we define a map from triangles of $\tilde{\tau}$ to $\mathbb{Q}(x_1, \ldots, x_X) \times \mathbb{Q}(x_1, \ldots, x_X)$ as described in Fig. 15.23.

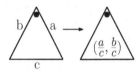

Fig. 15.23 $a, b, c \in \mathbb{Q}(x_1, \ldots x_X)$, which are associated to each edge of the triangulation τ

Chapter 16

The Three-dimensional Case

16.1 The group Γ_n^5

Consider a configuration of n points in general position in $\mathbb{R}^3$. We can think of these points as lying in a fixed tetrahedron $ABCD$. The points induce a unique Delaunay triangulation of the tetrahedron: four points form a simplex of the triangulation if and only if there are no other points inside the sphere circumscribed over these points. The triangulation transforms as soon as the points move in the space.

In order to avoid degenerate Delaunay triangulations we exclude configurations where four points lie on one circle (intersection of a plane and a sphere).

Transformations of the combinatorial structure of the Delaunay triangulation correspond to configurations of codimension 1 when five points lie on a sphere which does not contain any points inside. At this moment two simplices of the triangulation are replaced with three simplices as shown in Fig. 16.1 (or vice versa). This transformation is called a *2-3 Pachner move*.

If we want to trace the evolution of triangulations that correspond to a dynamics of the points, then we can mark any Pachner move with a letter with indices. For the move that replace the simplices $iklm, jklm$ with the simplices $ijkl, ijkm, ijlm$ in Fig. 16.1 we use the generator $a_{ij,klm}$. Note that

(1) we can split the indices into two subsets according to the combinatorics of the transformation;

(2) the generator $a_{ij,klm}$ is not expected to be involutive because it changes the number of simplices of the triangulation.

The relations on generators $a_{ij,klm}$ correspond to configurations of

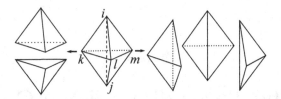

Fig. 16.1   A 2-3 Pachner move

codimension 2 which occurs when either

(1) six points lie on the same sphere with empty interior, or
(2) there are two spheres with five points on each of them, or
(3) five points on one sphere compose a codimension 1 configuration.

The last case means the convex hull of the five points has a quadrangular face (Fig. 16.2). The vertices of this face lie on one circle so we exclude this configuration.

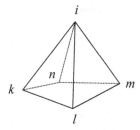

Fig. 16.2   A quadrangular pyramid

If there are two different spheres with five points on each of them, then the inscribed simplices for the first sphere and for the second sphere are different. We can suppose that the faces of the inscribed convex hull has only triangular face for each sphere; otherwise, the codimension should be greater than two. So the convex hulls can have one common face at most, and we can transform them independently. This gives us a commutation relation

$$a_{ij,klm}a_{i'j',k'l'm'} = a_{i'j',k'l'm'}a_{ij,klm},$$

where

$$|\{i,j,k,l,m\} \cap \{i',j',k',l',m'\}| < 4,$$

$$|\{i,j\} \cap \{i',j',k',l',m'\}| < 2,$$

and

$$|\{i',j'\} \cap \{i,j,k,l,m\}| < 2.$$

Consider now the case of six points on one sphere. The convex hull of these points is a convex polyhedron. The polyhedron must have only triangular faces; otherwise, there is an additional linear condition (four points lie on one plane) which raises the codimension beyond 2. There are two such polyhedra, see Fig. 16.3.

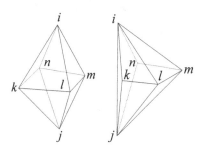

Fig. 16.3   Convex polyhedra with triangular faces and 6 vertices

For the octahedron on the left we specify the geometrical configuration assuming that the orthogonal projection along the direction ij maps the points i and j near the projection of the edge kl and the line ln is higher than km if you look from i to j (Fig. 16.4). In this case we have the six triangulations:

(1) $ijkl, ijkn, ijlm, ijmn;$
(2) $iklm, ikmn, jklm, jkmn;$
(3) $ikln, ilmn, jkln, jlmn;$
(4) $ijkl, ijkm, ijlm, ikmn, jkmn;$
(5) $ijkl, ijkn, ijln, ilmn, jlmn;$
(6) $ikln, ilmn, jklm, jkmn, klmn.$

The first three have four simplices each, the last three have five simplices each. The 2-3 Pachner moves between the triangulations are shown in Fig. 16.5. Thus, we have the relation

$$a_{km,ijn}a_{ij,klm}^{-1}a_{ln,ikm}a_{km,jln}^{-1}a_{ij,kln}a_{ln,ijm}^{-1} = 1.$$

In the case of the shifted octahedron (Fig. 16.3 right) we assume the line ln lies higher than the line km when one looks from the vertex i to the vertex j (Fig. 16.6). Then we get the six triangulations:

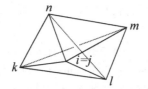

Fig. 16.4   An orthogonal projection of the octahedron

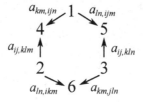

Fig. 16.5   Triangulation graph for the octahedron

(1) $ijkl, ijkm, ijmn$;

(2) $ijkl, ijln, ilmn, jlmn$;

(3) $ijkm, ijmn, iklm, jklm$;

(4) $ijkn, ikln, ilmn, jkln, jlmn$;

(5) $ijkn, iklm, ikmn, jklm, jkmn$;

(6) $ijkn, ikln, ilmn, jklm, jkmn, klmn$.

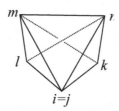

Fig. 16.6   An orthogonal projection of the shifted octahedron

The 2-3 Pachner moves between the triangulations are shown in Fig. 16.7. This gives us the relation

$$a_{ln,ijm}a_{kn,ijl}a_{km,jln} = a_{km,ijl}a_{kn,ijm}a_{ln,ikm}.$$

Now we can give the definition for Γ_n^5.

Definition 16.1. The group Γ_n^5 is the group with generators

$$\{a_{ij,klm} \mid \{i,j,k,l,m\} \subset \bar{n}, |\{i,j,k,l,m\}| = 5\}$$

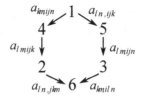

Fig. 16.7   Triangulation graph for the shifted octahedron

and relations

(1) $a_{ij,klm} = a_{ji,klm} = a_{ij,lkm} = a_{ij,kml}$,

(2) $a_{ij,klm} a_{i'j',k'l'm'} = a_{i'j',k'l'm'} a_{ij,klm}$, for $|\{i,j,k,l,m\} \cap \{i',j',k',l',m'\}| < 4$, $|\{i,j\} \cap \{i',j',k',l',m'\}| < 2$ and $|\{i',j'\} \cap \{i,j,k,l,m\}| < 2$,

(3) $a_{km,ijn} a_{ij,klm}^{-1} a_{ln,ikm} a_{km,jln}^{-1} a_{ij,kln} a_{ln,ijm}^{-1} = 1$, for distinct i,j,k,l,m,n,

(4) $a_{ln,ijm} a_{kn,ijl} a_{km,jln} = a_{km,ijl} a_{kn,ijm} a_{ln,ikm}$, for distinct i,j,k,l,m,n.

The group Γ_n^5 can be used to construct invariants of braids and dynamical systems like the groups G_n^k.

Let us consider the configuration space $\tilde{C}_n(\mathbb{R}^3)$ which consists of non-planar n-tuples $(x_1, x_2, \ldots, x_n)$ of points in $\mathbb{R}^3$ such that for all distinct i,j,k,l the points x_i, x_j, x_k, x_l do not lie on the same circle.

We construct a homomorphism from $\pi_1(\tilde{C}_n(\mathbb{R}^3))$ to Γ_n^5. Let

$$\alpha = (x_1(t), \ldots, x_n(t)), \quad t \in [0,1],$$

be a loop in $\tilde{C}_n(\mathbb{R}^3)$. For any t the set $\mathbf{x}(t) = (x_1(t), \ldots, x_n(t))$ determines a Delaunay triangulation $T(t)$ of the polytope $\operatorname{conv}(\mathbf{x}(t))$. If α is in general position, then there will be a finite number of moments $0 < t_1 < t_2 < \cdots < t_N < 1$ when the combinatorial structure of $T(t)$ changes, and for each p the transformation of the triangulation at the moment t_p will be the 2-3 Pachner move on simplices with vertices i_p, j_p, k_p, l_p, m_p. We assign to this move the generator $a_{i_p j_p, k_p l_p m_p}$ or $a_{i_p j_p, k_p l_p m_p}^{-1}$ and denote

$$\phi(\alpha) = \prod_{p=1}^{N} a_{i_p j_p, k_p l_p m_p} \in \Gamma_n^5.$$

Theorem 16.1. *The homomorphism* $\phi \colon \pi_1(\tilde{C}_n(\mathbb{R}^3)) \to \Gamma_n^5$ *is well defined.*

Proof. We need to show the element $\phi(\alpha)$ does not depend on the choice of the representative α in a given homotopy class. Given a homotopy in general position $\alpha(\tau), \tau \in [0,1]$, of loops in $\tilde{C}_n(\mathbb{R}^3)$, the transformations of the words $\phi(\alpha(\tau))$ correspond to point configurations of codimension 2, considered above the definition of Γ_n^5, and thus are counted by the relations in the group Γ_n^5. Therefore, the element $\phi(\alpha(\tau))$ of the group Γ_n^5 remains the same when τ changes. $\qquad\square$

16.2 The general strategy of defining Γ_n^k for arbitrary k

The groups Γ_n^4 and Γ_n^5 defined above admit extrapolation to groups $\Gamma_n^k, k \geq 4$, that we define in this section.

Consider the configuration space $\tilde{C}_n(\mathbb{R}^{k-2})$, $4 \leq k \leq n$, consisting of n-point configurations

$$\mathbf{x} = (x_1, x_2, \ldots, x_n)$$

in $\mathbb{R}^{k-2}$, such that $\dim \operatorname{conv} \mathbf{x} = k - 2$ and there are no $k - 1$ points which lie on one $(k - 4)$-dimensional sphere (intersection of a $(k - 3)$-sphere and a hyperplane in in $\mathbb{R}^{k-2}$).

A configuration $\mathbf{x} = (x_1, x_2, \ldots, x_n) \in \tilde{C}_n(\mathbb{R}^{k-2})$ determines a Delaunay triangulation of $\operatorname{conv} \mathbf{x}$ which is unique when $\mathbf{x}$ in general position. If the vertices $x_{i_1}, \ldots, x_{i_{k-1}}$ span a simplex of the Delaunay triangulation, then the interior of the circumscribed sphere over these points does not contain other points of the configuration. The inverse statement is true when $\mathbf{x}$ is generic. The condition that no $k - 1$ points which lie on one $(k - 4)$-dimensional sphere ensures that there are no degenerate simplices in the Delaunay triangulation.

Let $\alpha = (x_1(t), \ldots, x_n(t))$, $t \in [0, 1]$, be a loop in $\tilde{C}_n(\mathbb{R}^{k-2})$. For any t, the configuration $\mathbf{x}(t) = (x_1(t), \ldots, x_n(t))$ determines a Delaunay triangulation $T(t)$ of the polytope $\operatorname{conv}(\mathbf{x}(t))$. If α is generic, then there will be a finite number of moments $0 < t_1 < t_2 < \cdots < t_L < 1$ when the combinatorial structure of $T(t)$ changes.

For each singular configuration $\mathbf{x}(t_i)$, either one simplex degenerates and disappears on the boundary $\partial \operatorname{conv}(\mathbf{x}(t))$ or the Delaunay triangulation is not unique that means that there is a sphere in $\mathbb{R}^{k-2}$ which contains k points of $\mathbf{x}$ on it and no points of $\mathbf{x}$ inside. Assuming $\mathbf{x}$ is generic, the span of these k points is a simplicial polytope. Below we shall count only the latter type of singular configurations.

The simplicial polytopes in $\mathbb{R}^{k-2}$ with k vertices are described in [Grünbaum, 2003]. Each of them is the join $\Delta_P * \Delta_Q$ of simplices $\Delta_P = \operatorname{conv}(x_{p_1}, \ldots, x_{p_l})$ of dimension $l - 1 \geq 1$ and $\Delta_Q = \operatorname{conv}(x_{q_1}, \ldots, x_{q_{k-l}})$ of dimension $k - l - 1 \geq 1$ such that the intersection $\operatorname{relint}(x_{p_1}, \ldots, x_{p_l}) \cap \operatorname{relint}(x_{q_1}, \ldots, x_{q_{k-l}})$ consists of one point.

We recall that the *join* of two sets $X, Y \subset \mathbb{R}^m$ is defined as

$$X * Y = \{\lambda x + (1 - \lambda)y \mid x \in X, y \in Y, \lambda \in [0,1]\}$$

and the *relative interior* of a finite set $X \subset \mathbb{R}^m$ is defined as

$$\text{relint } X = \left\{ \sum_{x \in X} \lambda_x x \mid \forall x \; \lambda_x > 0, \sum_{x \in X} \lambda_x = 1 \right\}.$$

The polytope $\Delta_P * \Delta_Q$ has two triangulations:

$$T_P = \{\mathbf{x}_{P \cup Q} \setminus \{x_p\}\}_{p \in P} = \left\{ x_{p_1} \ldots x_{p_{i-1}} x_{p_{i+1}} \ldots x_{p_l} x_{q_1} \ldots x_{q_{k-l}} \right\}_{i=1}^{l}$$

and

$$T_Q = \{\mathbf{x}_{P \cup Q} \setminus \{x_q\}\}_{q \in Q} = \left\{ x_{p_1} \ldots x_{p_l} x_{q_1} \ldots x_{q_{i-1}} x_{q_{i+1}} \ldots x_{q_{k-l}} \right\}_{i=1}^{k-l}.$$

Here $P = \{p_1, \ldots, p_l\}$, $Q = \{q_1, \ldots, q_{k-l}\}$ and $\mathbf{x}_J = \{x_j\}_{j \in J}$ for any $J \subset \{1, \ldots, n\}$.

The condition $\text{relint}(\mathbf{x}_P) \cap \text{relint}(\mathbf{x}_Q) = \{z\}$ implies $P \cap Q = \emptyset$.

Thus, when the configuration $\mathbf{x}(t)$ goes over a singular value $t_i, i = 1, \ldots, L$, in the Delaunay triangulation simplices T_{P_i} are replaced by simplices T_{Q_i} for some subsets $P_i, Q_i \subset \{1, \ldots, n\}$, $P_i \cap Q_i = \emptyset$, $|P_i|, |Q_i| \geq 2$, $|P_i \cup Q_i| = k$. This transformation is called a *Pachner move*. We assign to the transformation the letter a_{P_i, Q_i}.

Hence, the loop α produces a word

$$\Phi(\alpha) = \prod_{i=1}^{L} a_{P_i, Q_i} \tag{16.1}$$

in the alphabet

$$\mathcal{A}_n^k = \{a_{P,Q} \mid P, Q \subset \{1, \ldots, n\}, P \cap Q = \emptyset, |P \cup Q| = k, |P|, |Q| \geq 2\}.$$

Now let us consider a generic homotopy $\alpha_s, s \in [0,1]$, between two generic loops α_0 and α_1. A loop $\alpha_s = \{\mathbf{x}(s,t)\}_{t \in [0,1]}$ can contain a configuration of codimension 2. This means that for some t the configuration $\mathbf{x}(s,t)$ has two different k-tuples of points, each of them lies on a sphere whose interior contains no points of $\mathbf{x}(s,t)$. If these spheres do not coincide, then their intersection contains at most $k-2$ points (the intersection cannot contain $k-1$ points because $\mathbf{x}(s,t) \in \tilde{C}_n(\mathbb{R}^{k-2})$). Hence, the simplices involved in one Pachner move can not be involved in the other one, so the Pachner moves can be performed in any order.

If the k-tuples of points lie on one sphere, then there is a sphere with $k + 1$ points of $\mathbf{x}(s,t)$ on it and its interior contains no points of $\mathbf{x}(s,t)$. These $k + 1$ points span a simplicial polytope in $\mathbb{R}^{k-2}$. Such polytopes are described in [Grünbaum, 2003]. The description uses the notion of the Gale transform.

Let $X = \{x_1, \ldots, x_n\}$ be a set of n points in $\mathbb{R}^d$ such that $\dim \operatorname{conv} X = d$. Then $n \geq d + 1$. Let $x_i = (x_{1i}, \ldots, x_{di}) \in \mathbb{R}^d$, $i = 1, \ldots, n$, be the coordinates of the points of X. The matrix

$$M = \begin{pmatrix} x_{11} & x_{12} & \cdots & x_{1n} \\ x_{21} & x_{22} & \cdots & x_{2n} \\ \cdots & \cdots & \cdots & \cdots \\ x_{d1} & x_{d2} & \cdots & x_{dn} \\ 1 & 1 & \cdots & 1 \end{pmatrix}$$

has rank $d+1$. Then the dimension of the space $\ker M = \{b \in \mathbb{R}^n \mid Mb = 0\}$ of dependencies between columns of M is equal to $n - (d+1)$. Take any basis $b_j = (b_{j1}, b_{j2}, \ldots, b_{jn})$, $j = 1, \ldots, n - d - 1$, of $\ker M$ and write it in matrix form

$$B = \begin{pmatrix} b_{11} & \cdots & b_{1n} \\ \cdots & \cdots & \cdots \\ b_{n-d-1,1} & \cdots & b_{n-d-1,n} \end{pmatrix}. \tag{16.2}$$

The columns of the matrix B form a set $Y = y_1, \ldots, y_n$, $y_i = (b_{1i}, \ldots, b_{n-d-1,i})$, in $\mathbb{R}^{n-d-1}$. The set Y is called a *Gale transform* of the point set X. Gale transforms which correspond to different bases of $\ker M$ are linearly equivalent. The vectors of the Gale transform Y may coincide.

Example 16.1. Let X be a pentagon in $\mathbb{R}^2$ with vertices $x_1 = (0, 2)$, $x_2 = (-2, 1)$, $x_3 = (-1, -1)$, $x_4 = (1, -1)$, $x_5 = (2, 1)$. Then

$$M = \begin{pmatrix} 0 & -2 & -1 & 1 & 2 \\ 2 & 1 & -1 & -1 & 1 \\ 1 & 1 & 1 & 1 & 1 \end{pmatrix}$$

and

$$B = \begin{pmatrix} -4 & 1 & 3 & -5 & 5 \\ -4 & 6 & -7 & 5 & 0 \end{pmatrix}.$$

The Gale transform Y consists of vectors $y_1 = (-4, -4)$, $y_2 = (1, 6)$, $y_3 = (3, -7)$, $y_1 = (-5, 5)$, $y_1 = (5, 0)$.

Let $Y = y_1, \ldots, y_n$ be a Gale transform of X. The set

$$\bar{Y} = \{\bar{y}_1, \ldots, \bar{y}_n\}, \quad \bar{y}_i = \begin{cases} \dfrac{y_i}{\|y_i\|}, & y_i \neq 0, \\ 0, & y_i = 0, \end{cases}$$

is called a *Gale diagram* of the point set X. It is a subset of $S^{n-d-2} \cup \{0\}$.

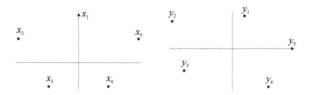

Fig. 16.8 A pentagon and its Gale transform

Denote $\bar{n} = \{1, \ldots, n\}$. Two subsets $\bar{Y} = \{\bar{y}_1, \ldots, \bar{y}_n\}$ and $\bar{Y}' = \{\bar{y}'_1, \ldots, \bar{y}'_n\}$ in $S^{n-d-2} \cup \{0\}$ are called *equivalent* if there is a permutation σ of $\bar{n}$ such that for any $J \subset \bar{n}$

$$0 \in \operatorname{relint} \bar{Y}_J \iff 0 \in \operatorname{relint} \bar{Y}'_{\sigma(J)}.$$

Here we denote $\bar{Y}_J = \{\bar{y}_i\}_{i \in J}$ and $\bar{Y}'_J = \{\bar{y}'_i\}_{i \in J}$.

The properties of Gale diagrams can be summarised as follows [Grünbaum, 2003].

Theorem 16.2.

(1) Let X be a set of n points which are vertices of some polytope P in $\mathbb{R}^d$ and $\bar{Y}$ be its Gale diagram. Then the set of indices $J \subset \bar{n}$ defines a face of P if and only if $0 \in \operatorname{relint} Y_{\bar{n} \setminus J}$.

(2) Let X and X' be sets of vertices of polytopes P and P', $|X| = |X'|$, and $\bar{Y}$ and $\bar{Y}'$ be their Gale diagrams. Then P and P' are combinatorially equivalent if and only if $\bar{Y}$ and $\bar{Y}'$ are equivalent.

(3) For any n-point set $\bar{Y} \in S^{n-d-2} \cup \{0\}$ such that $\bar{Y}$ spans $\mathbb{R}^{n-d-1}$ and 0 lies in the interior of $\operatorname{conv} \bar{Y}$, there is an n-point set X in $\mathbb{R}^d$ such that $\bar{Y}$ is a Gale diagram of X.

This theorem implies (see [Grünbaum, 2003]) that simplicial polytopes with $k + 1$ vertices in $\mathbb{R}^{k-2}$ are in a bijection with standard Gale diagrams in $\mathbb{R}^2$ (defined uniquely up to isometries of the plane).

A *standard Gale diagram* of order $l = k + 1$ is a subset $\bar{Y}$, $|\bar{Y}| = l$, of the vertices set $\{e^{\frac{\pi i p}{l}}\}_{p=0}^{2l-1}$ of the regular $2l$-gon inscribed in the unit circle S^1, such that:

(1) any diameter of S^1 contains at most one point of $\bar{Y}$;
(2) for any diameter of S^1, any open half-plane determined by it contains at least two points of $\bar{Y}$.

The first property means that the corresponding polytope is simplicial, the second means that any of the $k + 1$ vertices of the polytope is a face.

The number c_l of standard Gale diagrams of order l is equal to

$$c_l = 2^{\left[\frac{l-3}{2}\right]} - \left[\frac{l+1}{4}\right] + \frac{1}{4l} \sum_{h:\, 2 \nmid h \mid l} \varphi(h) \cdot 2^{\frac{l}{h}},$$

where $l = 2^{a_0} \prod_{i=1}^{t} p_i^{a_i}$ is the prime decomposition of l and φ is Euler's function. For small l the numbers are $c_5 = 1, c_6 = 2, c_7 = 5$, see Fig. 16.9.

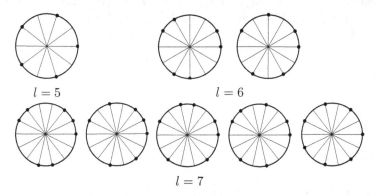

$l = 5$ $l = 6$

$l = 7$

Fig. 16.9   Standard Gale diagrams of small order

Let us describe the triangulations of the simplicial polytopes with $k + 1$ vertices in $\mathbb{R}^{k-2}$.

Let $X = \{x_1, \dots, x_n\}$ be a subset in $\mathbb{R}^d$, so $x_i = (x_{i1}, \dots, x_{id})$, $i = 1, \dots, n$. Let $P = \operatorname{conv} X$ be the convex hull of X, assume that $\dim P = d$. A triangulation T of the polytope P with vertices in X is called *regular* if there is a height function $h \colon X \to \mathbb{R}$ such that T is the projection of the lower convex hull of the lifting $X^h = \{x_1^h, \dots, x_n^h\} \subset \mathbb{R}^{d+1}$, where $x_i^h = (x_i, h(x_i))$. This means a set of indices $J \subset \{1, \dots, n\}$ determines a face of T if and only if there exists a linear functional ϕ on $\mathbb{R}^{d+1}$ such that $\phi(0, \dots, 0, 1) > 0$ and $J = \{i \mid \phi(x_i^h) = \min_{x^h \in X^h} \phi(x^h)\}$. In case T is regular we write $T = T(X, h)$. Any generic height function induces a regular triangulation.

The Delaunay triangulation of X is regular with the height function $h \colon \mathbb{R}^d \to \mathbb{R}$, $h(z) = \|z\|^2 = \sum_{i=1}^{d} z_i^2$ if $z = (z_1, \dots, z_d) \in \mathbb{R}^d$.

A height function $h \colon X \to \mathbb{R}$ can be regarded as a vector $h = (h_1, \dots, h_n) \in \mathbb{R}^n$ where $h_i = h(x_i)$. Denote $\beta(h) = Bh \in \mathbb{R}^{n-d-1}$ where B is the matrix (16.2) used to define a Gale transform of X. Let

$\bar{Y} = \{\bar{y}_1, \ldots, \bar{y}_n\}$ be a Gale diagram of X. Convex cones generated by the subsets of $\bar{Y}$ split the space $\mathbb{R}^{n-d-1}$ into a union of conic cells. A relation between the triangulation $T(X, h)$ and the Gale diagram $\bar{Y}$ can be described as follows.

Theorem 16.3.

(1) If $T(X, h)$ is a regular triangulation of X, then $\beta(h)$ belongs to a conic cell of maximal dimension in the splitting of $\mathbb{R}^{n-d-1}$ induced by $\bar{Y}$.

(2) Let $J \subset \bar{n} = \{1, \ldots, n\}$. The set X_J spans a cells of the triangulation $T(X, h)$ if and only if $\beta(h) \in \mathrm{concone}(\bar{Y}_{\bar{n} \setminus J})$, where $\mathrm{concone}(X_{\bar{n} \setminus J})$ is the convex cone spanned by the set $\bar{Y}_{\bar{n} \setminus J}$.

Let P be a simplicial polytopes with $l = k + 1$ vertices in $\mathbb{R}^{l-3}$ and $\bar{Y} = \{\bar{y}_1, \ldots, \bar{y}_l\}$ be the corresponding standard Gale diagram. By Theorem 16.3 there are l different regular triangulations which correspond to open sectors between the rays spanned by the vectors of $\bar{Y}$. The graph whose vertices are combinatorial classes of triangulations of P and the edges are Pachner moves, is a cycle. Let us find which Pachner moves appears in this cycle.

We change the order of vertices of P (and hence, the order of point of $\bar{Y}$) so that the points $\bar{y}_1, \ldots, \bar{y}_l$ appear in this sequence when one goes counterclockwise on the unit circle. For each i denote $R_{\bar{Y}}(i)$ ($L_{\bar{Y}}(i)$) be the set of indices j of vectors $\bar{y}_j$ that lie in the right (left) open half-plane incident to the oriented line spanned by the vector $\bar{y}_i$. Then the Pachner moves which occurs when the vector $\beta(h)$ passes $\bar{y}_i$ from right to left, will be marked with the letter $a_{R_{\bar{Y}}(i), L_{\bar{Y}}(i)} \in \mathcal{A}_l^{l-1}$. The Pachner moves of the whole cycle of triangulation give the word $w_{\bar{Y}} = \prod_{i=1}^{l} a_{R(i), L(i)}$.

Thus, we have the relation $w_{\bar{Y}} = 1$ that we should imply in order make the words like $\Phi(\alpha)$ independent on resolutions of configurations of codimension 2.

Example 16.2. Consider the standard Gale diagram of order 5, Fig. 16.10. Then we have $R(1) = \{4, 5\}$, $L(1) = \{2, 3\}$, $R(2) = \{1, 5\}$, $L(2) = \{3, 4\}$ etc. The corresponding word is equal to

$$w = a_{45,23} a_{15,34} a_{12,45} a_{23,15} a_{34,12}.$$

We can give now the definition of Γ_n^k groups.

Definition 16.2. Let $4 \leq k \leq n$. The group Γ_n^k is the group with the generators

$$\{a_{P,Q} \mid P, Q \in \{1, \ldots, n\}, P \cap Q = \emptyset, |P \cup Q| = k, |P|, |Q| \geq 2\},$$

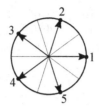

Fig. 16.10   Standard Gale diagram of order 5

and the relations:

(1) $a_{Q,P} = a_{P,Q}^{-1}$;
(2) (far commutativity) $a_{P,Q}a_{P',Q'} = a_{P',Q'}a_{P,Q}$ for each generators $a_{P,Q}$, $a_{P',Q'}$ such that

$$|P \cap (P' \cup Q')| < |P|, \qquad |Q \cap (P' \cup Q')| < |Q|,$$
$$|P' \cap (P \cup Q)| < |P'|, \qquad |Q' \cap (P \cup Q)| < |Q'|;$$

(3) $((k+1)$-gon relations) for any standard Gale diagram $\bar{Y}$ of order $k+1$ and any subset $M = \{m_1, \ldots, m_{k+1}\} \subset \{1, \ldots, n\}$

$$\prod_{i=1}^{k+1} a_{M_R(\bar{Y},i),M_L(\bar{Y},i)} = 1$$

where $M_R(\bar{Y}, i) = \{m_j\}_{j \in R_{\bar{Y}}(i)}$, $M_L(\bar{Y}, i) = \{m_j\}_{j \in L_{\bar{Y}}(i)}$.

Example 16.3. Let us write the $(k+1)$-gon relations in Γ_n^k for small k. The group Γ_n^4 has one pentagon relation

$$a_{m_4m_5,m_2m_3}a_{m_1m_5,m_3m_4}a_{m_1m_2,m_4m_5}a_{m_2m_3,m_1m_5}a_{m_3m_4,m_1m_2} = 1.$$

The group Γ_n^5 has two hexagon relations

$$a_{m_5m_6,m_2m_3m_4}a_{m_1m_5m_6,m_3m_4}a_{m_1m_2,m_4m_5m_6}a_{m_1m_2m_3,m_5m_6} \cdot$$
$$a_{m_3m_4,m_1m_2m_6}a_{m_3m_4m_5,m_1m_2} = 1,$$
$$a_{m_5m_6,m_2m_3m_4}a_{m_1m_5m_6,m_3m_4}a_{m_1m_2m_6,m_4m_5}a_{m_1m_2m_3,m_5m_6} \cdot$$
$$a_{m_3m_4,m_1m_2m_6}a_{m_4m_5,m_1m_2m_3} = 1.$$

The group Γ_n^6 has five heptagon relations

$$a_{m_6m_7,m_2m_3m_4m_5}a_{m_1m_6m_7,m_3m_4m_5}a_{m_1m_2m_6m_7,m_4m_5}a_{m_1m_2m_3,m_5m_6m_7}.$$
$$a_{m_1m_2m_3m_4,m_6m_7}a_{m_4m_5,m_1m_2m_3m_7}a_{m_4m_5m_6,m_1m_2m_3}=1,$$
$$a_{m_6m_7,m_2m_3m_4m_5}a_{m_1m_6m_7,m_3m_4m_5}a_{m_1m_2m_6m_7,m_4m_5}a_{m_1m_2m_3m_7,m_5m_6}.$$
$$a_{m_1m_2m_3m_4,m_6m_7}a_{m_4m_5,m_1m_2m_3m_7}a_{m_5m_6,m_1m_2m_3m_4}=1,$$
$$a_{m_6m_7,m_2m_3m_4m_5}a_{m_1m_6m_7,m_3m_4m_5}a_{m_1m_2m_7,m_4m_5m_6}a_{m_1m_2m_3m_7,m_5m_6}.$$
$$a_{m_1m_2m_3m_4,m_6m_7}a_{m_3m_4m_5,m_1m_2m_7}a_{m_5m_6,m_1m_2m_3m_4}=1,$$
$$a_{m_6m_7,m_2m_3m_4m_5}a_{m_1m_6m_7,m_3m_4m_5}a_{m_1m_2m_7,m_4m_5m_6}a_{m_1m_2m_3,m_5m_6m_7}.$$
$$a_{m_1m_2m_3m_4,m_6m_7}a_{m_3m_4m_5,m_1m_2m_7}a_{m_4m_5m_6,m_1m_2m_3}=1,$$
$$a_{m_5m_6m_7,m_2m_3m_4}a_{m_1m_6m_7,m_3m_4m_5}a_{m_1m_2m_7,m_4m_5m_6}a_{m_1m_2m_3,m_5m_6m_7}.$$
$$a_{m_2m_3m_4,m_1m_6m_7}a_{m_1m_4m_5,m_1m_2m_3m_7}a_{m_4m_5m_6,m_1m_2m_3}=1.$$

For groups Γ_n^k we have a generalisation of Theorem 16.1.

Theorem 16.4. *Formula* (16.1) *defines a correct homomorphism*
$$\Phi\colon \pi_1(\tilde{C}_n(\mathbb{R}^{k-2})) \to \Gamma_n^k.$$

The proof repeats the arguments of the proof of Theorem 16.1.

16.3 The groups $\tilde{\Gamma}_n^k$

We finish this part with a slight variation of the groups Γ_n^k. Geometrically speaking, here we consider oriented triangulations. Therefore, the indices of the generators of the groups are not independent and do not freely commute as it was seen, for example, in the groups Γ_n^5 (see Definition 16.1, first relation).

To be precise, we introduce the following:

Definition 16.3. Let $5 \leq k \leq n$. The group $\tilde{\Gamma}_n^k$ is the group with the generators

$$\{a_{P,Q} \mid P, Q \text{ — cyclically oriented subsets of } \{1, \ldots, n\},$$
$$P \cap Q = \emptyset, |P \cup Q| = k, |P|, |Q| \geq 2\},$$

and the relations:

(1) $a_{Q,P} = a_{Q',P'}$, where $Q = Q', P = P'$ as unordered sets, and as cyclically ordered sets Q differs from Q' by one transposition and P differs from P' by one transposition;

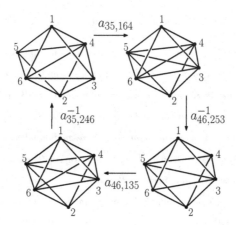

Fig. 16.11   A movement of a point around the configuration of four points on one circle

(2) $a_{Q,P} = a_{P,Q}^{-1}$;

(3) (far commutativity) $a_{P,Q} a_{P',Q'} = a_{P',Q'} a_{P,Q}$ for each generators $a_{P,Q}$, $a_{P',Q'}$ such that

$$|P \cap (P' \cup Q')| < |P|, \qquad |Q \cap (P' \cup Q')| < |Q|,$$
$$|P' \cap (P \cup Q)| < |P'|, \qquad |Q' \cap (P \cup Q)| < |Q'|;$$

(4) $((k+1)$-gon relations) for any standard Gale diagram $\bar{Y}$ of order $k+1$ and any subset $M = \{m_1, \ldots, m_{k+1}\} \subset \bar{n}$

$$\prod_{i=1}^{k+1} a_{M_R(\bar{Y},i), M_L(\bar{Y},i)} = 1,$$

where $M_R(\bar{Y}, i) = \{m_j\}_{j \in R_{\bar{Y}}(i)}$, $M_L(\bar{Y}, i) = \{m_j\}_{j \in L_{\bar{Y}}(i)}$.

These groups have the same connection to geometry and dynamics as the groups Γ defined above. To illustrate that, consider the following dynamical system.

Example 16.4. Let us have a dynamical system describing a movement of a point around the configuration of four points on one circle, see Fig. 16.11. Such system may be presented as the word in the group $\tilde{\Gamma}_6^5$

$$w = a_{35,164} a_{46,253}^{-1} a_{46,135} a_{35,246}^{-1}.$$

We can show that $\phi(\alpha)$ is nontrivial in the abelianisation $(\tilde{\Gamma}_6^5)_{ab} = \tilde{\Gamma}_6^5 / [\tilde{\Gamma}_6^5, \tilde{\Gamma}_6^5]$ of the group $\tilde{\Gamma}_6^5$. Computer calculations show the group $(\tilde{\Gamma})_{ab}$

can be presented as the factor of a free commutative group with 120 generators modulo $2 \cdot 6! = 1440$ relations that span a space of rank 90 if we work over $\mathbb{Z}_2$. Adding the element w to the relations increases the rank to 91, so the element w is nontrivial in $(\tilde{\Gamma}_6^5)_{ab}$, therefore it is nontrivial in $\tilde{\Gamma}_6^5$.

Hence, we have encountered a peculiar new effect in the behaviour of $\tilde{\Gamma}_n^5$ which is not the case of G_n^k. Certainly, the abelianisation of G_n^k is nontrivial and very easy to calculate since any generator enters each relation evenly many times. However, this happens not only for relations but for any words which come from braids. Thus, any abelianisations are trivial in interesting cases.

For $\tilde{\Gamma}_n^k$, it is an interesting new phenomenon, and the invariants we have demonstrated so far are just the tip of the iceberg to be investigated further.

Note that both for the groups Γ_n^k and $\tilde{\Gamma}_n^k$ we have many invariants since the corank of those groups is big.

PART 5
Unsolved Problems

Open Problems in the Groups G_n^k and Γ_n^k Theory and Related Fields

17.1 The groups G_n^k and Γ_n^k

17.1.1 Algebraic problems

Problem 17.1. Solve the word problem in the groups G_n^k for $n > k+1, k \neq 2$.

For the case of G_5^3 (private communication to Manturov) L. A. Bokut' constructed the *normal form* of the group (see [Bokut' and Kukin, 1994]). The word problem for G_5^4 was proved to be algorithmically solvable in the work [Fedoseev, Karpov, and Manturov, 2020]. The following lemmas were used to that end:

Lemma 17.1. *In the group* $G_5^4 = \langle a, b, c, d, e | a^2 = b^2 = \cdots = 1 = (abcde)^2 = (eadcb)^2 = \ldots \rangle$ *it can be algorithmically decided whether an element* $\beta \in G_5^4$ *belongs to the subgroup* H_4 *generated by* a, b, c, d.

Lemma 17.2. *The group* H_4 *allows the following presentation:*
$$H_4 = \langle a, b, c, d | a^2 = \cdots = 1, (abab)(cdcd) = (cdcd)(abab), \ldots \rangle.$$

The solvability of the word problem of G_5^4 follows from the solvability of the word problem of H_4.

Consider the following generalisation of the notion of the H_4 subgroup. Let $b_1, \ldots, b_{k+1}$ be generators of G_{k+1}^k. Let H_k be the subgroup of G_{k+1}^k generated by $\{b_1, \ldots, b_k\}$. The structure of the G_{k+1}^k groups can be partially understood through their H_k subgroups. In particular, the following theorem holds.

Theorem 17.1. *If the word problem in* H_k *is solvable, then the word problem in* G_{k+1}^k *is solvable. Moreover, the solution of the latter is constructive once the solution of the former is constructive.*

Conjecture 17.1. *An analogue of Lemma 17.2 holds for the groups H_k for any k. To be precise, for any group G_{k+1}^k its subgroup H_k admits the following presentation:*

$$H_k = \langle b_1, \ldots, b_k \mid b_1^2 = \cdots = b_k^2 = 1, [(b_i b_j)^2, (b_m b_l)^2] = 1 \rangle,$$

where $\{i, j, m, l\}$ go over all possible sets of distinct numbers from $\{1, \ldots, k\}$.

An important question is, whether the groups H_k have a linear presentation.

A powerful tool to study groups is the small cancellations theory. We can pose the following question: do the groups Γ_n^k satisfy the small cancellation condition? The answer, though, is in general negative due to commutativity. On the other hand, it can be checked, that, for example, the groups $\tilde{\Gamma}_{k+1}^k$ satisfy the small cancellation condition $C'(\frac{1}{6} + \varepsilon)$ for any $\varepsilon > 0$.

It is interesting to investigate special cases of the groups Γ and $\tilde{\Gamma}$ to find out if some of them satisfy the small cancellation conditions. Thus the following problem may be formulated:

Problem 17.2. To study the groups Γ_n^k and $\tilde{\Gamma}_n^k$ in context of small cancellation conditions for different particular case of n and k.

In addition, we can ask whether some conditions akin to small cancellations are satisfied. For example,

Problem 17.3. Are the groups Γ_{k+1}^k hyperbolic?

Problem 17.4. What is the boundary of the groups G_n^k?

Problem 17.5. Solve the conjugacy problem for the groups G_n^k and Γ_n^k.

Problem 17.6. Using the equality $\binom{n}{k} = \binom{n-1}{k} + \binom{n-1}{k-1}$ to consider elements of G_n^k as elements of type G_{n-1}^k with labels of type G_{n-1}^{k-1}.

Problem 17.7. The pentagon relation of the group Γ_n^4 expresses the associativity law. Which "higher associativities" correspond to the relations of the groups Γ_n^k?

Problem 17.8. Define an energy for the word of the groups G_n^k and Γ_n^k analogous to the energy function on braids. Investigate the gradient descent process induced by such an energy.

Problem 17.9. Which groups are embeddable in G_n^k?

We know that, for example, certain Coxeter groups, braid groups, mapping class groups may be embedded into G_n^k. The question is to find more examples and to understand the general problem of embeddability.

In particular, the following problem is an interesting question regarding the relationship between different G_n^k.

Problem 17.10. Does G_n^k embed into G_m^l when k is greater than l and n less than m? What are the conditions on the indices n, k, m, l?

Most likely the answer is negative for $G_m^l = G_m^2$, but considering G_m^3 may yield a positive answer.

Problem 17.11. Which homomorphisms from pure braid group or the fundamental group of configuration space to G_n^k are surjective or injective? What is the kernel of the homomorphism if it is surjective?

Problem 17.12. Is there any torsion in G_n^k except for two-torison?

Faithful representations were constructed for the group G_n^2 (see Chapter 10). The general problem, nevertheless, stands:

Problem 17.13. How to construct faithful representations of the groups G_n^k?

17.1.2 *Topological problems*

In Chapter 4 we studied the groups Γ_n^k which appear from triangulations of the $(k-2)$-dimensional space and their modification — groups $\tilde{\Gamma}_n^k$, which appear from oriented triangulations. At the same time, in Chapter 11 we studied the realisation of the groups G_n^k in the sense of construction of spaces where G_n^k (or a finite index subgroup of it) acts faithfully.

It seems natural to pose the same question in case of the Γ_n^k groups. In short it may be formulated in the following form:

Problem 17.14. Find realisations of the groups Γ_n^k. In other words, construct topological spaces where the groups Γ_n^k (or their finite index subgroups) act faithfully.

Problem 17.15. Study the connection between the groups Γ_n^k and the Manin–Shechtmann braids (see [Manin and Schechtmann, 1990]).

Problem 17.16. Do higher-dimensional braids detect smooth structures of higher-dimensional manifolds? For example, can we extract the Pontrjagin numbers of manifolds from Γ_n^k?

The codimension 1 properties which were used in the definition of the groups G_n^k and, to smaller extent, the groups Γ_n^k were *integral* in the sense that the forbidden configurations of points were fixed beforehand in the form of certain manifolds (spheres, planes, etc.). The following problem is big and important for both the G_n^k-groups theory and the Γ_n^k-groups theory:

Problem 17.17. Which good non-integral properties depending on k points out of n exist?

Problem 17.18. Which relations are there between the mapping class group of 2-manifolds and the groups G_n^k or Γ_n^k?

Problem 17.19. Construct invariants of 3-manifolds using the groups G_n^k and Γ_n^k.

Problem 17.20. Define Kontsevich-like invariants for braids in manifolds. A good starting point is to consider the group G_n^3.

It is known (see, e.g., the book [Conway and Sloane, 1998]) that the permutation group for 12 balls of radius 1 touching one ball of radius 1 is the whole symmetric group S_{12}. What is known about the corresponding braid group?

Problem 17.21. To study configurational spaces of balls by the means of the G_n^k and Γ_n^k groups.

Problem 17.22. To study knot analogues of the groups Γ_n^4 and G_n^k.

A possible approach to this problem is to obtain "free k-knots" as the closures of "free k-braids" — that is, of the elements of G_n^k. To that end we need to introduce the Markov moves and consider the group G_∞^k with those additional relations.

Problem 17.23. How to construct a fundamental domain so that G_n^k acts on it and generates a topological space?

Again, as in the case of faithful representations, the problem is solved for the case of G_n^2 but the general case is yet unsolved.

17.1.3   *Geometric problems*

Problem 17.24. Given a polyhedron, investigate the set of its triangulations using methods of G_n^k theory and Γ_n^k theory.

Problem 17.25. Describe spaces of packings and lattices (see [Conway and Sloane, 1998]) and their properties in terms of the groups G_n^k and Γ_n^k. Does the gradient descent method exist and work in that context?

Problem 17.26. How to describe the geometric difference between 1-dimensional homology group and fundamental group in terms of G_n^k?

17.2   *G*-braids

An important area of study is to construct a theory of braid groups GB_n. Let G be an abelian group. The goal is to construct a braid group GB_n such that if $G = \{1\}$, $GB_n = Br_n$.

Generators of the group should be of the form $\sigma_{i,g}$ where i stands for a usual index ranging from 1 to $n-1$, and g is an element of the group G. The problem which arises here is that if the group G is not a direct sum of several groups $\mathbb{Z}_2$, at each vertex we must choose which element we use: g or g^{-1}.

Consider a simple case: $G = \mathbb{Z}_2$. In that case the group GB_n has $2 \cdot (n-1)$ generators: $\sigma_{1,0}, \sigma_{1,1}, \ldots, \sigma_{(n-1),0}, \sigma_{(n-1),1}$. The relations are of the form:

$$\sigma_{i,k}\sigma_{j,l} = \sigma_{j,l}\sigma_{i,k} \text{ for } |i - j| \geq 2 \text{ and for any } k, l,$$

$$\sigma_{i,p}\sigma_{i+1,q}\sigma_{i,r} = \sigma_{i+1,r}\sigma_{i,q}\sigma_{i+1,p} \text{ for } p + q + r \equiv 0 \mod 2.$$

Note, that if we add the relations $\sigma_{i,1}^2 = 1$ for all $i = 1 \ldots, n-1$, we obtain the virtual braid group.

Our goal is to send the group $\mathbb{Z}_2 B_n$ to the classical braid group.

An immensely important question is to understand the following:

Problem 17.27. What is the dynamics behind the virtual braids?

Problem 17.28. How to embed the virtual braid group into a quotient group of the classical braid group? How to embed the virtual braid group into the group G_n^k?

Problem 17.29. Obtain a representation of the virtual braid group using the previous Problem.

Note, that the $\mathbb{Z}_2 B_n$ group may be connected to braids with dots. That leads to another important question: how to construct a "dot theory" for 2-dimensional knots?

General problem of the groups GB_n should be formulated as follows:

Problem 17.30. Can a group GB_n be included in a group B_{N+k} for some $N(n)$ and $k(G)$?

Note that this problem is related to Problem 17.10.

Another important area of study is the following:

Problem 17.31. Understand how braids act on braids.

17.3 Weavings

Weavings are a broad and important area of study. In particular, they can be used to study the balls collisions problems. The key observation is that a system of moving and non-colliding balls always has the same weaving associated with it, and this weaving changes at the moment of collision. Thus a study of the dynamics of a system of balls boils down to the study of the "regular" weavings and their bifurcations. Those bifurcations of weavings lead to a question about discriminants.

Problem 17.32. If there are many balls, many limitations on the weaving must appear. What are they?

Problem 17.33. How to understand the "future" of a weaving looking only at the weaving and the velocity of the balls?

The most obvious partial cases are the cases of three and four moving balls.

17.4 Free knot cobordisms

17.4.1 *Cobordism genera*

In Section 6.1 we discussed the coborism theory for free knots (decorated 4-valent graphs) and proved several sliceness criteria for certain families of knots. Therefore we got the techniques which allow us to study slice, or, in other words, knots, 0-cobordant to the trivial knot. An important question arises if one considers cobordisms of higher genus.

Problem 17.34. How to estimate the genus of a cobordism if it is known to be greater than 1?

The questions of the interconnections between the cobordism genera of different knots are also important. For example, consider the following:

Problem 17.35. If the connected sum of two knots is slice, those knots are cobordant of the same genus. On the other hand, the connected sum of two genus g cobordant knots is genus $g + 1$ slice. Can its slice genus be lowered? In other words, are there any pairs of g-cobordant knots such that their connected sum is $(g + 1)$-slice, but not g-slice?

17.5 Picture calculus

17.5.1 *Picture-valued solutions of the Yang–Baxter equations*

The problem of picture-valued solution to the Yang–Baxter equation may be tackled step-by-step. First it can be done for virtual braids. Then the groups G_n^2 should be considered. Finally, classical braids should be dealt with using the G_n^3 approach.

17.5.2 *Picture-valued classical knot invariants*

Problem 17.36. Construct picture-valued invariants, which are nontrivial for classical knots.

In Section 5.6.1 we studied the parity and picture-valued invariants of virtual knots: the *parity projection* and the *functorial mapping f*. By means of the parity projection it can be shown that classical knots are embedded into virtual knots. By means of the functorial mapping f the minimality of odd virtual (free) knots can be proved.

But those invariants are not fine enough for classical knots. More precisely, the value of functorial mapping f on classical knots is a trivial knot and the value of the parity projection on a classical knot is the knot itself.

On the other hand, in [Nelson, Orrison and Rivera, 2015] Nelson, Orrison and Rivera first suggested a bracket polynomial for virtual links with given biquandle colorings. In [Nelson, Oshiro, Shimizu and Yaguchi, 2017]

Nelson, Oshiro, Shimizu and Yaguchi suggested an enhancement of the bracket polynomial for virtual links with given biquandle colorings defined as follows. Let D be an oriented link diagram and C be its coloring by a biquandle $(X, \circ, *)$. Then the bracket polynomial f_C is defined as the sum of states on D as Fig. 17.1:

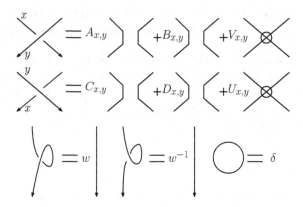

Fig. 17.1   Virtual states of a diagram

The coefficients A, B, C, D, V, U and δ, w are valued in a ring R with unity and they satisfy the axioms from [Nelson, Oshiro, Shimizu and Yaguchi, 2017, p. 5, Definition 3].

Proposition 17.1. [Nelson, Oshiro, Shimizu and Yaguchi, 2017] *The set* $\{f_C(D) \mid C\colon a\ coloring\ on\ D\ by\ X\}$ *is an invariant of virtual links.*

In [Ilyutko and Manturov, 2009] Manturov and Ilyutko suggested a bracket polynomial valued in pictures (framed 4-valent graphs) as follows: let D be an oriented diagram with coloring C by a biquandle $(X, \circ, *)$. Then the bracket polynomial g_C valued in pictures is defined by the sum of states on D as Fig. 17.2.

The coefficients A, B, C, D, E, F and δ, w are valued in a ring R with unity and they satisfy the axioms given in [Ilyutko and Manturov, 2009, p. 16, Definition 3.6]. We call the bracket $g_C(D)$ a *picture-valued bracket.*

Proposition 17.2. [Ilyutko and Manturov, 2009] *The set*

$$\{g_C(D) \mid C\colon a\ coloring\ on\ D\ by\ X\}$$

is an invariant of virtual links.

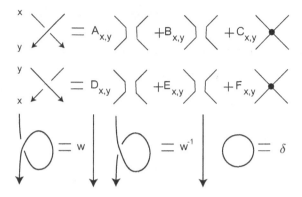

Fig. 17.2   States of a diagram used in the Ilyutko–Manturov invariant

Remark 17.1. The framed 4-valent graphs obtained as shown in Fig. 17.2 are considered as graphs modulo the second Reidemeister moves, not as flat or free links. This makes the bracket polynomial valued in pictures stronger than the Nelson bracket polynomial. Moreover, if $g_C(D)$ has a non-trivial 4-valent graph, then D is a non-trivial virtual link. On the other hand, if we define the bracket polynomial valued in (framed) 4-valent graphs modulo Reidemeister moves, then it is the same as the bracket polynomial defined by Nelson, Oshiro, Shimizu and Yaguchi.

We can define the bracket polynomial valued in (framed) 4-valent graphs modulo a subset of Reidemeister moves, for example, modulo Reidemeister moves 1 and 2. In other words, we can block some moves for (framed) 4-valent graphs as we want.

Modified problem: Find good examples of the use of the picture-valued biquandle bracket.

Unfortunately, for now the biquandle colorings which give a non-trivial value $g_C(K)$ for a classical knot K are not known.

17.5.3   *Categorification of polynomial invariants*

In Section 5.1 we defined the parity bracket for knots and links.

Problem 17.37. How to categorise the integer parity bracket?

17.6   Theory of secants

It is well known that any non-trivial knot admits a non-trivial quadrisecant. However, secants were not used for constructing knot invariants. The problem is to construct a 1-dimensional fromalism for the G_n^k groups. In particular, the following problem arises:

Problem 17.38. Is it possible to construct triod diagrams and define an analogue of parity theory for triods in a way similar to that for chord diagrams?

For a knot with "trisecants" each trisecant corresponds to trivalent vertex as described in Fig. 17.3.

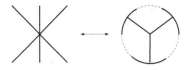

Fig. 17.3   Trivalent vertex on a chord diagram

It can be expected that a parity for trisecants can be obtained from the triod chord diagram by counting the number of intersections in the same way as the Gaußian parity was defined.

But there is no obvious parity of such sort since just two chord endpoints a and a' can change places as depicted in Fig. 17.4.

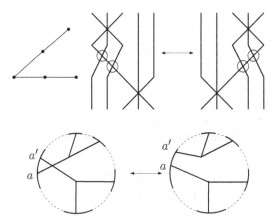

Fig. 17.4   Obstruction to the Gaußian parity

One can generalise the idea "*if two triples intersect then there is a quadruple intersection*".

Problem 17.39. Find the connections between quadrisecants and 2-dimensional knots.

In that manner we can study knots, braids and configurations of lines. Every object would have a *quadrisecant diagram* and *horizontal trisecant diagram*.

Resolution of singularities may be used in that theory: we can resolve a multiple point on a screen in which all quadrisecants cross. Then draw all appearing pictures on one plane to obtain a free knot (or braid). It is conjectured, that this free knot (or braid) is an invariant of the initial object.

Another direction of study is the use of braid-valued braid invariants to create classical knot picture-valued invariants.

17.7 Surface knots

In Section 4 free surface knots are introduced (Definition 4.10). Note that the notion of *abstract 2-knots* (Definition 4.5) corresponds to virtual knots (see Section 3.1). The notion of "free" surface knots is analogous to the 1-dimensional free knots.

Note that there exists another knot theory in one-dimensional case: the theory of flat knots. A *flat knot* is a knot without the over/under information. It is well-known that free knots are flat knots modulo *virtualisation*, see Fig. 3.17. Now we define *a flat surface knot* as a surface knot without the over/under information. Naturally the following question arises:

Problem 17.40. What is the difference between free and flat surface knots?

17.7.1 *Parity for surface knots*

In Chapter 2 we studied parities for virtual knots and many invariants of virtual knots, which were obtained from invariants of classical knots by means of parity and the parity projection from virtual knots to classical knots.

The simplest example of parity — *the Gaußian parity* is defined by Gauß diagrams, which correspond to spherical diagrams for 2-knots. From the above observation, since triple points of surface knots play a role similar to that of classical crossings of (virtual) knots, it can be expected that there are indices arising from triple points of surface knots such as the parity for virtual knots, one example of which is obtained from spherical diagrams. And the following questions follow:

Problem 17.41. Is it possible to prove "minimality" of free surface knots?

Problem 17.42. Is there a projection from 2-dimensional virtual (flat) knots to 2-dimensional classical (flat) knots?

Problem 17.43. Are the classical surface knots embedded in 2-dimensional virtual surface knots?

The series of the above questions should be solved step-by-step. Note that the "minimality" of free 1-dimensional knots can be studied by means of the picture-valued invariant — the functorial mapping f, which is obtained by parity of free knots. We can formulate the following question:

Problem 17.44. Is there a picture-valued invariant for (free) surface knots?

Problem 17.45. Can we study manifolds using multidimensional knot techniques?

17.8 Link homotopy

17.8.1 *Knots in $S_g \times S^1$*

In virtual knot theory there are many combinatorial formulas obtained by using "local information" on classical crossings, which we call *labels on the crossings*. An example of such labels is given by parities and "early/later" information for (virtual) long knots.

Proposition 17.3. *Let K be a knot in a thickened 2-surface trivial in $\mathbb{Z}_2$-homologies. Then the Khovanov homology theory for such knots may be constructed with additional gradings.*

We would like to introduce a new method of giving labels to classical crossings of knots in 3-dimensional space (see [Manturov and Rushworth, 2018]).

17.8.2 *Links in $S_g \times S^1$*

Definition 17.1. Let L and L' be two links in $S_g \times S^1$. If L' can be obtained from L by diffeomorphisms and stabilisations/destabilisations of $S_g \times S^1$, then the links L and L' are called *equivalent*.

By *destabilisation* for $S_g \times S^1$, we mean the following. Let C be a non-contractible circle on the surface S_g such that there exists a torus T homotopic to the torus $C \times S^1$ and not intersecting the link. Then the destabilisation is performed by cutting the manifold $S_g \times S^1$ along the torus $C \times S^1$ and pasting the two newborn components by $D \times S^1$.

Consider a point $x_0 \in S^1$ such that $(S_g \times \{x_0\}) \cap L(S^1)$ is a set of points without transversal points.

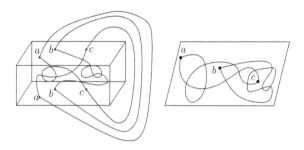

Fig. 17.5   Knots in $S_g \times S^1$ and the projection on the plane

Assume that the counterclockwise orientation is given on S^1. Then $cl(S_g \times S^1) \setminus (S_g \times \{x_0\}) \cong_h S_g \times [0,1]$ where h is an orientation preserving diffeomorphism. Let $M_L = \overline{(S_g \times S^1) \setminus (S_g \times \{x_0\})}$. Then L in M_L has a diagram on the surface S_g. The diagram of L in M_L has n arcs with n vertices and m circles. Two arcs near a vertex correspond to arcs near $S_g \times \{0\}$ and $S_g \times \{1\}$, respectively. We change a vertex to two small lines such that if one of the lines is corresponding to an arc which is close to $S_g \times (1 - \epsilon, 1]$, then the line is longer than the other one, see Fig. 17.6.

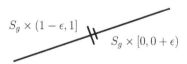

$S_g \times (1 - \epsilon, 1]$

$S_g \times [0, 0 + \epsilon)$

Fig. 17.6   The vertex notation

We shall call this a *diagram* of L in $S_g \times S^1$ on S_g. For any diagram with vertices on S_g we can easily construct a link L in $S_g \times S^1$.

A link L in M_L has a diagram on the plane as a virtual link. The following corollary also holds:

Theorem 17.2. *Let L and L' be two links in $S_g \times S^1$. Let D_L and $D_{L'}$ be diagrams of L and L' on the plane, respectively. Then L and L' are equivalent if and only if $D_{L'}$ can be obtained from D_L by applying the following moves, see Fig. 17.7.*

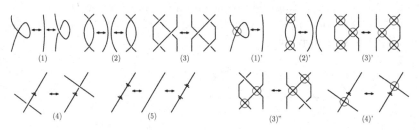

Fig. 17.7   Moves allowed

Now we can formulate a very important question:

Problem 17.46. How do we define "labels" for classical crossings by using the information from S^1?

17.8.3   *Degree of knots in $S_g \times S^1$*

Let K be a knot in $S_g \times S^1$. We can consider K as a mapping from $[0, 1]$ to $S_g \times S^1$ such that $K(0) = K(1) = (x, (1, 0))$ for some $x \in S_g$. Let p be the covering map from $S_g \times \mathbb{R}$ to $S_g \times S^1$. Let $\pi_2 \colon S_g \times S^1 \to S^1$ and $\pi_2' \colon S_g \times \mathbb{R} \to \mathbb{R}$ be the second projections of $S_g \times S^1$ and $S_g \times \mathbb{R}$, respectively. Then $\pi_2 \circ K \colon [0, 1] \to S^1$ is a loop on S^1 and there is the unique lifting $\overline{\pi_2 \circ K} \colon [0, 1] \to \mathbb{R}$ of K along exp where $\overline{\pi_2 \circ K}(0) = 0$. For a knot K in $S_g \times S^1$ define $deg(K) := deg(\pi_2 \circ K)$. From the definition we see that $deg(K)$ is an invariant for knots in $S_g \times S^1$.

Remark 17.2. If $deg(K) \neq 0$, then no liftings of K along $id \times exp$ are knots. In this case, we say that the covering space of $S_g \times S^1$ is $S_g \times \mathbb{R}/\mathbb{Z}_n$ where $n = deg(K)$. Then there is a lifting of K along the covering map into $S_g \times \mathbb{R}/\mathbb{Z}_n$ and this is a knot again.

Let $\tilde{K}$ be a lifting of K into $S_g \times \mathbb{R}/\mathbb{Z}_n$. For a line segment l of a diagram D of K, there is a line segment l' of $\tilde{K}$ in $S_g \times \mathbb{R}/\mathbb{Z}_n$. Let $q_2 : S_g \times \mathbb{R}/\mathbb{Z}_n \to \mathbb{R}/\mathbb{Z}_n$. If $q_2(l') \in [a, a+1)$, then give a label a to l. We consider the label a as an element of $\mathbb{Z}_n$.

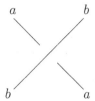

Fig. 17.8   Labeled crossing

Without loss of generality we may assume that the two under-arcs related to a classical crossing are labeled by the same element $a \in \mathbb{Z}_n$ and the label of the over-arc does not change in the classical crossing.

Remark 17.3. For labels a, b in the Fig. 17.9, $b = a + 1$.

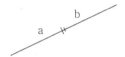

Fig. 17.9   Labels of an arc

For each crossing if the over-arc is labeled by b and the under-arc is labeled by a for some $a, b \in \mathbb{Z}_n$, then we give a label $i = b - a$ to the crossing, where $b - a$ is in $\mathbb{Z}_n$. Then we call D with this labelling for each classical crossing a *labeled diagram*.

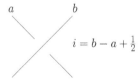

Fig. 17.10   Labelling for classical crossing of knots in $S_g \times S^1$

In other words, the label described above shows "how many times the diagram turns along S^1 before it comes back to the classical crossing".

17.8.4   *Questions*

As described previously, the labels of classical crossings of knots in $S_g \times S^1$ can be obtained. The first question is the following:

Problem 17.47. Is there an invariant of knots in $S_g \times S^1$, which is obtained from the labels of crossings?

There exists a covering map p from $S_g \times \mathbb{R}$ to $S_g \times S^1$. It is easy to see that for a knot $K \colon S^1 \to S_g \times S^1$ there exists a lift $\tilde{K} \colon S^1 \to S_g \times \mathbb{R}$ such that $p \circ \tilde{K} = K$. By means of the lift of K the multicomponent (infinite) link $L = \tilde{K}_1 \cup \ldots \tilde{K}_k$ in $S_g \times \mathbb{R}$ is constructed, such that $k = deg(K)$ and $p \circ \tilde{K}_i = K$ for each $i \in \{1, \ldots k\}$.

Problem 17.48. Is it possible to construct combinatorial invariant for knots in $S_g \times S^1$ by using the multicomponent link from the knot in $S_g \times S^1$?

Recall that virtual knots can be considered as knots in $S_g \times [0, 1]$. It is easy to see that there is a natural map from $S_g \times [0, 1]$ to $S_g \times S^1$ by attaching $S_g \times 0$ and $S_g \times 1$, and it follows that there is a natural map from virtual knots to knots in $S_g \times S^1$. But this map is not sufficient to study virtual knots by means of knots in $S_g \times S^1$, because every knot in $S_g \times S^1$ obtained from a knot in $S_g \times [0, 1]$ never turns around S^1 and, therefore, we cannot obtain anything for knots in $S_g \times [0, 1]$ by knots in $S_g \times S^1$.

Problem 17.49. Is there a reasonable way to embed virtual knots into $S_g \times S^1$?

Bibliography

M. Chrisman, V. O. Manturov, "Combinatorial formulae for finite-type invariant via parity", https://arxiv.org/abs/1002.0539.

D. A. Fedoseev, A. B. Karpov, V. O. Manturov, "Word and conjugacy problems in groups G_{k+1}^k", Lobachevskii Journal of Mathematics, **41**:02 (2020), pp. 176–193.

D. A. Fedoseev, V. O. Manturov, "A sliceness criterion for odd free knots", *Math. Sb.*, **210**:10 (2019), 161–178.

D. A. Fedoseev, V. O. Manturov, "Cobordisms of graphs. A sliceness criterion for stably odd free knots and related results on cobordisms", *J. Knot Theory Ramifications*, **27**:13 (2018), https://doi.org/10.1142/S0218216518420117.

D. A. Fedoseev, V. O. Manturov, "Parities on 2-knots and 2-links", *J. Knot Theory Ramifications*, **25**:14 (2016).

D. A. Fedoseev, V. O. Manturov, Z. Cheng, "On marked braid groups", *Journal of Knot Theory and Its Ramifications*, **24**:13 (2015).

A. A. Gaifullin, V. O. Manturov, "On the recognition of Braids", *J. Knot Theory Ramifications*, **11**:8 (2002), pp. 1193–1209.

D. P. Ilyutko, V. O. Manturov, "Introduction to graph-link theory", *J. Knot Theory Ramifications*, **18**:6, pp. 791–823.

D. P. Ilyutko, V. O. Manturov, "Virtual Knots: The State of the Art", *Series on Knots and Everything, vol. 51., World Scientific*, 2013.

D. P. Ilyutko, V. O. Manturov, "Parity and Patterns in Low-dimensional Topology", *Reviews in Mathematics and Mathematical Physics, Cambridge Scientific Publishers*, 2015.

D. P. Ilyutko, V. O. Manturov, I. M. Nikonov, "Parity in knot theory and graph-links", *M.: PFUR*, **41**:6 (2011), pp. 3–163.

D. P. Ilyutko, V. O. Manturov, I. M. Nikonov, "Virtual knot invariants arising from parities", *Banach Center Publ.*, **100** (2014), pp. 99–130.

S. Kim, V. O. Manturov, "The braid groups and imaginary generators", http://arxiv.org/abs/1612.03486v1.

V. A. Krasnov, V. O. Manturov, "Graph-valued invariants of virtual and classical links and minimality problem", *J. Knot Theory Ramifications*, **22**:12 (2013).

D. Yu. Krylov, V. O. Manturov, "Parity and relative parity in knot theory", https://arxiv.org/abs/1101.0128.

V. O. Manturov, "A note on a map from knots to 2-knots", http://arxiv.org/abs/1604.06597.

V. O. Manturov, "An elementary proof of the embeddability of classical braids into virtual braids", *Dokl. Akad. Nauk*, **469**:5 (2016), pp. 535–538 (in Russian).

V. O. Manturov, "Geometry and combinatorics of virtual knots", *Doctoral thesis*, 2007.

V. O. Manturov, "Invariants of classical braids valued in G_n^2", http://arxiv.org/abs/1611.07434v2.

V. O. Manturov, "Knot theory", 2nd ed., Boca Raton: CRC Press, 2018, https://doi.org/10.1201/9780203710920.

V. O. Manturov, "Multivariable polynomial invariants for virtual knots and links", *J. Knot Theory Ramifications*, **12**:8 (2003), pp. 1131–1144.

V. O. Manturov, "Non-Reidemeister knot theory and its applications in dynamical systems, geometry and topology", http://arxiv.org/abs/1501.05208.

V. O. Manturov, "O raspoznavanii virtual'nykh kos (On the Recognition of Virtual Braids)", *POMI Scientific Seminars*, **299**. Geometry and Topology, 8 (2003) (in Russian).

V. O. Manturov, "On free knots", preprint, https://arxiv.org/abs/0901.2214.

V. O. Manturov, "On the groups G_n^2 and Coxeter groups", http://arxiv.org/abs/1512.09273.

V. O. Manturov, "One-term parity bracket for braids", http://arxiv.org/abs/1501.00580.

V. O. Manturov, "Parity and cobordism of free knots", *Math. Sb.*, **203**:2 (2012), pp. 45–76.

V. O. Manturov, "Parity in knot theory", *Math. Sb..*, **201**:5 (2010), pp. 693–733.

V. O. Manturov, "The groups G_n^k and fundamental groups of configuration spaces", *J. Knot Theory Ramifications*, **26**:6 (2017).

V. O. Manturov, "Virtual crossing numbers for virtual knots", *J. Knot Theory Ramifications*, **21**:13 (2012).

V. O. Manturov, "The groups G_{k+1}^k and fundamental groups of configuration spaces", *Proceedings of the third international congress on algebra and combinatorics (E.I.Zelmanov, K.P.Shum, P.S.Kolesnikov, Editors)*, World Scientific, Singapore (2019), pp. 310–323.

V. O. Manturov, I. M. Nikonov, "On braids and groups G_n^k", *J. Knot Theory Ramifications*, **24**:13 (2015).

V. O. Manturov, H. Wang, "Markov theorem for free links", *J. Knot Theory Ramifications*, **21**:13 (2012).

V. O. Manturov, W. Rushworth, "Additional gradings on generalisations of Khovanov homology and invariants of embedded surfaces", *J. Knot Theory Ramifications*, **27**:09 (2018).

I. M. Nikonov, "Weak parities and functorial maps", *Topology*, SMFN, **51**, *M.: PFUR*, 2013, pp. 123–141.

D. M. Afanasiev, "Refining virtual knot invariants by means of parity", *Math. Sb.*, **201**:6 (2010), pp. 3–18.

E. Artin, "Theorie der Zöpfe", *Abh. Math. Sem. Univ. Hamburg*, **4** (1925), pp. 27–72.

E. Artin, "Theory of braids", *Math. Ann.* (2) **48** (1947), pp. 101–126.

V. G. Bardakov, "The virtual and universal braids", *Fundamenta Mathematicae*, **184** (2004), pp. 1–18.

A. J. Berrick, F. R. Cohen, Y. L. Wong, J. Wu, "Configurations, braids, and homotopy groups", *Journal AMS*, **19**:2 (2005), pp. 265–326.

S. Bigelow, "Braid groups are linear", *J. Amer. Math. Soc.*, **14** (2001), pp. 471–486

S. Bigelow, "Does the Jones polynomial detect the unknot", *J. Knot Theory Ramifications*, **11**:4 (2002), pp. 493–505.

J. S. Birman, "Braids, links and mapping class groups", *Princeton, NJ: Princeton Univ. Press*, 1974 (*Ann. Math. Stud.*, 1982).

J. S. Birman, "Studying links via closed braids", In: *Lecture Notes on the Ninth Kaist Mathematical Workshop*, **1** (1994), pp. 1–67.

J. S. Birman, W. Menasco, "Studying links via closed braids II: On a theorem of Bennequin", *Topology and Its Applications*, **40** (1991), pp. 71–82.

J. S. Birman, W. Menasco, "Studying links via closed braids III: Classifying links which are closed 3-braids", *Pacific Journal of Mathematics*, **161** (1993), pp. 25–113.

J. S. Birman, W. Menasco, "Studying links via closed braids IV: Composite links and split links", *Inventiones Mathematicae*, **102** (1990), pp. 115–139.

J. S. Birman, W. Menasco, "Studying links via closed braids V: The unlink", *Transactions of the AMS*, **329** (1992), pp. 585–606.

J. S. Birman, W. Menasco, "Studying links via closed braids VI: A non-finiteness theorem", *Pacific Journal of Mathematics*, **156** (1992), pp. 265–285.

W. A. Bogley, S. J. Pride, "Aspherical relative presentations", Proc. Edinburgh Math. Soc. No.35 (1992), pp. 1–39.

O. Bogopolski, A. Martino, O. Maslakova, E. Ventura, "Free-by-cyclic groups have solvable conjugacy problem", Bulletin of the London Mathematical Society, Volume 38, Issue 5, October 2006, Pages 787–794.

L. A. Bokut', G. P. Kukin, "Algorithmic and combinatorial algebra", *Springer-Science+Business Media, B. V.* (1994).

K. Borsuk, "On the k-independent subsets of the Euclidean space and of the Hilbert space", *Bull. Acad. Polon. Sci.*, Cl. III. 5 (1957), pp. 351–356, XXIX.

E. V. Brieskorn, "Die fundamentalgruppe des raumes der regulären orbits einer endlichen komplexen spiegelsgruppe", *Inventiones Mathematicae*, **12**:1 (1971), pp. 57–61.

E. V. Brieskorn, "Sur les groupes des tresses (d'après V.I.Arnol'd)", In: *Seminaire Bourbaki 1971-72*, **401**, *Lect. Notes Math.*, **317**, Springer, Berlin, 1973, pp. 21–44.

W. Burau, "Über zopfgruppen und gleichzeitig verdrillte verkettungen", *Abh. Math. Sem. Univ. Hamburg*, **11** (1936), pp. 179–186.

E. Cartan, "Leçons sur la géométrie des espaces de Riemann", Gauthiers-Villars, 1928.

J. S. Carter, "Reidemeister/Roseman-Type Moves to Embedded Foams in 4-Dimensional Space", *New Ideas in Low Dimensional Topology (Series on Knots and Everything)* (2015).

J. S. Carter, S. Kamada, M. Saito, "Surfaces in 4-Space", *Encyclopaedia of Mathematical Sciences (Book 142)*, Springer, 2004.

J.S. Cho, S. Yoon, C. K. Zickert, *On the Hikami-Inoue conjecture*, https://arxiv.org/abs/1805.11841v2.

O. Chterental, "Virtual braids and virtual curve diagrams", *J. Knot Theory Ramifications*, **24**:13 (2015).

O. Chterental, "Distinguishing virtual braids in polynomial time", https://arxiv.org/abs/1706.01273.

A.M. Cohen, Coxeter groups, Notes of a MasterMath course, Fall 2007, January 24, 2008.

R. H. Crowell, R. H. Fox, "Introduction to knot theory", *New York: Ginn & Co* (1963).

J. H. Conway, N. J. A. Sloane, "Sphere Packings, Lattices and Groups", Springer, New York, NY, 1998.

P. Dehornoy, "A fast method for comparing braids", *Adv. Math.*, **125**:2 (1997), pp. 200–235.

I.A. Dynnikov, "On a Yang–Baxter map and the Dehornoy ordering", Russian Mathematical Surveys(2002), 57(3):592.

P. I. Etingof, I. Frenkel, A. A. Kirillov, "Lectures on Representation Theory and Knizhnik-Zamolodchikov Equations", Mathematical Surveys and Monographs, 58 (1998), American Mathematical Society, ISBN 0821804960.

R. Fenn, M. Jordan, L. Kauffman, "Biracks, biquandles and virtual knots", *Topol. Appl.* **145**:1–3, (2004), pp. 157–175.

R. Fenn, P. Rimanyi, C. P. Rourke, "The braid-permutation group", *Topology*, **36**:1 (1997), pp. 123–135.

S. Fomin, A. N. Kirillov, "Quadratic Algebras, Dunkl Elements, and Schubert Calculus", 1999, *Advances in Geometry*, pp. 147–182.

I. M. Gelfand, M. M. Kapranov, A. V. Zelevinsky, "Discriminants, Resultants, and Multidimensional Determinants", Birkhäuser Boston, 1994.

A. Gibson, "Homotopy invariants of Gauss words", *Math. Ann.*, 349 (2011), pp. 871–887.

A. I. Goldberg, "O nevozmozhnosti usileniya nekotorykh resultatov Greendlingera i Lyndona", *UMN*, 1978, vol. 33, **6**, pp. 201–202.

W. Goldman, "Invariant functions on Lie groups and Hamiltonian flows of surface group representations", *Invent. Math.* **85** (1986), pp. 263–302.

M. Goussarov, M. Polyak, O. Ya. Viro, "Finite type invariants of classical and virtual knots", *Topology* **39** (2000), pp. 1045–1068.

M. Greendlinger, "An analogue of a theorem of Magnus", *Archiv der Mathematik*, vol 12 (1961), pp. 94–96.

B. Grünbaum, "Convex polytopes", 2nd ed., *Spriger-Verlag* (2003).

J. Hass, P. Scott, "Shortening curves on surfaces", *Topology* **33**:1 (1994), pp. 25–43.

K. Hikami and R. Inoue, "Braids, complex volume and cluster algebras", Algebraic and Geometric Topology, 15(4):2175–2194, 2015.

J. Howie, "The solution of length three equations over groups", Proc. Edinburgh Math. Soc. (2) No26 (1983), pp. 89–96.

D. Joyce, "A classifying invariant of knots, the knot quandle", *J. Pure Appl. Algebra* **23** (1982), pp. 37–65.

S. Kamada, "Braid presentation of virtual knots and welded knots", `https://arxiv.org/abs/math/0008092`.

N. Kamada, S. Kamada, "Abstract link diagrams and virtual knots", *J. Knot Theory Ramifications* **9**:1 (2000).

E. R. van Kampen, "On some lemmas in the theory of groups", *Amer. J. Math*, 1933, vol. 55, pp. 268–273.

M. Kapranov, V. Voevodsky, "Braided monoidal 2-categories and Manin-Schechtmann higher braid groups", *J. Pure and Applied Math.*, N. 92 (1994), pp. 241–167.

L. H. Kauffman, "Virtual Knots", *talks at MSRI Meeting in January 1997 and AMS meeting at University of Maryland, College Park in March 1997.*

L. H. Kauffman, S. Lambropoulou, "Hard unknots and collapsing tangles", *Introductory lectures on knot theory, Ser. Knots Everything*, **46** (2012), pp. 187–247.

A. Kolmogorov, "A remark on P. L. Chebyshev polynomials deviating least from zero", *Uspekhi Mat. Nauk* **3**:1 (1948), pp. 216–221.

G. Kuperberg, "What is a virtual link?", *Alg. Geom. Topol.*, **3** (2002), pp. 587–591.

E.-K. Lee, "A positive presentation for the pure braid group", *Journal of the Chungcheong mathematical society*, Vol. 23, No. 3, Sep., 2010.

J.A. De Loera, J. Rambau, F. Santos, "Triangulations: structures for algorithms and applications", *Springer*, (2010).

R. C. Lyndon, "On Dehn's algorithm", *Math. Ann.*, 1966, Bd 166, pp. 208–228.

R. C. Lyndon, P E. Schupp, "Combinatorial group theory", Reprint of the 1977 ed. - Berlin; Heidelberg; New York; New York; Barcelona; Hong Kong; London; Milan; Paris; Singapore; Tokyo; Springer, 2001.

G. A. Makanina, "Defining relation in the pure braid group", *Moscow Univ. Math. Bull.*, **3** (1992), pp. 14–19.

Y. I. Manin, V. V. Schechtmann, "Arrangements of hyperplanes, higher braid groups and higher Bruhat orders", *Adv. Stud. Pure Math.*, **17** (1990), pp. 289–308.

S. V. Matveev, "Distributive grouppoids in knot theory", *Mathematics of the USSR-Sbornik* **47**:1 (1984) 73.

J. Milnor, "Link Groups", *Ann. Math* **59**:2 (1954), pp. 177–195.

J. A. Moody, "The Burau representation of the braid group B_n is unfaithful for large n", *Bull. Amer. Math. Soc.*, **25** (1991), pp. 379–284.

A. Nabutovsky, "Fundamental group and contractible closed geodesics", Comm. Pure Appl. Math. 49(12) (1996), pp. 1257–1270.

A. Nabutovsky, "Physics Geometry of the Space of Triangulations of a Compact Manifold", Communications in Mathematical Physics, 181, pp. 303–330, 1996.

S. Nelson, M. E. Orrison, V. Rivera, "Quantum enhancements and biquandle brackets", preprint, https://arxiv.org/abs/1508.06573.

S. Nelson, K. Oshiro, A. Shimizu, Y. Yaguchi, "Biquandle Virtual Brackets", preprint, https://arxiv.org/abs/1701.03982v2.

M. H. A. Newman, "On theories with a combinatorial definition of equivalence", *Math. Ann.*, 1942, **43**, 2, pp. 223–243.

P. S. Novikov, "Ob algoritmicheskoy nerazreshimosti problemy tozhdestva slov v teorii grupp", *M.:Nauka*, 1955.

A. Yu. Olshanskii, "Geometry of defining relations in groups", *M.:Nauka*, 1989.

U. Pachner, "P.L. homeomorphic manifolds are equivalent by elementary shellings", European Journal of Combinatorics (1991), 12 (2): 129–145, https://doi.org/10.1016/S0195-6698(13)80080-7.

K. Reidemeister, "Knot Theory", *New York: Chelsea Publ. & Co.* (1948).

D. Roseman, "Elementary moves for higher dimensional moves", *Fund. Math.* **184** (2004), pp. 291–310.

D. Roseman, "Reidemeister-type moves for surfaces in four-dimensional space", *Knot Theory, Banach Center Publications*, Vol. 42 (Polish Academy of Sciences, Warsaw, 1998), pp. 347–380.

P. E. Schupp, "On Dehn's algorithm and the conjugacy problem", *Math. Ann.*, 1968, Bd 178, **2**, pp. 119–130.

V. G. Turaev, "Problem list from workshop on low dimensional topology", *Luminy* (1989).

V. G. Turaev, "Skein quantization of Poisson algebras of loops on surfaces", *Ann. Sci. École Norm. Sup.* **4**, 24 (1991), pp. 635–704.

V. G. Turaev, "Topology of words", *Proc. Lond. Math. Soc.*, **95**:3 (2007), pp. 360–412.

V. Vershinin, "On homology of virtual braids and Burau representation", *Journal of Knot Theory and Its Ramifications*, **18**:5 (2001), pp. 795–812.

O. Ya. Viro, "Topological problems concerning lines and points of three-dimensional space", *Dokl. Akad. Nauk SSSR*, 284, 5 (1985), pp. 1049–1052; English transl.: *Soviet Math. Dokl.*, 32, 528–531 (1985).

B. K. Winter, "Virtual links in arbitrary dimensions", *J. Knot Theory Ramifications* **24**:14 (2015).

Index

SERIES ON KNOTS AND EVERYTHING

ISSN: 0219-9769

Editor-in-charge: Louis H. Kauffman *(Univ. of Illinois, Chicago)*

The Series on Knots and Everything: is a book series polarized around the theory of knots. Volume 1 in the series is Louis H Kauffman's Knots and Physics.

One purpose of this series is to continue the exploration of many of the themes indicated in Volume 1. These themes reach out beyond knot theory into physics, mathematics, logic, linguistics, philosophy, biology and practical experience. All of these outreaches have relations with knot theory when knot theory is regarded as a pivot or meeting place for apparently separate ideas. Knots act as such a pivotal place. We do not fully understand why this is so. The series represents stages in the exploration of this nexus.

Details of the titles in this series to date give a picture of the enterprise.

Published:

More information on this series can also be found at http://www.worldscientific.com/series/skae